# The Boundary Integral Equation Method for Porous Media Flow

OF RELATED INTEREST

*Boundary element methods in solid mechanics*

S. L. Crouch and A. M. Starfield

# The Boundary Integral Equation Method for Porous Media Flow

James A. Liggett
Philip L-F. Liu

*School of Civil and Environmental Engineering, Cornell University, Ithaca, New York, USA*

London
GEORGE ALLEN & UNWIN
Boston Sydney

**George Allen & Unwin (Publishers) Ltd,
40 Museum Street, London WC1A 1LU, UK**

George Allen & Unwin (Publishers) Ltd,
Park Lane, Hemel Hempstead, Herts HP2 4TE, UK

Allen & Unwin Inc.,
9 Winchester Terrace, Winchester, Mass 01890, USA

George Allen & Unwin Australia Pty Ltd,
8 Napier Street, North Sydney, NSW 2060, Australia

First published in 1983

---

**British Library Cataloguing in Publication Data**

Liggett, James A.
The boundary integral equation method for porous media flow.
1. Hydrogeology 2. Integral equations
I. Title II. Liu, Philip L.-F.
551.49'01'5145 GB1197.7
ISBN 0-04-620011-8

---

**Library of Congress Cataloging in Publication Data**

Liggett, James A.
The boundary integral equation method for porous media flow.
Bibliography: p.
Includes index.
1. Groundwater flow–Mathematics. 2. Integral equations. 3. Boundary value problems.
I. Liu, Philip L. F. II. Title.
GB1197.7.L53 1983 551.49'01'51 82-8795
ISBN 0-04-620011-8 AACR2

---

Set in 10 on 12 point Times by Pintail Studios Ltd, Ringwood, Hampshire and printed in Great Britain by Mackays of Chatham

# Preface

Due to the complexities of geometry and of inhomogeneous and/or anisotropic media there are few analytical solutions to porous media problems. Numerical solutions (e.g. the Theis method) have long played an important role in groundwater calculations. Finite differences were used extensively in the 1960s (e.g. Remson *et al.* 1971). In the 1970s finite elements began to replace finite differences as the dominant technique (e.g. Pinder & Gray 1977). The advantages of finite elements are: (1) it is easier to conform to the physical geometry whereas with finite differences the geometry is usually 'adjusted' to fit the grid spacing; (2) the boundary conditions are easier to apply for geometrical reasons; (3) it is easier to use variable grid spacing whereby small elements are used in regions of interest or rapidly changing flow and large elements are used in regions of little interest and slowly changing flow, and (4) it is easier to write universal programs which apply to any geometry and a large number of physically different situations without having to change the code for each individual case. The primary disadvantages of finite element calculations are: (1) the time consuming data input of the grid (although grid generation programs may substantially shorten this process); (2) long computing times, especially for complex problems and three-dimensional problems; and (3) the requirement that the analyst must find an element size such that the governing differential equation is satisfactorily approximated everywhere.

Lately, the Boundary Integral Equation Method (BIEM) has emerged as a contender in the computation Derby. It carries all the above-mentioned advantages of the finite element method while considerably reducing the disadvantages. It does not, however, replace the finite element method in all cases. The BIEM is not as powerful as the finite element method in that it cannot handle all of the physical complications. (The previous statement is not literally true as the BIEM could be broken into subregions as small as elements. If such a division proves necessary, however, the finite element method would be a better choice.) An example is unsaturated flow, in which case the BIEM is unsuited but can be combined with finite elements as explained in Chapter 9. In the great majority of problems the BIEM can provide a solution and, even if some approximation is necessary, the method is far more accurate than justified by knowledge of the physical constants such as permeability, storativity, etc.

In one important aspect the BIEM is more powerful than finite elements. Although three-dimensional finite element solutions are entirely possible and conceptually easy, the work of creating the grid and the excessive computer requirements of storage and computer time make the method frequently impractical. In contrast, the BIEM requires only a two-dimensional grid for the solution of three-dimensional problems and both the computing time and computer storage requirements are much less. Thus large problems can be solved routinely instead of on a research basis or for one time only.

The BIEM has the virtue of reducing the domain of calculation by one dimension. It is a surface method (in three dimensions) or a line method (in two dimensions) as opposed to a volume method (3-D) or an area method (2-D). Two-dimensional problems are solved by integration over a line; three-dimensional problems are solved by integration over a surface. The lack of necessity of discretizing the entire domain leads, in general, to a great gain in efficiency. Moreover, the governing differential equation, or at least the differential operator, is solved exactly; all approximations are confined to the boundaries. The user need not be concerned with approximations to derivatives. He needs only to choose boundary elements such that the variables are approximated accurately on the boundary. This choice can be made from the user's physical, as opposed to mathematical, knowledge of the problem.

It is usual in a book of this type to cite the history of the subject. That is particularly difficult to do in the case of the BIEM. It is an integral equation method and thus the history should include the development of integral equations since Abel (1823). To do so would obviously be superfluous, but then it is unclear at what point the BIEM emerged from the general development of integral equations and their numerical solution. It is clear that the BIEM, or the closely related technique of source distributions, has been available for many years (Prager 1928). However, such methods are largely impractical without the use of digital computers. Thus the modern emergence of the BIEM for the solution of engineering problems stems from the 1960s (e.g. Kupradze 1965, Jaswon & Symm 1977). The method was used to solve linear water wave problems (Shaw 1975) and flow in porous media (Liggett 1977) in the 1970s. It is expected that in the 1980s the BIEM will become a major tool of engineering analysis for groundwater computation.

In the chapters that follow the equations for porous media are presented, and both the theory and numerical algorithms of the BIEM are developed. The BIEM is then applied to two-dimensional problems with fixed boundaries. In Chapter 3 special elements are considered whereby the solution accuracy can be improved. Chapter 4 treats problems with nonhomogeneous and anisotropic permeability. Free surface problems and transients are considered in Chapter 5. Chapters 6 and 7 extend the solutions to problems with axial symmetry and to three-dimensional cases. Problems of two or more fluids with sharp interfaces between them are considered in Chapter 8. In Chapter 9 the BIEM is combined with the finite element method in order to solve efficiently those problems in which one method is advantageous in one region and the other method is best applied to other regions.

In Chapter 10 many of the previously described methods are combined to treat complex groundwater situations. In that chapter it is shown that some fundamentally three-dimensional problems can be solved using only line integration. Finally, three of the basic programs are presented in Chapter 11. These illustrate how the general method can be adapted to automatic computation and will give the reader a quick start on BIEM calculation.

Jim Liggett
Phil Liu
February 1982


## Acknowledgements

We gratefully acknowledge the work of our students and associates who have contributed to both the theory and computational aspects of the BIEM. Dr Gerard Lennon has made numerous contributions to all aspects of the work and he is directly responsible for the axisymmetric and three-dimensional programs. Dr Olurinde Lafe has written programs for flow in zoned media and is largely responsible for the computations in complex groundwater basins. Dr James Salmon has written several BIEM programs. Dr Alexander Cheng has contributed to the studies on singular elements and to the multiphase flow program. Mr Melih Erdemir is largely responsible for the combination of the BIEM with the finite element method. Without these contributions the groundwater applications of the BIEM would be very limited.

We also gratefully acknowledge the American Geophysical Union, who have given us permission to use illustrations from the following journal papers:

Liu, P. L-F. and J. A. Liggett 1978. *Water Resources Research* **14**(3), June; Lennon, G. P. *et al.* 1979. Ibid. **15** (2 papers); Liggett, J. A. and P. L-F. Liu 1979. Ibid. **15**(2), also **15**(4), 861–6; Lennon, G. P. *et al.* 1979. Ibid. **15**(5), 1102–6 and 1107–15; Liu, P. L-F. and J. A. Liggett 1980. Ibid. **16**(4); Lafe, O. E. *et al.* 1981. Ibid. **17**(5), 1431–44; Liu, P. L-F. *et al.* 1981. Ibid. **17**(5), 1445–52.

# Contents

# 1 Model equations for groundwater flow problems

In this chapter we review briefly the governing equations and boundary conditions which describe various types of groundwater flow problems. Using appropriate assumptions and approximations, the general equations are reduced to the simplified model equations that will be used in the remainder of this book. All but the simpler examples require numerical methods for their solution. Odd geometrical shapes and the complexities of an aquifer form the primary difficulties. For free surface problems the non-linear nature of the kinematic boundary condition and the fact that the location of the free surface is part of the solution make the problem challenging.

## 1.1 Conservation of mass

In studying problems concerning flows in porous media, we are usually interested in the macroscopic flow patterns but not in the microscopic details. We can then define the specific discharge (or seepage velocity) as the volume flow rate per unit total area (pores plus particles) normal to the flow. Assuming that the averaged flow field can be considered as a continuum and applying the principle of mass conservation, we can derive the following equation (Eagleson 1970):

$$-n\rho\frac{\partial S}{\partial t} - n\rho S\beta\frac{\partial p}{\partial t} + S\rho\alpha\frac{\partial \sigma_z}{\partial t} = \nabla \cdot (\rho \mathbf{q}) \tag{1.1}$$

where

$\mathbf{q}$ = the specific discharge (or the seepage velocity)
$p$ = pore-water pressure intensity
$\sigma_z$ = the normal intergranular stress on a horizontal plane
$S$ = degree of saturation (volume of fluid/volume of voids)
$\rho$ = density of the fluid
$n$ = porosity
$\beta$ = compressibility of the fluid (reciprocal of bulk modulus of elasticity of the fluid)
$\alpha$ = the vertical compressibility of the granular skeleton.

The relationship between the normal stress $\sigma_z$ and the pore-water pressure depends on the nature of the medium and the fluid.

For saturated flows, $S = 1$ and $\partial S/\partial t = 0$, (1.1) can be reduced to the following form:

$$-n\rho\beta \frac{\partial p}{\partial t} + \rho\alpha \frac{\partial \sigma_z}{\partial t} = \nabla \cdot (\rho \mathbf{q}) \tag{1.2}$$

A brief discussion on unsaturated flows is presented in Section 1.8. To simplify (1.2) further we must consider the properties of the medium and the fluid.

In the case of a confined aquifer where the permeable layer is bounded by rigid and impermeable material on both top and bottom, a change in the pore-water pressure must spontaneously produce an equal but opposite change in the normal intergranular stress, i.e. $\sigma_z = -p$. Therefore, (1.2) becomes

$$-\rho(\alpha + n\beta) \frac{\partial p}{\partial t} = \nabla \cdot (\rho \mathbf{q}) \tag{1.3}$$

We may introduce the specific storage,

$$S_e = \rho g(\alpha + n\beta) \tag{1.4}$$

and substitute it into (1.3) to get

$$-\frac{1}{g} S_e \frac{\partial p}{\partial t} = \nabla \cdot (\rho \mathbf{q}) \tag{1.5}$$

In the case of an unconfined aquifer where a free surface appears, the compressibility of the medium and of the fluid are relatively insignificant as compared with the changes of fluid volume due to the changes in the free surface elevation. Hence, we can assume that $\alpha = \beta = |\nabla \rho| = 0$, and equation (1.2) becomes

$$\nabla \cdot \mathbf{q} = 0 \tag{1.6}$$

## 1.2 Darcy's law

In order to complete the description of flow in porous media a relationship between the pore pressure, elevation, and seepage velocity is needed. With simplifying assumptions such a relationship can be derived from the

Navier–Stokes equations (Bear 1972, Eagleson 1970). However, the result would contain one or more unknown factors that stem from an incomplete description of the exact geometry of the soil particles and pores in the medium. Darcy (1856) arrived at the same result through experiments of saturated flow of water through a column of soil. He demonstrated that the head difference is inversely proportional to the length of the column. This one-dimensional empirical relationship is written in three dimensions as

$$\mathbf{q} = -K\,\nabla\Phi \tag{1.7}$$

in which $K$ is the permeability or hydraulic conductivity and $\Phi$ is the piezometric head,

$$\Phi = \frac{p}{\rho g} + z \tag{1.8}$$

Assuming that the compressibility of the fluid and the soil are small enough to be neglected, (1.7) is substituted into (1.6) to give

$$\nabla \cdot (K\nabla\Phi) = 0 \tag{1.9}$$

If the medium is homogeneous (i.e., $K$ is a constant),

$$\nabla^2\Phi = \frac{\partial^2\Phi}{\partial x^2} + \frac{\partial^2\Phi}{\partial y^2} + \frac{\partial^2\Phi}{\partial z^2} = 0 \tag{1.10}$$

which is Laplace's equation. This relationship is valid for steady flow in a confined or unconfined aquifer and is an excellent approximation to unsteady flow in an unconfined aquifer. When combined with proper boundary conditions, (1.10) furnishes an adequate description of many problems in porous media.

If the assumption of small compressibility is not sufficiently accurate, (1.5) must be used. Using (1.8) and (1.7) in (1.5), we obtain

$$S_e\rho\frac{\partial\Phi}{\partial t} = \nabla\cdot(\rho K\,\nabla\Phi) \tag{1.11}$$

Usually the density of the fluid can be assumed constant even though the medium may be compressible. Thus

$$S_e\frac{\partial\Phi}{\partial t} = \nabla\cdot(K\,\nabla\Phi) \tag{1.12}$$

The additional assumption of homogeneity gives

$$S_e \frac{\partial \Phi}{\partial t} = K \nabla^2 \Phi \tag{1.13}$$

which is the equation that governs heat flow in solids.

## 1.3 Sources and sinks

In a regional (large scale) groundwater problem a recharge or pumping well is often idealized as a source or sink. In such a case water is injected into or pumped from a point in the medium and the equation of conservation of mass, (1.9), must be modified to

$$\nabla \cdot (K \nabla \Phi) = -\sum_{k=1}^{np} Q_k \,\delta(x - x_k)\,\delta(y - y_k)\,\delta(z - z_k) \tag{1.14}$$

in which $n_p$ is the number of sources and sinks, $Q_k$ is the volume rate of flow (positive for a source, negative for a sink) of the $k$th source, $(x_k, y_k, z_k)$ is the co-ordinate of the $k$th source or sink, and $\delta(p)$ is the Dirac delta function, which is zero for $p \neq 0$ and unity for $p = 0$. For a homogeneous medium (1.14) becomes Poisson's equation

$$\nabla^2 \Phi = -\sum_{k=1}^{np} \frac{Q_k}{K} \,\delta(x - x_k)\,\delta(y - y_k)\,\delta(z - z_k) \tag{1.15}$$

which is comparable to (1.10). Similarly (1.13) becomes

$$S_e \frac{\partial \Phi}{\partial t} = K \nabla^2 \Phi + \sum_{k=1}^{np} Q_k \,\delta(x - x_k)\,\delta(y - y_k)\,\delta(z - z_k) \tag{1.16}$$

## 1.4 Non-Darcy flow

The use of Darcy's law, (1.7), assumes that the Reynolds number (based on pore size) is small enough for inertial effects to be unimportant. The Reynolds number for most groundwater flows is well into the Darcy range. However, flow with large gradients and/or flow where the soil particles are large (coarse sand) may exhibit either general or localized non-Darcy properties.

Forchheimer (1901) suggested the nonlinear relationship

$$\nabla \Phi = -(a + b|\mathbf{q}|)\mathbf{q} \tag{1.17}$$

in which $a$ and $b$ are constants to be empirically determined for any medium. Actually $a$ and $b$ are not strictly constant but should be functions of the Reynolds number (Stark & Volker 1967); however, they seem to vary little over a wide range of Reynolds numbers. Others have proposed a similar equation

$$\nabla\Phi = -c|\mathbf{q}|^{m-1}\mathbf{q} \tag{1.18}$$

in which $c$ and $m$ are now the empirical constants. Equation (1.17) reduces to Darcy's law for $b = 0$ and $a = K$; (1.18) reduces to Darcy's law for $m = 1$ and $c = K$.

In either case it is common to define an 'effective' permeability

$$K = 1/(a + b|\mathbf{q}|) \tag{1.19}$$

or

$$K = 1/(c|\mathbf{q}|^{m-1}) \tag{1.20}$$

in which case (1.9) is valid. A number of the relationships have been proposed to describe non-Darcy flow. For a more thorough treatment see Bear (1972).

The discussion presented in the rest of this chapter and the book are confined to Darcy flows with the exception of a portion of Chapter 9.

## 1.5 Anisotropic media

In stating Darcy's law, (1.7), we assumed that the constant of proportionality, the permeability, is the same in all directions. A more general formulation is

$$\mathbf{q} = -\boldsymbol{K} \cdot \nabla\Phi \tag{1.21}$$

in which $\boldsymbol{K}$ is a permeability tensor which has nine components. For Cartesian co-ordinates $(x, y, z)$ (1.21) becomes

$$\begin{Bmatrix} u \\ v \\ w \end{Bmatrix} = -\begin{bmatrix} K_{xx} & K_{xy} & K_{xz} \\ K_{yx} & K_{yy} & K_{yz} \\ K_{zx} & K_{zy} & K_{zz} \end{bmatrix} \begin{Bmatrix} \partial\Phi/\partial x \\ \partial\Phi/\partial y \\ \partial\Phi/\partial z \end{Bmatrix} \tag{1.22}$$

in which $(u, v, w)$ are the velocity components in the $(x, y, z)$ directions, respectively, i.e. $\mathbf{q} = (u, v, w)$. Anisotropic media occur frequently where the soil has been deposited in layers. Often the compaction is more severe in one direction, the soil grains have a preferred orientation, and/or there are seams, any of which may create a situation where water flows more easily in one direction than another.

Of the nine components of $\boldsymbol{K}$ there are only six different permeability coefficients since the $\boldsymbol{K}$ matrix is symmetric. That is $K_{xy} = K_{yx}$, $K_{zx} = K_{xz}$, and $K_{yz} = K_{zy}$ since for all common soils the flow in any direction is equal and opposite under an equal and opposite pressure gradient. For such a matrix we can find a Cartesian co-ordinate system in which the off-diagonal elements are zero. The axes of this new co-ordinate system are in the directions of the eigenvectors of the permeability matrix (e.g. Long 1961). We write

$$\boldsymbol{K} \cdot \mathbf{A} = \lambda \mathbf{A} \tag{1.23}$$

in which $\lambda$ is a scalar and $\mathbf{A}$ is an unknown vector. Equation (1.23) is equivalent to the simultaneous scalar equations

$$\begin{aligned} (K_{xx} - \lambda)A_1 + K_{xy}A_2 + K_{xz}A_3 &= 0 \\ K_{yx}A_1 + (K_{yy} - \lambda)A_2 + K_{yz}A_3 &= 0 \\ K_{zx}A_1 + K_{zy}A_2 + (K_{zz} - \lambda)A_3 &= 0 \end{aligned} \tag{1.24}$$

In order that $A_1 = A_2 = A_3 = 0$ be not the only solution, the determinant of the coefficient matrix must be zero, i.e.,

$$\begin{vmatrix} K_{xx} - \lambda & K_{xy} & K_{xz} \\ K_{yx} & K_{yy} - \lambda & K_{yz} \\ K_{zx} & K_{zy} & K_{zz} - \lambda \end{vmatrix} = 0 \tag{1.25}$$

Equation (1.25) forms a cubic equation in the eigenvalues, $\lambda$, of the permeability tensor and so we expect three values for $\lambda$. After solving for these eigenvalues we can find the direction of three eigenvectors, $\mathbf{A}_1$, $\mathbf{A}_2$, $\mathbf{A}_3$, corresponding to each of the three eigenvalues. These three directions form the co-ordinate system in which the permeability tensor is diagonal. (Proof of the previous statement is rather long and since it is not central to our theme the reader is referred to any book on tensor analysis or to Long (1961).) The resultant is

$$K_{i_* j_*} = \sum_{i=1}^{3} \sum_{j=1}^{3} a_{i_* i} a_{j_* j} K_{ij} \qquad i_* = 1, 2, 3; \qquad j_* = 1, 2, 3 \tag{1.26}$$

in which $i_*$ and $j_*$ take on the values $x_*, y_*, z_*$ where the starred co-ordinate system consists of the principal axes of the permeability tensor and $a_{i_* i}$ and $a_{j_* j}$ are the angle cosines between the two co-ordinate systems (e.g. $a_{x_* y} = \cos\theta_{x_* y}$ in which $\theta_{x_* y}$ is angle between the $x_*$ co-ordinate and the $y$ co-ordinate).

Equation (1.22) becomes

$$\begin{Bmatrix} u_* \\ v_* \\ w_* \end{Bmatrix} = -\begin{bmatrix} K_{x_*x_*} & 0 & 0 \\ 0 & K_{y_*y_*} & 0 \\ 0 & 0 & K_{z_*z_*} \end{bmatrix} \begin{Bmatrix} \partial\Phi/\partial x_* \\ \partial\Phi/\partial y_* \\ \partial\Phi/\partial z_* \end{Bmatrix} \tag{1.27}$$

Using (1.6),

$$\frac{\partial}{\partial x_*}\left(K_{x_*x_*}\frac{\partial\Phi}{\partial x_*}\right) + \frac{\partial}{\partial y_*}\left(K_{y_*y_*}\frac{\partial\Phi}{\partial y_*}\right) + \frac{\partial}{\partial z_*}\left(K_{z_*z_*}\frac{\partial\Phi}{\partial z_*}\right) = 0 \tag{1.28}$$

If we assume homogeneity

$$K_{x_*x_*}\frac{\partial^2\Phi}{\partial x_*^2} + K_{y_*y_*}\frac{\partial^2\Phi}{\partial y_*^2} + K_{z_*z_*}\frac{\partial^2\Phi}{\partial z_*^2} = 0 \tag{1.29}$$

which can be reduced to Laplace's equation by the transformation

$$x_{**} = x_* \quad y_{**} = y_*(K_{x_*x_*}/K_{y_*y_*})^{\frac{1}{2}} \quad z_{**} = z_*(K_{x_*x_*}/K_{z_*z_*})^{\frac{1}{2}} \tag{1.30}$$

Then

$$\frac{\partial^2\Phi}{\partial x_{**}^2} + \frac{\partial^2\Phi}{\partial y_{**}^2} + \frac{\partial^2\Phi}{\partial z_{**}^2} = 0 \tag{1.31}$$

Thus the non-isotropic problem can be solved as an isotropic problem by a rotation and stretching of co-ordinates.

## 1.6 Boundary conditions for saturated flow

For most steady state groundwater problems the boundary conditions can be classified into two types: (1) the Dirichlet condition

$$\Phi = \Phi_b \qquad \text{on } \Omega_1 \in \Omega \tag{1.32}$$

where $\Phi_b$ is the prescribed potential along $\Omega_1$ which is a portion of the boundary $\Omega$, and (2) the Neumann condition

$$\frac{\partial\Phi}{\partial n} = -q_b \qquad \text{on } \Omega_2 \in \Omega \tag{1.33}$$

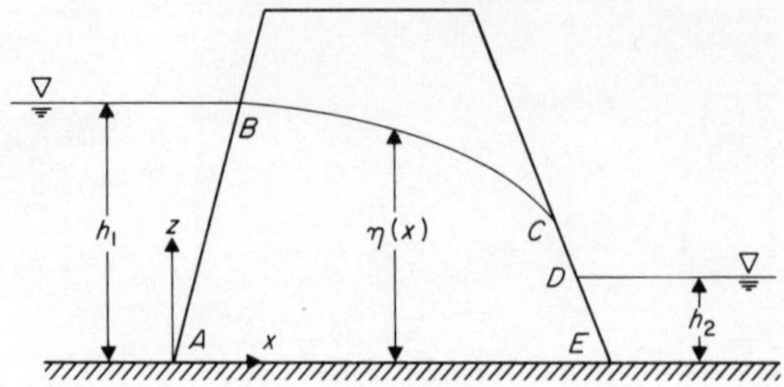


**Figure 1.1** A steady state seepage flow through an earth dam.

where $\partial\Phi/\partial n$ is the normal derivative of the potential (i.e. $\partial\Phi/\partial n = \nabla\Phi \cdot \mathbf{n}$, where $\mathbf{n}$ is the unit outward normal from the boundary). In addition, the free surface boundary conditions are used on unsteady problems in which the flow is unconfined. Seepage flow through an earth dam (Fig. 1.1) is an example in which all the types of boundary conditions must be applied.

### 1.6.1 Fixed boundaries

For the earth dam of Figure 1.1, the potentials on the upstream surface $AB$ and the downstream surface $DE$ are given as

$$\Phi = h_1 \qquad \text{along } AB \tag{1.34a}$$

$$\Phi = h_2 \qquad \text{along } DE \tag{1.34b}$$

Along the seepage surface $CD$, the potential is the same as the vertical elevation,

$$\Phi = z \qquad \text{along } CD \tag{1.34c}$$

since the atmospheric pressure is assumed to be zero. Along the impervious boundary $AE$, the normal flux is zero, thus

$$\frac{\partial\Phi}{\partial n} = 0 \qquad \text{along } AE \tag{1.34d}$$

Finally, along the free surface the normal flux also vanishes because the free surface is a streamline for steady flow. Thus

$$\frac{\partial\Phi}{\partial n} = 0 \qquad \text{along } BC \tag{1.34e}$$

$$\Phi = \eta \qquad \text{along } BC \tag{1.34f}$$

where $z = \eta(x)$ is the free surface elevation. If the earth dam is assumed to be a homogeneous medium, we can use (1.10) as the governing equation of the flow. Equations (1.34) provide the necessary boundary conditions.

### *1.6.2 Unsteady free surface boundary conditions*

For unconfined, saturated groundwater flow problems the free surface elevation may be represented by $z = \eta(x, y, t)$. The free surface (e.g. line $BC$, Fig. 1.1) is taken to be the line of constant atmospheric pressure even though a capillary fringe and/or unsaturated zone may be present above this line. This constant can be chosen as zero without loss of generality. From the definition of the velocity potential, (1.8), the dynamic free surface boundary condition becomes

$$\Phi = \eta \qquad \text{on } z = \eta \tag{1.35}$$

Since the free surface is a material surface (in other words, the free surface always consists of the same fluid particles), the rate of change of the position of the free surface must be equal to the vertical velocity on the surface. Hence

$$\frac{\mathrm{D}}{\mathrm{D}t}(z-\eta) = -\frac{\partial\eta}{\partial t} + \frac{\mathbf{q}}{n}\cdot\nabla(z-\eta) = 0 \qquad \text{on } z = \eta \tag{1.36}$$

in which $\mathbf{q}/n$ represents the velocity of the fluid particles on the free surface. If the soil moisture above the free surface is non-zero, the effective porosity, $n_e$, should be used in (1.36) in place of $n$. Employing Darcy's law we can rearrange (1.36) into the following form:

$$\frac{\partial\eta}{\partial t} = \frac{K}{n}\left[\nabla_2\Phi\cdot\nabla_2\eta - \frac{\partial\Phi}{\partial z}\right] \qquad \text{on } z = \eta \tag{1.37}$$

in which $\nabla_2$ denotes the horizontal gradient operator. The kinematic free surface boundary condition (1.37) is clearly a non-linear condition. Furthermore, both free surface boundary conditions are applied on the free surface, the position of which is unknown *a priori*.

Consider an isotropic and homogeneous medium, $K = \text{constant}$. We can introduce the following dimensionless variables:

$$t^* = tK/nL \qquad (\Phi^*, \eta^*, x^*, y^*, z^*) = (\Phi, \eta, x, y, z)/L \tag{1.38}$$

in which $L$ is a characteristic length scale. All quantities with an asterisk are non-dimensional quantities. However, for brevity all asterisks will be dropped in the rest of this section. Substitution of (1.38) into (1.35) and (1.37) yields the dimensionless free surface boundary conditions

$$\Phi = \eta \tag{1.39a}$$

$$\text{on } z = \eta$$

$$\frac{\partial \eta}{\partial t} = \nabla_2 \Phi \cdot \nabla_2 \eta - \frac{\partial \Phi}{\partial z} \tag{1.39b}$$

The unit normal vector on the free surface can be defined as

$$\mathbf{n} = \frac{\nabla(z - \eta)}{|\nabla(z - \eta)|} \tag{1.40}$$

We can also define the normal derivative of the potential on the free surface as follows:

$$\frac{\partial \Phi}{\partial n} = \mathbf{n} \cdot \nabla \Phi = (|\nabla(z - \eta)|)^{-1} \nabla(z - \eta) \cdot \nabla \Phi$$

$$= (1 + |\nabla_2 \eta|^2)^{-\frac{1}{2}} \left( \frac{\partial \Phi}{\partial z} - \nabla_2 \eta \cdot \nabla_2 \Phi \right) \tag{1.41}$$

in which (1.40) has been used. Substituting (1.41) into the free surface boundary conditions (1.39), we obtain

$$\Phi = \eta \tag{1.42a}$$

$$\text{on } z = \eta$$

$$\frac{\partial \eta}{\partial t} = -(1 + |\nabla_2 \eta|^2)^{\frac{1}{2}} \frac{\partial \Phi}{\partial n} \tag{1.42b}$$

In the case that the free surface movement is affected by vertical recharge, the dynamic free surface boundary condition (1.42b) should be modified to

$$\frac{\partial \eta}{\partial t} = -(1 + |\nabla_2 \eta|^2)^{\frac{1}{2}} \frac{\partial \Phi}{\partial n} + W \qquad \text{on } z = \eta \tag{1.43}$$

where $W$ represents the dimensionless (vertical) recharge intensity per unit area and is, in general, a function of the horizontal co-ordinates and time. The recharge intensity, $W$, which is always positive, causes the free surface elevation to rise.

## 1.7 Two-dimensional (Dupuit) flow in confined and unconfined aquifers

In dealing with large scale (regional) groundwater flow problems, where the horizontal length scale is much greater than the thickness of the aquifer, one can simplify the governing equations by assuming the vertical flow is small compared to the horizontal flow. The result is a reduction in the dimensionality of the equations, thus reducing the computational effort for a solution.

### *1.7.1 Confined aquifers*

In the case of a confined aquifer bounded by upper and lower confining strata, $z = h_t$ and $z = h_b$, respectively (Fig. 1.2), the continuity equation

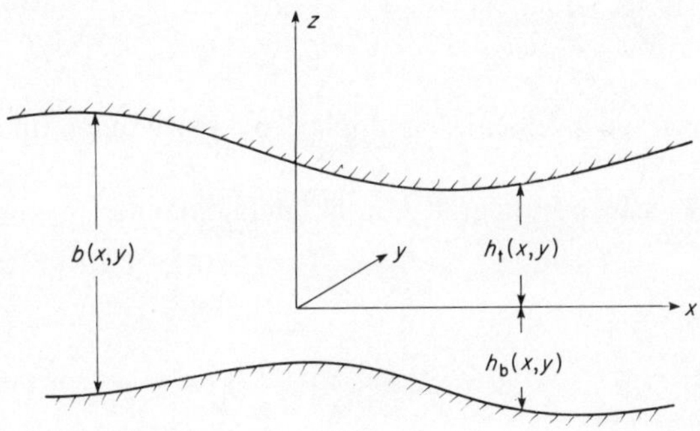


**Figure 1.2** Flow in a confined aquifer.

(1.6) can be integrated vertically to give

$$\int_{h_b}^{h_t} \frac{\partial u}{\partial x}\, dz + \int_{h_b}^{h_t} \frac{\partial v}{\partial y}\, dz + \int_{h_b}^{h_t} \frac{\partial w}{\partial z}\, dz = 0 \tag{1.44}$$

Making use of Leibniz' rule, (1.44) is written as

$$\begin{aligned} &\frac{\partial}{\partial x}\int_{h_b}^{h_t} u\, dz - u(x, y, h_t)\frac{\partial h_t}{\partial x} + u(x, y, h_b)\frac{\partial h_b}{\partial x} \\ &+ \frac{\partial}{\partial y}\int_{h_b}^{h_t} v\, dz - v(x, y, h_t)\frac{\partial h_t}{\partial y} + v(x, y, h_b)\frac{\partial h_b}{\partial y} \\ &+ w(x, y, h_t) - w(x, y, h_b) = 0 \end{aligned} \tag{1.45}$$

in which $u(x, y, h_t)$ indicates that $u$ is to be evaluated at the upper confining surface. The condition that there is no flow across the upper confining surface is

$$u(x, y, h_t)\frac{\partial h_t}{\partial x} + v(x, y, h_t)\frac{\partial h_t}{\partial y} = w(x, y, h_t) \tag{1.46}$$

with a similar condition for the lower confining surface. Thus equation (1.45) simplifies without approximation to

$$\frac{\partial \bar{u}}{\partial x} + \frac{\partial \bar{v}}{\partial y} = 0 \tag{1.47}$$

where

$$\bar{u} = \int_{h_b}^{h_t} u \, dz \qquad \bar{v} = \int_{h_b}^{h_t} v \, dz \tag{1.48}$$

are the vertically averaged velocities multiplied by the aquifer thickness, $b(x, y) = h_t - h_b$.

Darcy's law, (1.7), can be integrated in a similar manner to give

$$\int_{h_b}^{h_t} u \, dz = -K \int_{h_b}^{h_t} \frac{\partial \Phi}{\partial x} dz \tag{1.49a}$$

$$\int_{h_b}^{h_t} v \, dz = -K \int_{h_b}^{h_t} \frac{\partial \Phi}{\partial y} dz \tag{1.49b}$$

We now make the approximation that the horizontal potential gradient is not a function of the vertical co-ordinate,

$$\bar{u} = -K(h_t - h_b)\frac{\partial \Phi}{\partial x} = -Kb\frac{\partial \Phi}{\partial x} \tag{1.50a}$$

$$\bar{v} = -K(h_t - h_b)\frac{\partial \Phi}{\partial y} = -Kb\frac{\partial \Phi}{\partial y} \tag{1.50b}$$

The approximation in (1.50) can be seen clearly by comparing (1.50a) to a more exact integration of (1.49a):

$$\bar{u} = -K\left[\frac{\partial}{\partial x}\int_{h_b}^{h_t} \Phi \, dz - \Phi(x, y, h_t)\frac{\partial h_t}{\partial x} + \Phi(x, y, h_b)\frac{\partial h_b}{\partial x}\right] \tag{1.51}$$

Thus the approximation is that $\Phi$ does not change in the vertical and that the aquifer thickness does not vary sharply.

The product of the permeability and aquifer thickness is called the transmissivity,

$$T = Kb \tag{1.52}$$

Using (1.50) and (1.47) gives

$$\frac{\partial}{\partial x}\left(T\frac{\partial \Phi}{\partial x}\right) + \frac{\partial}{\partial y}\left(T\frac{\partial \Phi}{\partial y}\right) = 0 \tag{1.53}$$

which is analogous to (1.9). The assumption of constant transmissivity produces Laplace's equation. On the other hand the transmissivity may vary directionally, in which case a procedure similar to that given in Section 1.5 is useful.

### 1.7.2 *Unconfined aquifers*

For unconfined aquifers it is convenient to take the $x$–$y$ co-ordinate system in the aquifer basement (Fig. 1.3). Although this system is not Cartesian, we assume that the vertical variation in the lower confining aquifer is small in agreement with the assumptions made above. Also the approximation that $\Phi$ does not change in the vertical means that we can take $\Phi$ as equivalent to the depth of fluid

$$\Phi(x, y) = h(x, y) \tag{1.54}$$

From (1.52) and (1.53)

$$\frac{\partial}{\partial x}\left(K\Phi\frac{\partial \Phi}{\partial x}\right) + \frac{\partial}{\partial y}\left(K\Phi\frac{\partial \Phi}{\partial y}\right) = 0 \tag{1.55}$$

If $K$ is constant with $x$ and $y$,

$$\frac{\partial^2 \Phi^2}{\partial x^2} + \frac{\partial^2 \Phi^2}{\partial y^2} = 0 \tag{1.56}$$

Thus the governing equation is still Laplace's equation but with the square of the potential instead of the potential itself.

Boundary conditions for (1.56) are those of Section 1.6, except that the free surface conditions do not apply as we have suppressed the vertical co-ordinate.

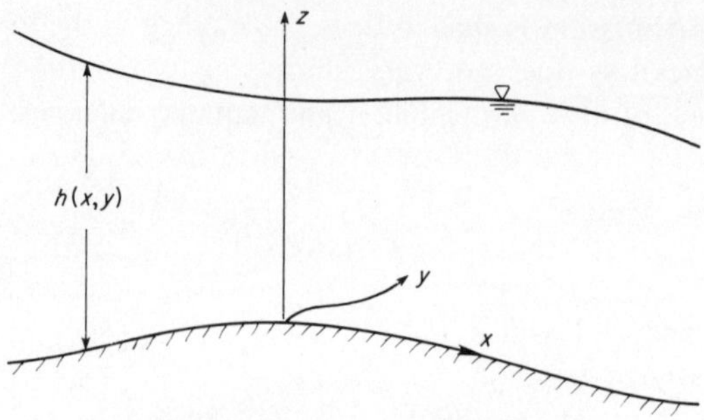


**Figure 1.3** Flow in an unconfined aquifer.

### 1.7.3 *Leaky aquifers*

In aquifers where the approximations of the last section are otherwise valid, it often occurs that the confining layers (aquitards) are not entirely impermeable. Instead they 'leak' or allow some transmission of the fluid into or out of the aquifer. We assume that the leak rate is proportional to the potential (head) difference across the confining layer (Fig. 1.4):

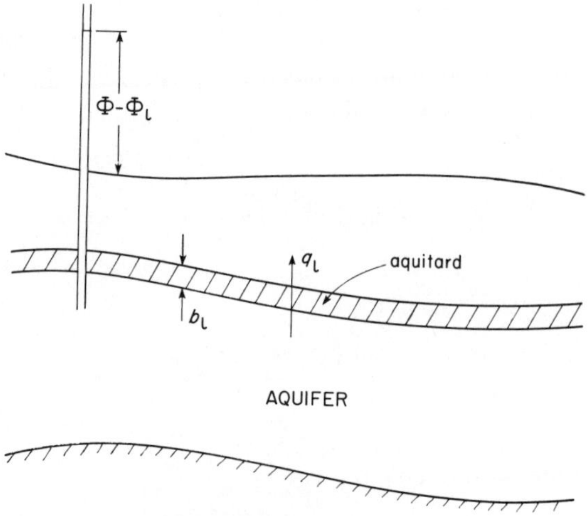


**Figure 1.4** A leaky aquifer.

$$q_l = \frac{K_l}{b_l}(\Phi - \Phi_l) \tag{1.57}$$

in which $K_l$ is the permeability of the aquitard, $\Phi_l$ is the head outside of the confining layer, and $b_l$ is the thickness of the confining layer. The equation of continuity, (1.47), becomes

$$\frac{\partial \bar{u}}{\partial x} + \frac{\partial \bar{v}}{\partial y} = \frac{K_1}{b_1}(\Phi_1 - \Phi) \tag{1.58}$$

and (1.53) becomes

$$\frac{\partial}{\partial x}\left(T\frac{\partial \Phi}{\partial x}\right) + \frac{\partial}{\partial y}\left(T\frac{\partial \Phi}{\partial y}\right) + \frac{K_1}{b_1}(\Phi_1 - \Phi) = 0 \tag{1.59}$$

We now define

$$\tilde{\Phi} = \Phi - \Phi_1 \tag{1.60}$$

and assume that $\Phi_1$ and $T$ are constant, giving

$$\frac{\partial^2 \tilde{\Phi}}{\partial x^2} + \frac{\partial^2 \tilde{\Phi}}{\partial y^2} - \frac{K_1}{b_1 T}\tilde{\Phi} = 0 \tag{1.61}$$

Sources/sinks representing wells can be added in the same way as in Section 1.3.

## 1.8 Unsaturated flow

When air fills parts of the pores in a porous medium, the flow can be very complex. Such flows have long challenged soil scientists and engineers to explain the physics adequately for accurate calculations to be made. The surface tension of water causes the pore pressure–moisture content curve to be hysteretic; it has different values for wetting than for draining. We confine the discussion herein to a very elementary level in order to arrive at the equations to be used in Chapter 9. Bear (1972) has a more complete discussion.

Neglecting the compressibility of the fluid and of the granular medium, (1.1) becomes

$$n\rho\frac{\partial S}{\partial t} + \nabla \cdot (\rho \mathbf{q}) = 0 \tag{1.62}$$

In order to use as much of the saturated theory as possible, we continue to use Darcy's law (1.7) but written as

$$\mathbf{q} = -K(C)\,\nabla\Phi \tag{1.63}$$

in which the permeability is written as a function of the moisture content $C$. Since the medium contains air and the flow is confined to the space occupied by the liquid, the available flow area is dependent on the degree of saturation. This area is not, however, proportional to the degree of saturation and it may be different in different media for the same degree of saturation. Since the permeability already reflects the available flow area for a given medium, it is natural to allow this quantity to vary with the water content also.

Taking the density of the fluid as a constant, (1.62) and (1.63) become

$$\frac{\partial C}{\partial t} - \mathbf{\nabla} \cdot (K\, \mathbf{\nabla}\Phi) = 0 \tag{1.64}$$

in which $C = nS$ and the medium has been considered to be nondeformable ($n$ = constant). In order to solve (1.64), $C$ needs to be expressed in terms of $\Phi$. Usually $C$ is taken as a function of the capillary pressure head,

$$\begin{aligned} h_c &= \Phi - z \qquad \text{for } \Phi \leqslant z \\ h_c &= 0 \qquad \text{for } \Phi \geqslant z \end{aligned} \tag{1.65}$$

The moisture content function, $C(h_c)$, is usually an empirical relationship which is specific for a given soil.

The solution of (1.64) also depends on finding a relationship between $K$ and $\Phi$. As in the case of $C$, it is usually empirical and specific for a given case. It is also expressed in terms of the capillary pressure head and should go to the saturated permeability as $h_c \rightarrow 0$. Examples are found in Chapter 9.

# 2 Development of the Boundary Integral Equation Method

The Boundary Integral Equation Method (BIEM) is formulated in this chapter for both two-dimensional and three-dimensional problems. The fluid flow is governed by the Laplace equation. Similar formulations for other types of equations which appear in groundwater problems are in Chapter 10.

For two-dimensional problems we show in this chapter how to discretize the boundaries by linear segments and how to perform the integrations by using linear interpolation functions for $\Phi$ and $\partial\Phi/\partial n$ on each segment. A simple numerical example is given so that the reader may easily confirm the general formulation. The basic BIEM is shown in this chapter for three-dimensional problems, but the detailed development is in Chapter 7.

## 2.1 Integral equations in two- and three-dimensional forms

An expression of continuity in a volume is the divergence theorem (Lass 1950, Franklin 1944)

$$\int_D (\nabla \cdot \mathbf{v})\, \mathrm{d}\tau = \int_\Omega \mathbf{v} \cdot \mathbf{n}\, \mathrm{d}A \tag{2.1}$$

in which $\nabla$ is the vector operator, $\mathbf{v}$ is any differentiable vector, $D$ is the domain of integration (a volume in three dimensions, an area in two dimensions), $\Omega$ is the boundary of $D$, and $\mathbf{n}$ is the unit outward normal to $D$ on $\Omega$. First, $\mathbf{v}$ is defined as $U\,\nabla V$, where $U$ and $V$ are any two functions, twice differentiable, in $D$. Thus

$$\int_D (\nabla U \cdot \nabla V + U\,\nabla^2 V)\, \mathrm{d}\tau = \int_\Omega U\,\nabla V \cdot \mathbf{n}\, \mathrm{d}A \tag{2.2}$$

Secondly, $\mathbf{v}$ is defined as $V\,\nabla U$, so that

$$\int_D (\nabla V \cdot \nabla U + V\,\nabla^2 U)\, \mathrm{d}\tau = \int_\Omega V\,\nabla U \cdot \mathbf{n}\, \mathrm{d}A \tag{2.3}$$

Subtracting (2.3) from (2.2) yields

$$\int_D (U\nabla^2 V - V\nabla^2 U)\,\mathrm{d}\tau = \int_\Omega (U\nabla V - V\nabla U)\cdot\mathbf{n}\,\mathrm{d}A \qquad (2.4)$$

Equation (2.4) is Green's second identity. It forms the basis for much of BIEM contained in this book. In most of what follows we adopt a shorthand notation for the normal component of a gradient with respect to the boundary

$$\nabla V\cdot\mathbf{n} = \frac{\partial V}{\partial n} \qquad \nabla U\cdot\mathbf{n} = \frac{\partial U}{\partial n} \qquad (2.5)$$

Thus Green's second identity is written

$$\int_D (U\nabla^2 V - V\nabla^2 U)\,\mathrm{d}\tau = \int_\Omega \left(U\frac{\partial V}{\partial n} - V\frac{\partial U}{\partial n}\right)\mathrm{d}A \qquad (2.6)$$

If $U$ and $V$ are both chosen such that they satisfy Laplace's equation, $\nabla^2 U = \nabla^2 V = 0$, then

$$\int_\Omega \left(U\frac{\partial V}{\partial n} - V\frac{\partial U}{\partial n}\right)\mathrm{d}A = 0 \qquad (2.7)$$

The present application makes direct use of (2.7) in which $U$ is chosen as a velocity potential, $\Phi$, and $V$ is chosen as a 'free space Green function' which satisfies

$$\nabla^2 V = 0 \qquad (2.8)$$

everywhere in $D$ except that it goes to infinity at a 'singular point', $P(\mathbf{x}_i)$. The free space Green function, $V$, has different forms for two-dimensional and three-dimensional problems. The resulting integral equations for these two classes of problems are discussed in the following sections separately.

### 2.1.1 Two-dimensional problems

The free space Green function satisfying (2.8) is (Greenberg 1971, see Appendix A.3 in this volume)

$$V = \ln r \qquad (2.9)$$

in which $r$ denotes the distance between the singular point $P$ (where $r = 0$)

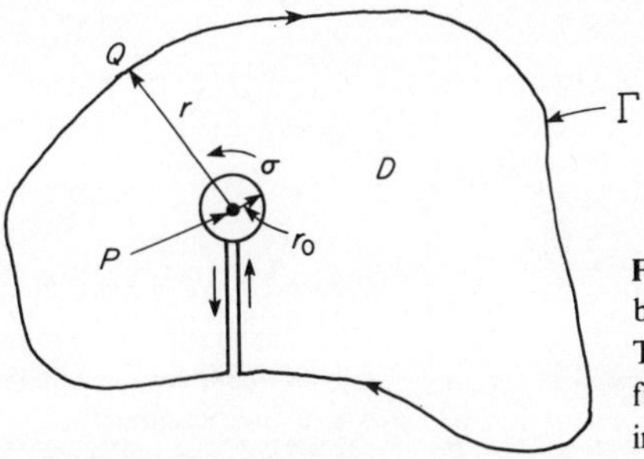


**Figure 2.1** The domain $D$ surrounded by the boundary curve $\Gamma$. The singular point $P$ is separated from $D$ by the circle $\sigma$. The arrows indicate the direction of integration.

and another point $Q$ on the boundary of the two-dimensional domain. In order to apply (2.7) we need to exclude the point $P$ by a small circle of radius $r_0$ as shown in Figure 2.1. Thus the integral (2.7) becomes

$$\int_\Gamma \left[\Phi \frac{\partial}{\partial n}(\ln r) - \ln r \frac{\partial \Phi}{\partial n}\right] ds + \lim_{r_0 \to 0} \int_\sigma \left[\Phi \frac{\partial}{\partial n}(\ln r) - \ln r \frac{\partial \Phi}{\partial n}\right] ds = 0 \tag{2.10}$$

in which $\Gamma$ is the bounding contour of $D$ and $\sigma$ is the circle of radius $r_0$ around $P$. The portions of the integral along the two lines connecting the circle to $\Gamma$ will cancel and thus they do not appear in (2.10).

On the circle $\sigma$ the outward normal from $D$ points inward toward $P$. Thus

$$\frac{\partial}{\partial n}(\ln r) = \frac{1}{r}\frac{\partial r}{\partial n} = \frac{1}{r}(\nabla r \cdot \mathbf{n}) = -\frac{1}{r} \tag{2.11}$$

and the integral around the circle becomes

$$\lim_{r_0 \to 0} \int_0^{2\pi} \left(-\frac{\Phi}{r_0} + \ln r_0 \frac{\partial \Phi}{\partial r}\right) r_0 \, d\theta = -2\pi\Phi(P) \tag{2.12}$$

The potential at any point $P$ is defined in terms of a boundary integral

$$2\pi\Phi(P) = \int_\Gamma \left[\Phi(Q)\frac{\partial}{\partial n}(\ln r) - \ln r \frac{\partial}{\partial n}\Phi(Q)\right] ds \tag{2.13}$$

If we know $\Phi$ and $\partial\Phi/\partial n$ everywhere on $\Gamma$, then we can find the solution for $\Phi$ at any interior point by a line integration. In a well posed problem, however, both $\Phi$ and $\partial\Phi/\partial n$ are not known on $\Gamma$; instead, either $\Phi$ or $\partial\Phi/\partial n$ or a relation between them is known at all points of the boundary. The integral equation can be used to find the 'missing data'.

To complete the boundary data the point $P$ can be moved to the

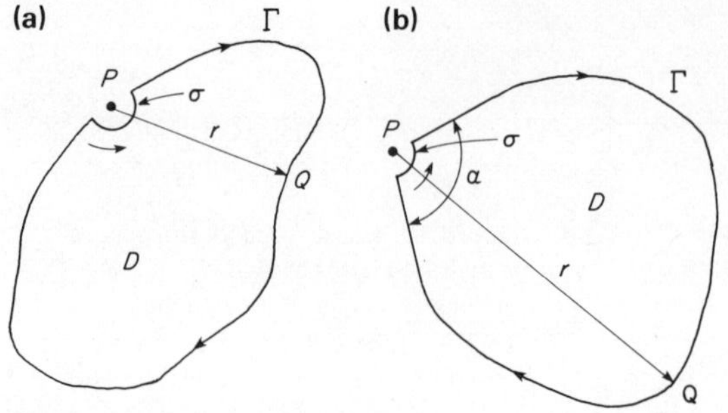


**Figure 2.2** The point $P$ on the boundary (a) at a smooth part of the boundary and (b) where the boundary contour forms an angle.

boundary as shown in Figure 2.2. It is still excluded from $D$ by a circular arc. The same considerations hold as before except that the integration indicated in (2.12) takes place over an angle which is less than $2\pi$. For the case in which $P$ is on a smooth part of the boundary (Fig. 2.2a)

$$\pi\Phi(P) = \int_{\Gamma} \left( \frac{\Phi}{r}\frac{\partial r}{\partial n} - \ln r \frac{\partial \Phi}{\partial n} \right) \mathrm{d}s \tag{2.14}$$

and in the general case (Fig. 2.2b)

$$\alpha\Phi(P) = \int_{\Gamma} \left( \frac{\Phi}{r}\frac{\partial r}{\partial n} - \ln r \frac{\partial \Phi}{\partial n} \right) \mathrm{d}s \tag{2.15}$$

in which $\alpha$ is the angle between the boundary segments at $P$. Equation (2.15) can be discretized and solved numerically to complete the boundary data, as is shown subsequently.

An alternative, but equivalent, point of view comes about from writing (2.7) as simply

$$\int_{\Gamma} \left( \frac{\Phi}{r}\frac{\partial r}{\partial n} - \ln r \frac{\partial \Phi}{\partial n} \right) \mathrm{d}s = 0 \tag{2.16}$$

in which $r$ is the distance between a fixed 'base point' on the boundary and a variable 'field point' which is also on the boundary. Since we can choose the base point anywhere on the boundary, (2.16) can provide a multiplicity of equations which relate $\Phi$ and $\partial\Phi/\partial n$; that is, there is one equation for each choice of the base point. By defining $\Phi$ and $\partial\Phi/\partial n$ at a number of 'nodes' (Fig. 2.3) we can make each node, in turn, a base point which will yield $N$ equations (2.16), one equation for each choice of the base point. Thus there are $N$ equations which relate $\Phi$ and $\partial\Phi/\partial n$ at each node. Since we already know either $\Phi$ or $\partial\Phi/\partial n$ at each node from the given boundary conditions, these are sufficient equations to solve for the unknown boundary values at the nodes.

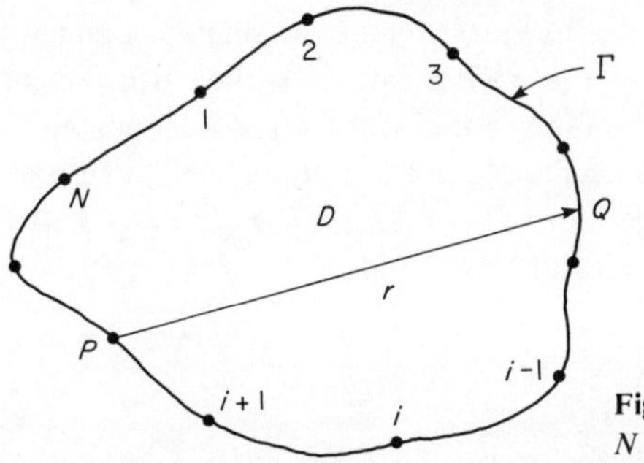


**Figure 2.3** The boundary with $N$ node points defined.

In integrating (2.16) there is one point of difficulty. For any chosen base point, as the field point moves around the contour Γ it must at one stage coincide with the base point, leading to $r = 0$. The second term of (2.16) gives no problem as ln $r$ is an integrable singularity. For the first term we define a circular segment around the base point as in Figure 2.2. Then

$$\int_{\sigma} \frac{\Phi}{r} \frac{\partial r}{\partial n} \, \mathrm{d}s = \int_{0}^{\alpha} \frac{\Phi_i}{r_0} (-1) r_0 \, \mathrm{d}\theta = -\alpha \Phi_i \tag{2.17}$$

which is the result of (2.15). If the boundary is smooth at the base point, $\alpha = \pi$.

In all the foregoing development no approximations have been made. Laplace's equation is satisfied exactly everywhere in the domain $D$. (Note, however, that sources and sinks are not allowed in $D$ at this point; such a case is treated in Sec. 3.6.) The approximations will come by defining $\Phi$ and $\partial\Phi/\partial n$ between nodes. Even then all the approximations will be confined to the boundary; the solution in the interior is exact in terms of the approximate boundary conditions.

### 2.1.2 *Three-dimensional problems*

For three-dimensional problems, the free space Green function $V$ satisfying (2.8) is (Greenberg 1971, see Appendix A.3 in this volume)

$$V = 1/R \tag{2.18}$$

where $R$ is the distance between the singular point $P(\mathbf{x}_i)$ and another point $Q(\mathbf{x})$ in space. We can substitute (2.18) into (2.7) and require that the point $Q(\mathbf{x})$ is on the boundary $\Omega$ and that the singular point $P$ is either inside the domain $D$ or on the boundary $\Omega$. Thus

$$\iint_{\Omega+\sigma} \left[ \Phi \frac{\partial}{\partial n} \left( \frac{1}{R} \right) - \frac{1}{R} \frac{\partial \Phi}{\partial n} \right] \mathrm{d}A = 0 \tag{2.19}$$

where σ denotes the small surface used to exclude the singular point $P$ from the domain $D$. If the point $P$ is an interior point, the surface σ is the surface of a sphere of radius $R_0$. On the other hand, if the point $P$ is located on the boundary Ω, the surface σ represents the portion of the surface of a sphere which is interior to $D$ (Fig. 2.4). For either case, (2.19) can be rewritten as

$$\iint_{\Omega}\left[\Phi\frac{\partial}{\partial n}\left(\frac{1}{R}\right)-\frac{1}{R}\frac{\partial\Phi}{\partial n}\right]\mathrm{d}A$$

$$+\lim_{R_0\to 0}\iint_{\sigma}\left[\Phi\left(-\frac{1}{R_0^2}\right)\frac{\partial R}{\partial n}-\frac{1}{R_0}\frac{\partial\Phi}{\partial n}\right]\mathrm{d}A=0 \quad (2.20)$$

The unit normal $\mathbf{n}$ is defined as an outward normal from $D$, thus $(\partial R/\partial n)_{R=R_0}=-1$. Furthermore, the differential surface area of a sphere is proportional to the square of the radius. Therefore, in the limit as $R_0\to 0$, the contribution from the integral in the term $\partial\Phi/\partial n$ vanishes. Thus, (2.20) may be reduced to

$$-\alpha\Phi(P)=\iint_{\Omega}\left[\Phi\frac{\partial}{\partial n}\left(\frac{1}{R}\right)-\frac{1}{R}\frac{\partial\Phi}{\partial n}\right]\mathrm{d}A \quad (2.21)$$

with

$$\alpha=\lim_{R_0\to 0}A_s/R_0^2 \quad (2.22)$$

where $\Phi(P)$ is the potential at the singular point $P$ and $A_s$ is the surface area of the portion of the sphere which lies inside the domain $D$. Equation (2.22) is the definition of a solid angle. If the singular point $P$ is an interior point, the area $A_s=4\pi R_0^2$ and (2.21) becomes

$$-4\pi\Phi(P)=\iint_{\Omega}\left[\Phi\frac{\partial}{\partial n}\left(\frac{1}{R}\right)-\frac{1}{R}\frac{\partial\Phi}{\partial n}\right]\mathrm{d}A \quad (2.23)$$

Therefore, when both $\Phi$ and $\partial\Phi/\partial n$ are known everywhere on the boundary, the preceding equation gives the potential at the point $P$.

As in the two-dimensional case $\Phi$ and $\partial\Phi/\partial n$ are not both known everywhere on the boundary, but instead either $\Phi$ or $\partial\Phi/\partial n$ or a relation between them is known on any given portion of the boundary. Equation (2.21) can be used as an integral equation for the unknown $\Phi$ or $\partial\Phi/\partial n$ along the boundary, provided that the point $P$ is located on the boundary. If the boundary Ω is smooth at the point $P$, the surface area in (2.22) becomes $2\pi R_0^2$ and the solid angle α is $2\pi$.

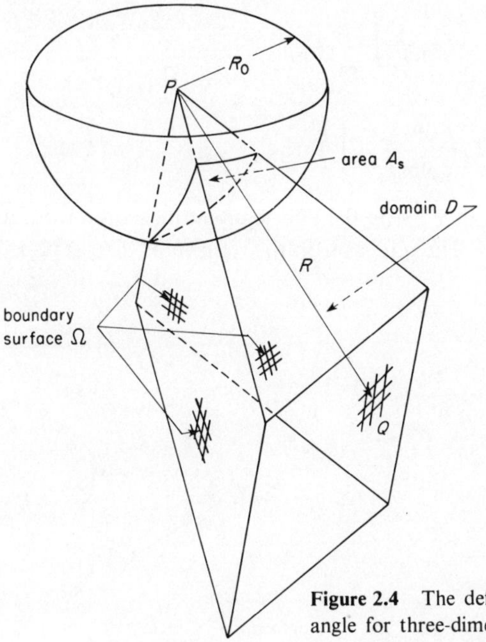


**Figure 2.4** The definition of the solid angle for three-dimensional problems.

## 2.2 Discretization of the two-dimensional boundary into linear elements

Equation (2.15) or (2.16) can be solved by choosing a finite number of points on the boundary and numerically performing the integration. In this section, only the two-dimensional problems are discussed. The axisymmetric and three-dimensional problems are presented in Chapters 6 and 7 respectively.

The numerical procedure involves the selection of a finite number of points, $P_j$ ($j = 1, 2, 3, \ldots, N$), on the boundary $\Gamma$. The contour integration is performed using the points $P_i$ ($i = 1, 2, 3, \ldots, N$) successively as origin for the $\ln r$ term. In other words, the points $P_i$ become sources ($\ln r$) and dipoles ($\partial(\ln r)/\partial n$), which are properly weighted by $\partial\Phi/\partial n$ and $\Phi$, respectively, so as to satisfy (2.15). For simplicity linear elements and linear interpolation functions are used in this chapter. Various orders of curved segments and higher order interpolation can be used and are discussed in a later chapter.

Over a small element $\Gamma_j$ between a pair of node points (Fig. 2.5), $P_j$ and $P_{j+1}$, the potential and its normal derivatives are approximated as

$$\Phi = [(\Phi_{j+1} - \Phi_j)\xi + (\xi_{j+1}\Phi_j - \xi_j\Phi_{j+1})]/(\xi_{j+1} - \xi_j) \qquad \xi_j \leqslant \xi \leqslant \xi_{j+1} \tag{2.24}$$

$$\frac{\partial\Phi}{\partial n}=\left\{\left[\left(\frac{\partial\Phi}{\partial n}\right)_{j+1}-\left(\frac{\partial\Phi}{\partial n}\right)_{j}\right]\xi\right.$$

$$\left.+\left[\xi_{j+1}\left(\frac{\partial\Phi}{\partial n}\right)_{j}-\xi_{j}\left(\frac{\partial\Phi}{\partial n}\right)_{j+1}\right]\right\}/(\xi_{j+1}-\xi_{j})\qquad \xi_j\leqslant\xi\leqslant\xi_{j+1}\quad (2.25)$$

where $\xi$ represents distance along the line segment measured from an origin as defined in Figure 2.5. The integrals on the right-hand side of (2.15) can be

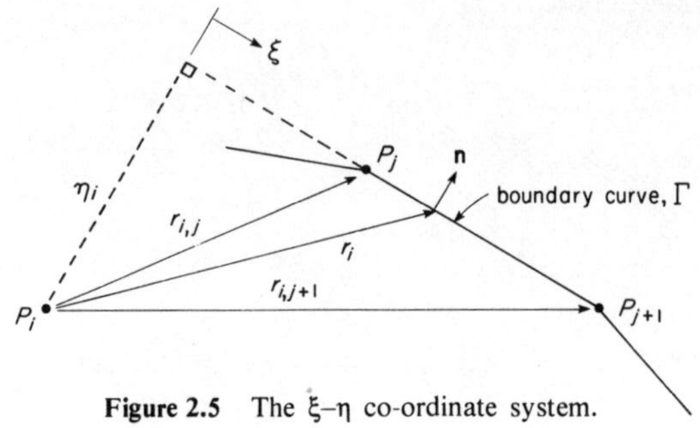


**Figure 2.5** The $\xi$–$\eta$ co-ordinate system.

integrated over the segment between $P_j$ and $P_{j+1}$ with reference to the singular point $P_i$ and they become

$$I_e=\int_{\xi_j}^{\xi_{j+1}}\left(\frac{\Phi}{r_i}\frac{\partial r_i}{\partial n}-\frac{\partial\Phi}{\partial n}\ln r_i\right)\mathrm{d}\xi$$

$$=\lfloor K_1^e\rfloor\left\{\begin{matrix}\Phi_j\\ \Phi_{j+1}\end{matrix}\right\}-\lfloor K_2^e\rfloor\left\{\begin{matrix}(\partial\Phi/\partial n)_j\\ (\partial\Phi/\partial n)_{j+1}\end{matrix}\right\}\qquad (2.26)$$

with (see Appendix A.2 for an explanation of the matrix notation)

$$\lfloor K_1^e\rfloor=\lfloor (K_1^e)_{i,j}\quad (K_1^e)_{i,j+1}\rfloor=\lfloor -I_{11}+\xi_{j+1}I_{12}\quad I_{11}-\xi_jI_{12}\rfloor$$

$$\lfloor K_2^e\rfloor=\lfloor (K_2^e)_{i,j}\quad (K_2^e)_{i,j+1}\rfloor=\lfloor -I_{21}+\xi_{j+1}I_{22}\quad I_{21}-\xi_jI_{22}\rfloor\qquad (2.27)$$

in which (see Appendix A.4 for additional detail)

$$I_{11}=\frac{1}{(\xi_{j+1}-\xi_j)}\int_{\xi_j}^{\xi_{j+1}}\frac{1}{r_i}\frac{\partial r_i}{\partial n}\,\xi\,\mathrm{d}\xi=\tfrac{1}{2}\eta_i\ln\left[\frac{\eta_i^2+\xi_{j+1}^2}{\eta_i^2+\xi_j^2}\right]/(\xi_{j+1}-\xi_j)\qquad (2.28)$$

$$I_{12} = \frac{1}{(\xi_{j+1} - \xi_j)} \int_{\xi_j}^{\xi_{j+1}} \frac{1}{r_i} \frac{\partial r_i}{\partial n} \, d\xi$$

$$= \left[ \tan^{-1}\left(\frac{\xi_{j+1}}{\eta_i}\right) - \tan^{-1}\left(\frac{\xi_j}{\eta_i}\right) \right] / (\xi_{j+1} - \xi_j) \quad (2.29)$$

$$I_{21} = \frac{1}{(\xi_{j+1} - \xi_j)} \int_{\xi_j}^{\xi_{j+1}} \ln r_i \xi \, d\xi = \tfrac{1}{4} \{(\eta_i^2 + \xi_{j+1}^2)[\ln(\eta_i^2 + \xi_{j+1}^2) - 1] -$$

$$- (\eta_i^2 + \xi_j^2)[\ln(\eta_i^2 + \xi_j^2) - 1]\} / (\xi_{j+1} - \xi_j) \quad (2.30)$$

$$I_{22} = \frac{1}{(\xi_{j+1} - \xi_j)} \int_{\xi_j}^{\xi_{j+1}} \ln r_i \, d\xi = \tfrac{1}{2} \{\xi_{j+1} \ln(\eta_i^2 + \xi_{j+1}^2) - \xi_j \ln(\eta_i^2 + \xi_j^2) -$$

$$- 2(\xi_{j+1} - \xi_j) + 2\eta_i [\tan^{-1}\left(\frac{\xi_{j+1}}{\eta_i}\right) - \tan^{-1}\left(\frac{\xi_j}{\eta_i}\right)]\} / (\xi_{j+1} - \xi_j) \quad (2.31)$$

The local $\xi$–$\eta$ co-ordinate system, as defined in Figure 2.5, has been used in the integrations shown above. Notice that

$$r_i = (\eta_i^2 + \xi^2)^{\frac{1}{2}} \quad (2.32)$$

and

$$\frac{\partial r_i}{\partial n} = \frac{\partial r_i}{\partial \eta_i} = \frac{\eta_i}{r_i} \quad (2.33)$$

have also been employed in the integration.

A system of algebraic equations for $\Phi_j$ and $(\partial\Phi/\partial n)_j$ can be obtained by summing the contributions from all boundary segments and by using the $P_i$ $(i = 1, 2, \ldots, N)$ successively. Thus

$$\sum_{j=1}^{N} R_{i,j} \Phi_j = \sum_{j=1}^{N} L_{i,j} \left(\frac{\partial \Phi}{\partial n}\right)_j \qquad i = 1, 2, 3, \ldots, N \quad (2.34)$$

where

$$R_{i,j} = [(K_1^e)_{i,j} - \delta_{i,j}\alpha_i] \quad \delta_{i,j} = \begin{cases} 0 & \text{if } i \neq j \\ 1 & \text{if } i = j \end{cases} \quad (2.35)$$

$$L_{i,j} = (K_2^e)_{i,j} \tag{2.36}$$

The summations in (2.34) range over all segments on the boundary. The numerical coefficients $R_{i,j}$ and $L_{i,j}$ depend only on the geometry of the boundary. Once the boundary conditions are used in (2.34), we can solve the system of linear algebraic equations for the unknown $\Phi$ or $\partial\Phi/\partial n$ on the boundary. The potential function in the interior of the region can be obtained by performing the integration along the boundary indicated by (2.13).

## 2.3 A numerical example – flow through a rectangular region

To demonstrate the numerical scheme developed in the previous section, we consider a potential flow in a rectangular region as shown in Figure 2.6. The region is two units long and one unit wide. The upper and the lower boundaries are solid walls such that the normal flux is zero, i.e.

$$\frac{\partial\Phi}{\partial n} = 0 \quad \text{on} \quad \begin{array}{ll} y = 0 & \text{for } 0 < x < 2 \\ y = 1 & \text{for } 0 < x < 2 \end{array} \tag{2.37}$$

Along the vertical boundaries the potentials are specified as follows:

$$\begin{array}{lll} \Phi = -1 & \text{for } x = 0 & \text{and } 0 \leqslant y \leqslant 1 \\ \Phi = 1 & \text{for } x = 2 & \text{and } 0 \leqslant y \leqslant 1 \end{array} \tag{2.38}$$

The analytical solution of this simple problem is

$$\Phi = x \tag{2.39}$$

Since the boundary consists of four linear segments, and the boundary conditions also vary linearly, a total four nodes, one at each corner, are sufficient for the discretization of the boundary and for the numerical integrations. The co-ordinates of the nodal points and the associated boundary conditions are shown in Figure 2.6.

The singular point $P$ is located at each nodal point successively. For each new location of $P$ the integral equation (2.15) is evaluated once in terms of the $\Phi$ and $\partial\Phi/\partial n$ at the nodal points. Four linear algebraic equations are obtained.

Consider the case that the singular point $P$ is located at node 1. The interior angle is 90°, $\alpha = \pi/2$, and the integral equation (2.15) can be rewritten as follows:

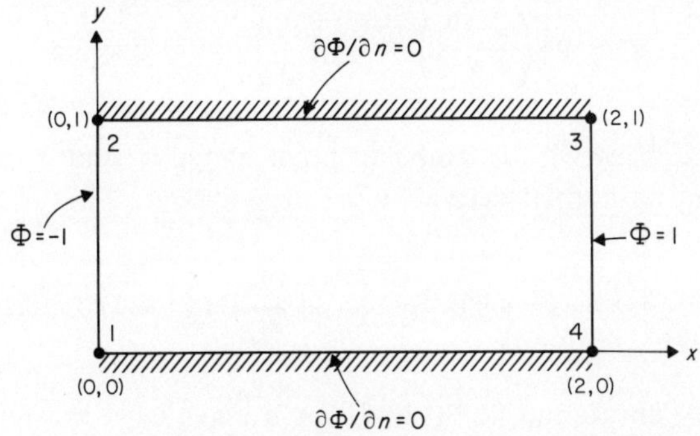


**Figure 2.6** Geometry and boundary conditions of the numerical example with the four-point grid system.

$$\frac{\pi}{2}\Phi_1 = \int_\Gamma \left[\Phi \frac{1}{r}\frac{\partial r}{\partial n} - \ln r \frac{\partial \Phi}{\partial n}\right] \mathrm{d}s, \qquad r = (x^2 + y^2)^{\frac{1}{2}} \tag{2.40}$$

where $\Phi_1$ is the potential at the point 1. The first integral on the right-hand side of the above equation can be divided into four integrals,

$$\int_\Gamma \Phi \frac{1}{r}\frac{\partial r}{\partial n} \mathrm{d}s = I_{1,1} + I_{1,2} + I_{1,3} + I_{1,4} \tag{2.41}$$

where

$$I_{1,1} = \int_{\text{pt. 1}}^{\text{pt. 2}} \Phi \frac{1}{r}\frac{\partial r}{\partial n} \mathrm{d}s = -\int_0^1 \Phi \frac{1}{y}\frac{\partial y}{\partial x} \mathrm{d}y = 0 \tag{2.42}$$

$$I_{1,2} = \int_{\text{pt. 2}}^{\text{pt. 3}} \Phi \frac{1}{r}\frac{\partial r}{\partial n} \mathrm{d}s = \int_0^2 \Phi \frac{1}{r}\frac{\partial r}{\partial y} \mathrm{d}x \tag{2.43}$$

$$I_{1,3} = \int_{\text{pt. 3}}^{\text{pt. 4}} \Phi \frac{1}{r}\frac{\partial r}{\partial n} \mathrm{d}s = \int_1^0 \Phi \frac{1}{r}\frac{\partial r}{\partial x} (-\mathrm{d}y) \tag{2.44}$$

$$I_{1,4} = \int_{\text{pt. 4}}^{\text{pt. 1}} \Phi \frac{1}{r}\frac{\partial r}{\partial n} \mathrm{d}s = -\int_2^0 \Phi \frac{1}{x}\frac{\partial x}{\partial y} (-\mathrm{d}x) = 0 \tag{2.45}$$

The integrals $I_{1,1}$ and $I_{1,4}$ are zero because the normal derivatives of the distance $r$ are zero in both cases. Now we consider the integral $I_{1,2}$. Along the segment between point 2 and point 3, the potential is linearly interpolated,

$$\Phi = \Phi_2\left(\frac{2-x}{2}\right) + \Phi_3\left(\frac{x}{2}\right) \qquad 0 \leqslant x \leqslant 2 \tag{2.46}$$

The distance between the singular point (node 1) and a point on the segment, and its normal derivative is

$$r = (x^2 + 1)^{\frac{1}{2}} \qquad \frac{\partial r}{\partial n} = \frac{\partial r}{\partial y} = (x^2 + 1)^{-\frac{1}{2}} \tag{2.47}$$

respectively. Substituting (2.46) and (2.47) into (2.43) we obtain

$$\begin{aligned} I_{1,2} &= \int_0^2 \left[\Phi_2\left(\frac{2-x}{2}\right) + \Phi_3\left(\frac{x}{2}\right)\right](x^2 + 1)^{-1}\,\mathrm{d}x \\ &= \Phi_2[\tan^{-1} x - \tfrac{1}{4}\ln(x^2 + 1)]\Big|_{x=0}^{x=2} + \Phi_3[\tfrac{1}{4}\ln(x^2 + 1)]\Big|_{x=0}^{x=2} \\ &= 0.70479\Phi_2 + 0.40236\Phi_3 \end{aligned} \tag{2.48}$$

Along the segment between node 3 and node 4 the potential is approximated by

$$\Phi = \Phi_3 y + \Phi_4(1 - y) \qquad 0 \leqslant y \leqslant 1 \tag{2.49}$$

In the integral $I_{1,3}$ of (2.44), $r$ and $\partial r/\partial x$ can be expressed in the following forms:

$$r = (4 + y^2)^{\frac{1}{2}} \qquad \frac{\partial r}{\partial n} = \frac{\partial r}{\partial x} = (4 + y^2)^{-\frac{1}{2}} \tag{2.50}$$

Using (2.49) and (2.50), we can readily integrate (2.44). Thus

$$\begin{aligned} I_{1,3} &= \int_0^1 [\Phi_3 y + \Phi_4(1 - y)]\,2(4 + y^2)^{-1}\,\mathrm{d}y \\ &= \Phi_3\left[\ln(4 + y^2)\right]\Big|_{y=0}^{y=1} + \Phi_4\left[\tan^{-1}\frac{y}{2} - \ln(4 + y^2)\right]\Big|_{y=0}^{y=1} \\ &= 0.22314\Phi_3 + 0.24050\Phi_4 \end{aligned} \tag{2.51}$$

The second integral in (2.40) can also be divided into four integrals with reference to the singular point 1 as follows:

$$-\int_{\Gamma} \ln r \frac{\partial \Phi}{\partial n} \, ds = -I'_{1,1} - I'_{1,2} - I'_{1,3} - I'_{1,4} \tag{2.52}$$

with

$$I'_{1,1} = \int_{\text{pt. 1}}^{\text{pt. 2}} \ln r \frac{\partial \Phi}{\partial n} \, ds = \int_{0}^{1} \ln y \frac{\partial \Phi}{\partial n} \, dy \tag{2.53}$$

$$I'_{1,2} = \int_{\text{pt. 2}}^{\text{pt. 3}} \ln r \frac{\partial \Phi}{\partial n} \, ds = \int_{0}^{2} \ln r \frac{\partial \Phi}{\partial n} \, dx = 0 \tag{2.54}$$

$$I'_{1,3} = \int_{\text{pt. 3}}^{\text{pt. 4}} \ln r \frac{\partial \Phi}{\partial n} \, ds = \int_{1}^{0} \ln r \frac{\partial \Phi}{\partial n} (-dy) \tag{2.55}$$

$$I'_{1,4} = \int_{\text{pt. 4}}^{\text{pt. 1}} \ln r \frac{\partial \Phi}{\partial n} \, ds = \int_{2}^{0} \ln x \frac{\partial \Phi}{\partial n} (-dx) = 0 \tag{2.56}$$

The integrals $I'_{1,2}$ and $I'_{1,4}$ are zero since $\partial\Phi/\partial n = 0$ as required by the boundary conditions (2.37). Along the segment between node 1 and node 2 the normal derivative of the potential is linearly interpolated as follows:

$$\frac{\partial \Phi}{\partial n} = \left(\frac{\partial \Phi}{\partial n}\right)_1 (1-y) + \left(\frac{\partial \Phi}{\partial n}\right)_2 y \tag{2.57}$$

Use of (2.57) in (2.53) gives the integral $I'_{1,1}$

$$\begin{aligned} I'_{1,1} &= \int_{0}^{1} \ln y \left[ \left(\frac{\partial \Phi}{\partial n}\right)_1 (1-y) + \left(\frac{\partial \Phi}{\partial n}\right)_2 y \right] dy \\ &= \left(\frac{\partial \Phi}{\partial n}\right)_1 \left[ y \ln y - y - \frac{y^2}{2} (\ln y - \tfrac{1}{2}) \right] \Bigg|_{y=0}^{y=1} \\ &\quad + \left(\frac{\partial \Phi}{\partial n}\right)_2 \left[ \frac{y^2}{2} \ln y - \frac{y^2}{4} \right] \Bigg|_{y=0}^{y=1} \\ &= -0.75 \left(\frac{\partial \Phi}{\partial n}\right)_1 - 0.25 \left(\frac{\partial \Phi}{\partial n}\right)_2 \end{aligned} \tag{2.58}$$

To evaluate the integral $I'_{1,3}$, the normal derivative of the potential between point 3 and point 4 is approximated by

$$\frac{\partial\Phi}{\partial n} = \left(\frac{\partial\Phi}{\partial n}\right)_3 y + \left(\frac{\partial\Phi}{\partial n}\right)_4 (1-y) \tag{2.59}$$

The distance between the singular point and a point on the segment is

$$r = (4 + y^2)^{\frac{1}{2}} \tag{2.60}$$

Equation (2.55) can now be integrated explicitly:

$$I'_{1,3} = \int_1^0 \ln (4 + y^2)^{\frac{1}{2}} \left[\left(\frac{\partial\Phi}{\partial n}\right)_3 y + \left(\frac{\partial\Phi}{\partial n}\right)_4 (1-y)\right](-\mathrm{d}y)$$

$$= \left(\frac{\partial\Phi}{\partial n}\right)_3 \{\tfrac{1}{4}(4 + y^2)[\ln (4 + y^2) - 1]\}\Big|_{y=0}^{y=1} + \left(\frac{\partial\Phi}{\partial n}\right)_4 \Big\{\tfrac{1}{2}\Big[y \ln (4 + y^2)$$

$$- 2y + 4 \tan^{-1}\left(\frac{y}{2}\right)\Big] - \tfrac{1}{4}(4 + y^2)[\ln (4 + y^2) - 1]\Big\}\Big|_{y=0}^{y=1}$$

$$= 0.37550\left(\frac{\partial\Phi}{\partial n}\right)_3 + 0.35651\left(\frac{\partial\Phi}{\partial n}\right)_4 \tag{2.61}$$

We can now substitute (2.41) and (2.52) into (2.40) with all the evaluated integrals to obtain an algebraic equation for the $\Phi$ and $(\partial\Phi/\partial n)$:

$$-1.57080\Phi_1 + 0.70479\Phi_2 + 0.62550\Phi_3 + 0.24050\Phi_4 + 0.75000(\partial\Phi/\partial n)_1$$

$$+ 0.25000(\partial\Phi/\partial n)_2 - 0.37550(\partial\Phi/\partial n)_3 - 0.35651(\partial\Phi/\partial n)_4 = 0 \tag{2.62}$$

We reiterate here that (2.62) is obtained by positioning the singular point at node 1. Three additional algebraic equations can be derived by locating the singular point at nodes 2, 3 and 4, successively. Combining these equations in a matrix form, we have

$$\begin{bmatrix} 0.75000 & 0.25000 & -0.37550 & -0.35651 \\ 0.25000 & 0.75000 & -0.35651 & -0.37550 \\ -0.37550 & -0.35651 & 0.75000 & 0.25000 \\ -0.35651 & -0.37550 & 0.25000 & 0.75000 \end{bmatrix} \begin{Bmatrix} (\partial\Phi/\partial n)_1 \\ (\partial\Phi/\partial n)_2 \\ (\partial\Phi/\partial n)_3 \\ (\partial\Phi/\partial n)_4 \end{Bmatrix}$$

$$= \begin{bmatrix} 1.57080 & -0.70479 & -0.62550 & -0.24050 \\ -0.70479 & 1.57080 & -0.24050 & -0.62550 \\ -0.62550 & -0.24050 & 1.57080 & -0.70479 \\ -0.24050 & -0.62550 & -0.70479 & 1.57080 \end{bmatrix} \begin{Bmatrix} \Phi_1 \\ \Phi_2 \\ \Phi_3 \\ \Phi_4 \end{Bmatrix} \tag{2.63}$$

Substitution of the boundary conditions (2.38) for the $\Phi_i$ leads to a system of equations for the unknown $(\partial\Phi/\partial n)_i$. Thus

$$\begin{bmatrix} 0.75000 & 0.25000 & -0.37550 & -0.35651 \\ 0.25000 & 0.75000 & -0.35651 & -0.37550 \\ -0.37550 & -0.35651 & 0.75000 & 0.25000 \\ -0.35651 & -0.37550 & 0.25000 & 0.75000 \end{bmatrix} \begin{Bmatrix} (\partial\Phi/\partial n)_1 \\ (\partial\Phi/\partial n)_2 \\ (\partial\Phi/\partial n)_3 \\ (\partial\Phi/\partial n)_4 \end{Bmatrix}$$

$$= \begin{Bmatrix} -1.73201 \\ -1.73201 \\ -1.73201 \\ -1.73201 \end{Bmatrix} \tag{2.64}$$

These equations can be solved by direct Gaussian elimination. The solutions are

$$\left(\frac{\partial\Phi}{\partial n}\right)_1 = -1.00000 \qquad \left(\frac{\partial\Phi}{\partial n}\right)_2 = -1.00000$$

$$\left(\frac{\partial\Phi}{\partial n}\right)_3 = 1.00000 \qquad \left(\frac{\partial\Phi}{\partial n}\right)_4 = 1.00000 \tag{2.65}$$

which are the exact solutions.

## 2.4 Interior solutions

Equation (2.13) provides a straightforward method of obtaining the potential at any interior point once both $\Phi$ and $\partial\Phi/\partial n$ are known everywhere on the boundary. The only important caveat in using (2.13) is that in the numerical method the accuracy will suffer if the point $P$ is too close to the boundary. Since all of the approximation is confined to the boundary, and since there is a jump of $\pi$ on the coefficient of $\Phi(P)$ as $P$ moves from the interior to the boundary, there exists a 'boundary layer' where the numerical accuracy suffers. The thickness of this boundary layer depends on the problem. It may be greater near the center of an element than near a node. As a rule of thumb a degradation of accuracy may occur at distances less than one element length from a boundary. Thus if greater accuracy is required, more shorter elements can be used.

In porous media problems the velocities are often needed as well as the head (potential). The velocities are related to the derivatives of the potential as in (1.7):

$$q_x = -K\frac{\partial \Phi}{\partial x} \qquad q_y = -K\frac{\partial \Phi}{\partial y} \tag{2.66}$$

The derivatives of $\Phi$ can be found by differentiating (2.13):

$$2\pi\frac{\partial}{\partial x}\Phi(P) = \int_\Gamma \left[\Phi(Q)\frac{\partial}{\partial x}\left(\frac{1}{r}\frac{\partial r}{\partial n}\right) - \frac{\partial}{\partial x}(\ln r)\frac{\partial}{\partial n}\Phi(Q)\right] ds \tag{2.67}$$

Only $\Phi$ at the base point $P$ is differentiated since that is where we want to find the derivative. The field point $Q$ is on the boundary and thus the $\Phi$ inside the integral is not differentiated. We now use $\partial r/\partial n = \eta/r$ and obtain

$$2\pi\frac{\partial \Phi}{\partial x} = \int_\Gamma \left(\frac{\Phi}{r^2}\frac{\partial \eta}{\partial x} - 2\frac{\Phi}{r^3}\eta\frac{\partial r}{\partial x} - \frac{1}{r}\frac{\partial r}{\partial x}\frac{\partial \Phi}{\partial n}\right) ds \tag{2.68}$$

Consider the integration from a base point $P$ along an element extending from $\xi_j$ to $\xi_{j+1}$ as shown in Figure 2.7. At the point $Q(\xi, \eta)$

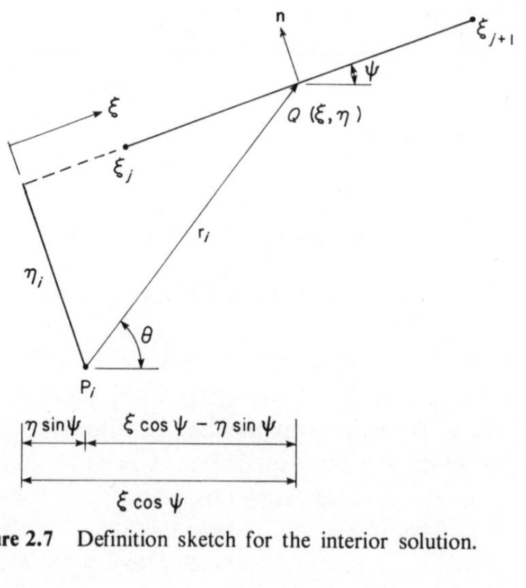


**Figure 2.7** Definition sketch for the interior solution.

$$\frac{\partial \eta}{\partial x} = \sin \psi \tag{2.69}$$

$$\frac{\partial r}{\partial x} = -\frac{\xi \cos \psi - \eta \sin \psi}{r} \tag{2.70}$$

Using these values in (2.68) produces

$$\frac{\partial}{\partial x}\Phi(P) = \frac{1}{2\pi}\int_\Gamma \left[\frac{\Phi \sin \psi}{r^2} + \frac{2\Phi\eta(\xi \cos \psi - \eta \sin \psi)}{r^4} + \frac{\xi \cos \psi - \eta \sin \psi}{r^2}\frac{\partial \Phi}{\partial n}\right] ds \tag{2.71}$$

Similarly,

$$\frac{\partial}{\partial y}\Phi(P) = \frac{1}{2\pi}\int_\Gamma \left[-\frac{\Phi \cos \psi}{r^2} + \frac{2\Phi\eta(\eta \cos \psi + \xi \sin \psi)}{r^4} + \frac{\eta \cos \psi + \xi \sin \psi}{r^2}\frac{\partial \Phi}{\partial n}\right] ds \tag{2.72}$$

Unlike (2.13) the point $P$ cannot be moved to the boundary to find the derivatives. The singularities that occur when the base point and field point coincide are not integrable.

The integrals involved in (2.71) and (2.72) are shown in Appendix A.4. In solving all the above integral equations care should be taken to obtain the proper sign on $\eta$. If $\eta$ and $\mathbf{n}$ are in the same direction as shown in Figures 2.5 and 2.7, $\eta$ is positive. If $\eta$ and $\mathbf{n}$ are in opposite directions, as may be the case if the vector $\mathbf{r}$ from $P$ to the point $Q$ passes outside of the region, the sign on $\eta$ is negative.

# 3 Special elements

Normally BIEM programs use linear interpolation between the nodal values. If, however, the analyst knows the functional behavior of the potential or its normal derivative, he can substitute a special interpolation function which more closely approximates the exact result. Such special functions are usually used in the neighborhood of a singularity or used to extend the solution to infinity. Several such examples are presented in this chapter.

## 3.1 Singularities and singular elements

In solving Laplace's equation the analyst often has to deal with singularities. In numerical solutions, especially using finite differences, singular behavior is often ignored in the expectation that the error will be limited to the vicinity of the singularity. Sometimes the statement is made that these singularities do not occur in nature, and thus one does not make any greater error in ignoring them than in the original approximation. It is true that the mathematical model is only an approximation to the real system. However, if an error is made in the solution of the governing equations, there is no reason to believe that this error is to some extent canceled by the fact that the model itself is an approximation. In order to achieve accurate results the numerical method must accurately solve the mathematical problem.

Special elements are most easily constructed from the finite element point of view of interpolation functions. Using linear interpolation as in (2.24) and (2.25) the value of the potential at any point on the element is

$$\Phi = \Phi_j N_1 + \Phi_{j+1} N_2 = \Phi_j + \frac{\xi}{L}(\Phi_{j+1} - \Phi_j) \tag{3.1}$$

The interpolating functions $N_1$ and $N_2$ are constructed so that $N_1$ is unity at the first node and zero at the last node (zero at all other nodes in a multinoded element), and $N_2$ is zero at the first node and unity at the second node. The boundary integrals are

$$I_1 = \int_j^{j+1} \frac{\Phi}{r}\frac{\partial r}{\partial n}\,\mathrm{d}s = \Phi_j \int_j^{j+1} \frac{N_1}{r}\frac{\partial r}{\partial n}\,\mathrm{d}s + \Phi_{j+1}\int_j^{j+1} \frac{N_2}{r}\frac{\partial r}{\partial n}\,\mathrm{d}s$$

$$I_2 = \int_j^{j+1} \frac{\partial \Phi}{\partial n} \ln r\,\mathrm{d}s = \left(\frac{\partial \Phi}{\partial n}\right)_j \int_j^{j+1} N_1 \ln r\,\mathrm{d}s + \left(\frac{\partial \Phi}{\partial n}\right)_{j+1} \int_j^{j+1} N_2 \ln r\,\mathrm{d}s \tag{3.2}$$

Thus the integration of the interpolating function furnishes the coefficient of $\Phi_j$ or $(\partial\Phi/\partial n)_j$. (Note that there is also a contribution to this coefficient from the previous element.) Equations (3.2) are limited to two-noded elements, but within that limitation the interpolation functions are arbitrary. As examples consider the following on an element one unit long with the origin of the $\xi$-axis at the first node:

(a) Constant

$$\begin{aligned} N_1 = 1, \quad N_2 = 0 \qquad & 0 \leqslant \xi < 0.5 \\ N_1 = 0, \quad N_2 = 1 \qquad & 0.5 \leqslant \xi < 1 \end{aligned} \tag{3.3}$$

Thus $\Phi$ and $\partial\Phi/\partial n$ are taken as constants for a distance of one element length on each side of the node. The method is equivalent to using a one-noded element with constant variation of $\Phi$ and $\partial\Phi/\partial n$ over the element.

(b) Linear

$$N_1 = 1 - \xi, \quad N_2 = \xi \tag{3.4}$$

The linear variation is shown in Figure 3.1 and is the approximation of (2.24) and (2.25).

(c) Power

$$N_1 = 1 - \xi^m, \quad N_2 = \xi^m \tag{3.5a}$$

$$N_1 = (1 - \xi)^m, \quad N_2 = 1 - (1 - \xi)^m \tag{3.5b}$$

$$N_1 = 1 - \xi^m, \quad N_2 = 1 - (1 - \xi)^m \tag{3.5c}$$

$$N_1 = (1 - \xi)^m, \quad N_2 = \xi^m \tag{3.5d}$$

These interpolation functions are shown in Figure 3.2. There is nothing to recommend one over another or to recommend quadratic variation over linear unless the analyst has reason to believe the variables in an actual problem behave in such a manner.

The integrals of (3.2) can be integrated in closed form when the interpolation functions are linear as was done in Chapter 2. Although some other

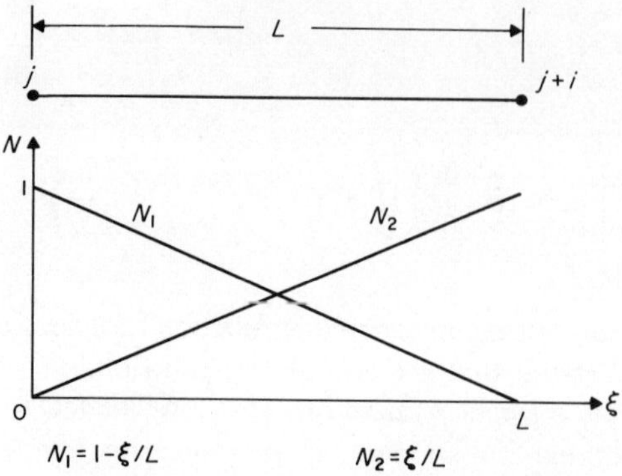


**Figure 3.1** Linear interpolation functions.

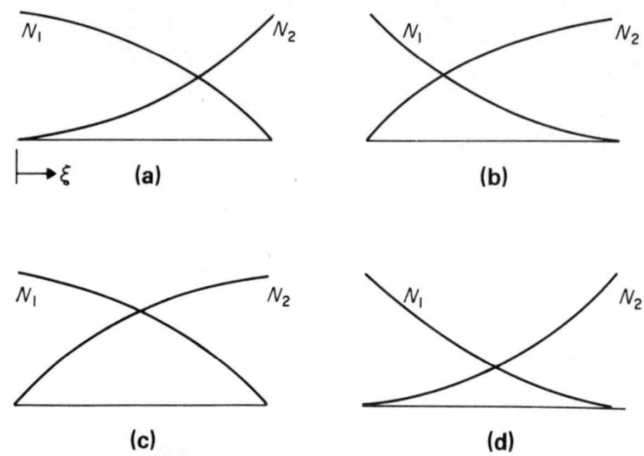


**Figure 3.2** Power law interpolation functions (see (3.5)).

interpolation functions can be integrated analytically, many cannot. In such cases it is usual to resort to Gaussian quadrature as the most efficient numerical integration method for a desired accuracy. Gaussian quadrature is explained in most books on numerical methods (e.g. Hornbeck 1975). The same books reference sources for the quadrature weights and zeros (e.g. Stroud & Secrest 1966, Abramowitz & Stegun 1974: see App. A.5).

The second of (3.2) contains a logarithmic singularity when $r$ is measured from one of the nodes $j$ or $j + 1$. Although this singularity presented no difficulty when using exact integration of Chapter 2, it can be troublesome when using numerical integration. In general special Gaussian quadrature formulas should be used which automatically account for the logarithmic singularity. If, however, the special interpolation function also contains an integrable singularity, great care must be taken to maintain accuracy.

The examples that follow illustrate how special interpolation functions can be used to approximate the actual behavior of the potential and normal derivative, thus increasing the accuracy of the calculation. Also illustrated is a closely related procedure whereby the BIEM can be combined with a locally accurate analytic solution, the combination of which improves the overall accuracy.

## 3.2 Flow around a right-angled corner

Consider as the first example the flow over a right-angled wedge as shown in Figure 3.3. Point 6 represents a singularity in that the velocity goes to infinity at that point. Since we are dealing only with boundary values and point 6 is on a solid boundary ($\partial\Phi/\partial n = 0$), we are concerned in the initial solution with the behavior of the potential function in the vicinity of the singularity. From theoretical hydrodynamics

$$\Phi \sim r^{2/3} \tag{3.6}$$

in which $r$ is the distance from the corner. Using special interpolation functions on elements 5–6 and 6–7 this behavior is reproduced as follows: for element 5–6

$$N_1 = (1 - \xi/L)^{2/3} \qquad N_2 = 1 - (1 - \xi/L)^{2/3} \tag{3.7}$$

where $\xi$ is the distance from point 5. For element 6–7

$$N_1 = 1 - (\xi/L)^{2/3} \qquad N_2 = (\xi/L)^{2/3} \tag{3.8}$$

in which $\xi$ is the distance from point 6. Table 3.1 shows the results of the example with both singular and linear elements. In this case the linear elements also give a reasonably good result. For stronger singularities the use of special elements is more important, whereas the linear elements give satisfactory results for weaker singularities.

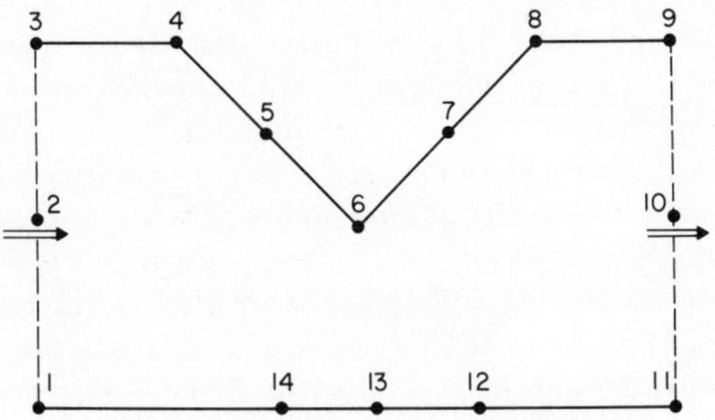


**Figure 3.3** Flow around a right-angled corner.

**Table 3.1** Solution for the 90° wedge.

| | $x=0$ $y=0$ | $x=0.25$ $y=0.25$ | $x=0.5$ $y=0.5$ |
|---|---|---|---|
| potential without singular elements | 0 | 0.505 | 0.720 |
| potential with singular elements | 0 | 0.520 | 0.728 |
| exact potential | 0 | 0.521 | 0.729 |

## 3.3 The sheetpile cut-off wall

If the angle in Figure 3.3 is sharpened to the limit, it becomes 180°, as on a sheetpile cut-off wall commonly used on the underside of a dam (Fig. 3.4). The BIEM cannot solve this problem directly because points must be placed on the left-hand side of the cut-off wall proceeding downward and on the right-hand side proceeding upward. Since the wall is infinitely thin, the left-hand side points occupy the same positions as the right-hand side points. The result is that the coefficient matrix has identical rows, and is therefore singular. The matrix is 'correct' in the sense that the correct solution would satisfy the BIEM equations; however, the singular matrix leads to an infinite number of solutions, and the correct solution is only one of them. Displacing the left-hand side and the right-hand side points vertically so that they do not occupy the same point but are collinear does not improve the situation. This dilemma is solved in Chapter 4 where zoned problems are treated. For the present, however, we can take advantage of the symmetry of the problem and solve only one half, say the left half. The boundary conditions are shown in Figure 3.5.

The solution using linear elements and from 2 to 9 points on the cut-off wall is shown in Table 3.2, columns 1–6. The solution was forced to give the correct value of the potential at the tip so the maximum error occurs at the first node point adjacent to the tip. Using more elements improves the solution.

The normal derivative (velocity) below the cut-off wall is shown in Figure 3.6 for various numbers of elements which are spaced exactly as those on the wall and given in Table 3.2. The normal derivative is highly oscillatory and inaccurate near the tip, although the use of more elements improves the solution.

The flow rate (column 6, Table 3.2) was calculated at the end of the box (line $AE$, Fig. 3.5). The quality of this number generally improves with more points. It appears fortuitous that the flow error is slightly less in the case of 6 points than it is for 9 points. We note that it is still a respectable number with only 2 points on the wall. This latter calculation was done using a total of only 5 points, the minimum required to define the geometry.

The singularity can be treated much more accurately through the use of

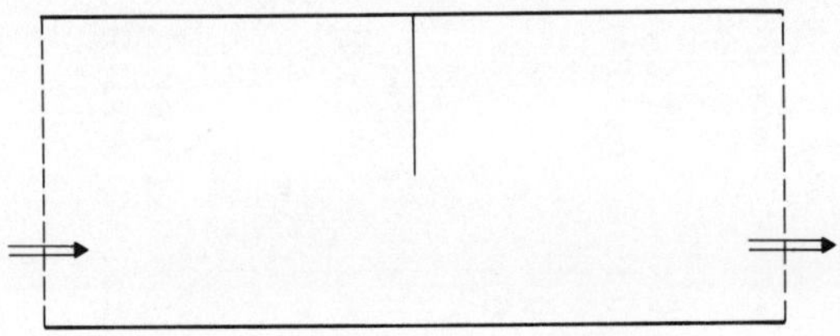

**Figure 3.4** The sheetpile cut-off wall.

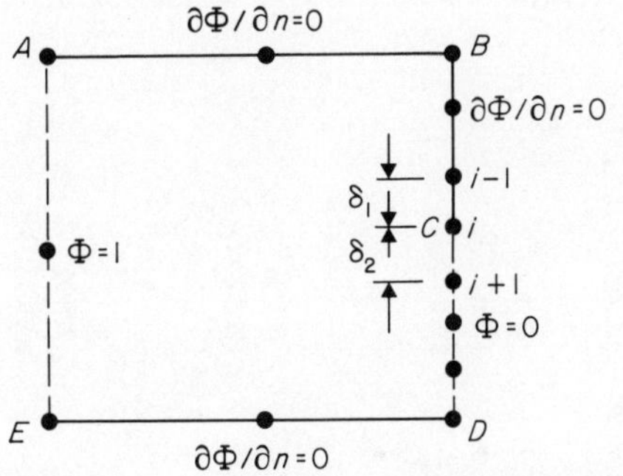


**Figure 3.5** Discretization of the left half of the cut-off wall problem.

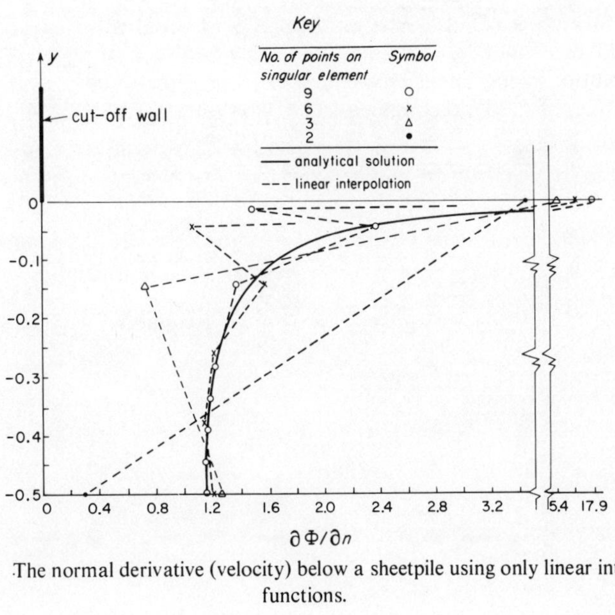


**Figure 3.6** The normal derivative (velocity) below a sheetpile using only linear interpolation functions.

**Table 3.2** Solution to the cut-off wall problem.

| | | | *Without using singular elements* | | | *Using singular elements* | | |
|---|---|---|---|---|---|---|---|---|
| *1* | *2* | *3* | *4* | *5* | *6* | *7* | *8* | *9* |
| *Number of points on cut-off wall* | *Position of the node* $y$ | *Exact solution* $\Phi$ | *Numerical solution* $\Phi$ | *% error* | *Flow error (exact = 0.8190)* | *Numerical solution* $\Phi$ | *% error* | *Flow erro* |
| | 0 | 0 | 0† | | | 0† | | |
| 9* | 0.01414 | 0.1096 | 0.1005 | –8.3 | 0.4% | 0.1091 | –0.4 | 0.2% |
| | 0.04243 | 0.1869 | 0.1803 | –3.5 | (0.8223) | 0.1859 | –0.5 | (0.8204) |
| | 0.14142 | 0.3225 | 0.3117 | –1.5 | (6 points) | 0.3210 | –0.5 | |
| | 0.28284 | 0.4159 | 0.4119 | –1.0 | | 0.4144 | –0.4 | |
| | 0.5 | 0.4614 | 0.4555 | –1.3 | | 0.4576 | –0.8 | |
| | 0 | 0 | 0 | | | 0 | | |
| 6 | 0.04243 | 0.1869 | 0.1739 | –7.0 | 0.2% | 0.1873 | 0.2 | 0.2% |
| | 0.14142 | 0.3225 | 0.3162 | –2.0 | (0.8207) | 0.3242 | 0.5 | (0.8207) |
| | 0.2609 | 0.4057 | 0.4029 | –0.7 | (3 points) | 0.4090 | 0.8 | |
| | 0.5 | 0.4614 | 0.4500 | –2.5 | | 0.4552 | –1.3 | |
| | 0 | 0 | 0 | | | 0 | | |
| 3 | 0.14142 | 0.3225 | 0.2986 | –7.4 | 2.8% | 0.3165 | –1.7 | 1.0% |
| | 0.5 | 0.4614 | 0.4364 | –5.4 | (0.8417) | 0.4500 | –2.5 | (0.8275) |
| | | | | | (3 points) | | | |
| | 0 | 0 | 0 | | | 0 | | |
| 2 | 0.5 | 0.4614 | 0.4063 | –11.9 | 5.8% | 0.4457 | –3.4 | 3.3% |
| | | | | | (2 points) | | | |

*Only 6 of the 9 points are shown with the $y$ co-ordinate as indicated in column 2.
†The zero value at the cut-off wall results from the conditions of the problem (symmetry) and is not truly a numerical result.

special elements. Near the tip of the wall the potential is approximated by the function

$$\Phi = (r/\delta_1)^{\frac{1}{2}}\Phi_{i-1} + (1 - (r/\delta_1)^{\frac{1}{2}})\Phi_i \tag{3.9}$$

where $\delta_1$ is defined on Figure 3.5. Below the cut-off wall the normal derivative behaves as

$$\frac{\partial\Phi}{\partial n} = (\delta_2/r)^{\frac{1}{2}}\left(\frac{\partial\Phi}{\partial n}\right)_{i+1} \tag{3.10}$$

that is, it goes to infinity as $r^{-\frac{1}{2}}$. Using (3.9) and (3.10) as interpolating functions produces the result shown in Table 3.2, columns 7, 8 and 9 and Figure 3.7. The effect was to reduce all error and, perhaps more importantly, to make the error more uniform. The maximum error using special elements no longer occurs at the point nearest the singularity but at the stagnation point. Presumably that error would also be reduced if a special element were used in the vicinity so that $\Phi \rightarrow r^{\frac{3}{2}}$ as $r \rightarrow 0$ where $r$ is now measured from the stagnation point.

Table 3.2 also indicates that use of singular elements reduces the error in the flow rate even though the error was computed some distance from the singularity. This fact indicates that the proper treatment of the singularity improves the overall solution, but the extent of the improvement is unknown and is probably strongly dependent on the geometry of the problem.

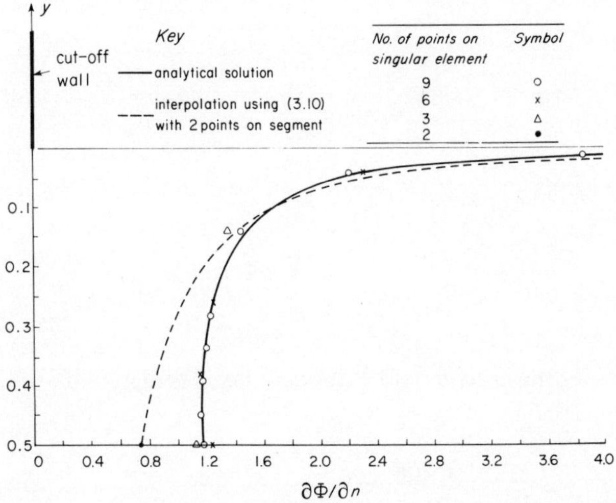


**Figure 3.7** The normal derivative (velocity) below a sheetpile using linear interpolation functions except near the tip where a special singular element was used.

## 3.4 The tailwater singularity

Figure 3.8 shows flow through an earth dam with a finite depth of tail water. The intersection of the tail water with the downstream face of the dam, point $P$, is a singular point in that the velocity goes to infinity at $P$. For simplicity the downstream face of the dam is vertical, although the same considerations apply to any slope (Polubarinova-Kochina 1962). A solution in the vicinity of point $P$ without reference to the remainder of the geometry should have the following properties:

$$\Phi = \Phi_0 + z, \quad \frac{\partial \Phi}{\partial z} = 1, \quad z > 0; \qquad \Phi = \Phi_0, \quad \frac{\partial \Phi}{\partial z} = 0, \quad z < 0 \tag{3.11}$$

where the co-ordinate system is defined in Figure 3.8. These properties are satisfied by

$$\Phi = \int_0^a m \tan^{-1} \frac{z - \eta}{x} \, d\eta \tag{3.12}$$

in which $m$ is a constant to be determined, and $a$ is an arbitrarily large number.

Integrating (3.12) yields

$$\frac{\Phi}{m} = z \tan^{-1} \frac{z}{x} - (a - z) \tan^{-1} \frac{a - z}{x} - \frac{x}{2} \ln \frac{1 + (z/x)^2}{1 + [(a - z)/x]^2} \tag{3.13}$$

The derivative with respect to $z$ is

$$\frac{1}{m} \frac{\partial \Phi}{\partial z} = \tan^{-1} \frac{z}{x} + \tan^{-1} \frac{a - z}{x} \tag{3.14}$$

For $a > z$ (3.14) gives

$$\frac{1}{m} \frac{\partial \Phi}{\partial z} = \pi \quad \text{for } z > 0, x = 0; \qquad \frac{1}{m} \frac{\partial \Phi}{\partial z} = 0 \quad \text{for } z < 0, x = 0 \tag{3.15}$$

which satisfies the criteria of (3.11) if $m = 1/\pi$. The derivative of (3.13) with respect to $x$ is

$$\frac{1}{m} \frac{\partial \Phi}{\partial x} = -\tfrac{1}{2} \ln \frac{x^2 + z^2}{x^2 + (z - a)^2} \tag{3.16}$$

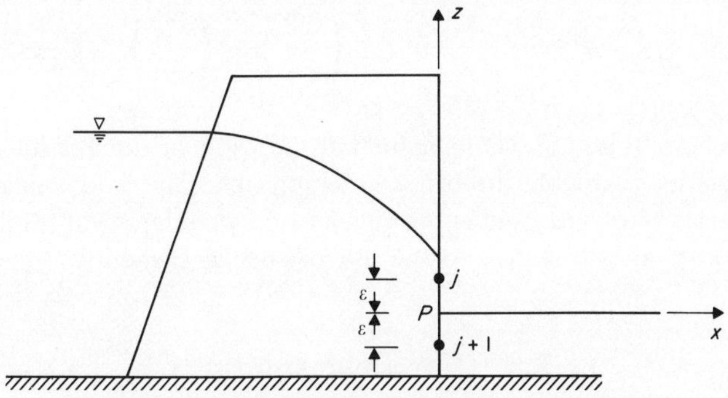


**Figure 3.8** Flow through an earth dam with tail water.

Setting $x = 0$, (3.16) shows that $\partial\Phi/\partial x$ tends to infinity as $\ln(1/|z|)$ near the dam–tail water intersection.

The singular boundary integral is

$$I = \int_j^{j+1} \frac{\partial \Phi}{\partial n} \ln r \, \mathrm{d}s \tag{3.17}$$

For small ε and for the base point $i$ not equal to $j$ or $j + 1$, $r$ is nearly unchanged over the integration and to a good approximation can be removed from the integrand, yielding

$$I = \ln r \left[ \int_\varepsilon^0 \left(\frac{\partial \Phi}{\partial n}\right)_j \ln \frac{e\varepsilon}{z} \, \mathrm{d}z + \int_0^{-\varepsilon} \left(\frac{\partial \Phi}{\partial n}\right)_{j+1} \ln\left(-\frac{e\varepsilon}{z}\right) \mathrm{d}z \right]$$

$$= -2\varepsilon \left[ \left(\frac{\partial \Phi}{\partial n}\right)_j + \left(\frac{\partial \Phi}{\partial n}\right)_{j+1} \right] \ln r \tag{3.18}$$

When the base point is located at $j$ or $j + 1$ the integral in (3.17) can be evaluated directly. For $i = j$ the integral from point $j$ to the origin is

$$I_j = \left(\frac{\partial \Phi}{\partial n}\right)_j \int_\varepsilon^0 \ln(\varepsilon - z) \ln \frac{e\varepsilon}{z} \, \mathrm{d}z \tag{3.19}$$

which is integrated to

$$I_j = \varepsilon \left(\frac{\partial \Phi}{\partial n}\right)_j \left[ -2 \ln \varepsilon + \sum_{n=1}^{\infty} \frac{1}{n(n+1)} + \sum_{n=1}^{\infty} \frac{1}{n(n+1)^2} \right] \tag{3.20}$$

A similar result is obtained for the integral between the origin and point $j + 1$. After evaluating the series and adding the two integrals the result is

$$I_j = \varepsilon(1.3550 - 2 \ln \varepsilon) \left[ \left( \frac{\partial \Phi}{\partial n} \right)_j + \left( \frac{\partial \Phi}{\partial n} \right)_{j+1} \right] \tag{3.21}$$

Equations (3.20) and (3.21) indicate that the value of the line integral near the singularity is double the result of simply ignoring the singularity. The 'pre-integration' of integrals in the vicinity of a singular point, as illustrated above, forms an alternative to the use of special elements.

## 3.5 Far field solutions

Many problems lack definite boundaries in one or more directions; that is, the solution domain extends to infinity. With any numerical solution technique it is difficult to know where to terminate the calculation. On one hand the calculation should be extended far enough so that any artificial boundaries introduced by the method do not seriously affect the solution. On the other hand the larger the domain, the larger the computational cost. Much of the computation may be devoted to a region which has little interest for the analyst.

The traditional answer to this dilemma is to begin with a limited computational region. The problem is solved, and then the artificial boundaries are moved outward. If there is no change in the region of interest, we assume that our domain is large enough and that we have arrived at the correct solution. If there is a change, the domain is again enlarged. This method works well for many problems, but in a few problems there are two difficulties:

(1) The method is likely to be expensive since each change of boundary requires a complete new solution. Also for unsteady problems a small solution domain may suffice for short times, whereas a greatly enlarged domain is necessary for larger times. Thus each new trial should be carried to completion.
(2) In many problems each change of domain creates small changes in the solution at the points of interest. Although these changes may be within the acceptable accuracy, the analyst may be uncertain as to the total of such changes. This process is analogous to summing an infinite series where each term has a smaller contribution than the last, but the sum of the remaining terms may be large.

A better technique is to determine in general terms how the far field solution behaves and then to use that behavior to extend the numerical solution to infinity. The far field behavior is usually independent of the details of the near field solution so that it can be determined by replacing the near field by elementary functions and/or linearized solutions. The idea is to produce an interpolation function, a special element, which is a much better approximation than the polynomials used in the remainder of the problem.

The general technique of using special elements for far field solutions is illustrated by a groundwater recharge example and the problem of flow through and under a dam which sits on a porous foundation.

### 3.5.1 Recharge

The problem of two-dimensional recharge into an infinite aquifer is shown in Figure 3.9. The region of interest is in the vicinity of the recharge, but the water spreads out to infinity. If the solution region is limited and no water is allowed to flow out, it will sooner or later begin to fill, causing the water table elevation in the entire region to rise.

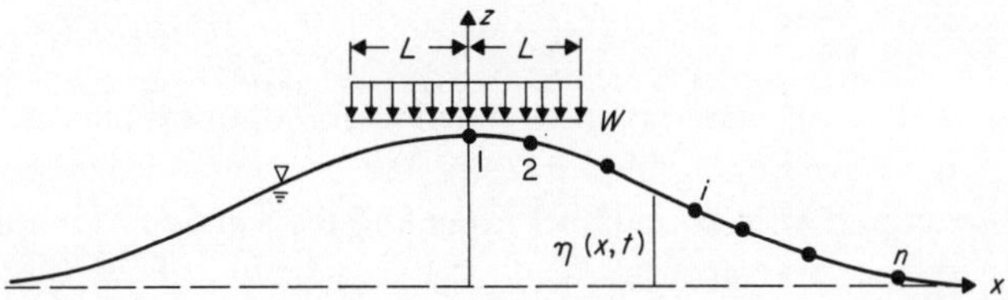


**Figure 3.9** Recharge into a water table of infinite horizontal and vertical extent.

For this problem, Dagan (1967a) gives the solution to the linearized equations as

$$\Phi = \eta = -\frac{W}{\pi}\left[\frac{x+1}{2}\ln\frac{(x+1)^2}{(x+1)^2+t^2} + \frac{x-1}{2}\ln\frac{(x-1)^2+t^2}{(x-1)^2} - t\tan^{-1}\frac{2t}{1-x^2-t^2}\right] \tag{3.22}$$

in which $W$ is the recharge rate and $t$ ($=t^*K/nL$) is the dimensionless time ($n$ is the effective porosity, $2L$ is the recharge width), and $x$ ($=x^*/L$) is the horizontal co-ordinate. Equation (3.22) can be simplified for $x \gg 1$ to

$$\Phi = \eta = \frac{Wt}{\pi}\tan^{-1}\frac{2t}{1-x^2-t^2} \qquad z = \eta \tag{3.23}$$

which represents the solution of the linearized equations for a point source at the origin with discharge $W$. Thus the details of the recharge (e.g. non-uniformity) can be neglected when considering the far field.

Our special element comes from the right-hand side of (3.23). This function serves as an interpolation function from the last calculated node, $x_n$, to infinity so that

$$\Phi = \frac{\Phi_n \tan^{-1}[2t/(1-x^2-t^2)]}{\tan^{-1}[2t/(1-x_n^2-t^2)]} \qquad |x| > x_n, z = \eta \tag{3.24}$$

Since the movement of the free surface is $\partial\eta/\partial t = -(\partial\Phi/\partial n)/\cos\beta$, the derivative $\partial\Phi/\partial n$ can be found by differentiating (3.24) with respect to $t$ and taking $\beta = 0$ (a consistent approximation). (For details of the development see Ch. 5.) The result is

$$\frac{\partial\Phi}{\partial n} \approx \frac{\partial\eta}{\partial t} = \left(\frac{\partial\Phi}{\partial n}\right)_n \frac{(1-x^2+t^2)[4t^2+(1-x_n^2-t^2)^2]}{(1-x_n^2+t^2)[4t^2+(1-x^2-t^2)^2]} \qquad |x| > x_n, z = \eta \tag{3.25}$$

which is used as the interpolation function for the normal derivative for $|x| > x_n$. The relevant integrals are then evaluated (numerically) between the last grid point and infinity, in effect extending the solution to infinity. Aside from providing a more accurate solution this technique allows the analyst to use fewer points.

Figure 3.10 shows recharge results where BIEM calculations are compared using far field solutions and terminating arbitrarily the calculation at

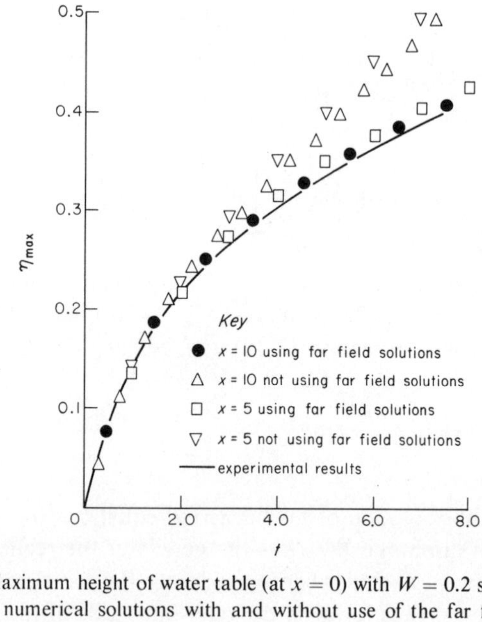


**Figure 3.10** Maximum height of water table (at $x = 0$) with $W = 0.2$ showing experimental data and numerical solutions with and without use of the far field solutions.

some point. For short time, when all the movement is near the origin, the use of far field interpolation is not necessary. As time goes on it becomes important to make use of the far field elements. For large enough time the two solutions would differ, even for small recharge and a large solution region.

### 3.5.2 *Dike on a permeable foundation*

Flow through an earth dike or dam is treated in detail in Chapter 5. The basic problem is used in this section as an illustration of infinite elements when the dike is placed on a foundation of the same permeability (although if the foundation were of different permeability, there would be little added complexity). Thus there is flow under as well as through the dike (Fig. 3.11).

When viewed from a large distance, the dike appears to cause a sudden change in potential on the foundation. Instead of a finite width dike this change could take place suddenly across a thin line and the effect of the far field would be the same. However, the boundary condition that the normal derivative is zero on the impermeable boundary must be maintained. These conditions can be simulated by replacing the dike by an infinite series of vortices (Fig. 3.12) which have the impermeable boundary as a line of symmetry, thus maintaining that boundary condition. The far field potential is

$$\Phi = \Phi_2 - \frac{\Phi_1 - \Phi_2}{\pi}\left[\tan^{-1}\frac{z}{x} + \sum_{k=1}^{\infty}(-1)^k\left(\tan^{-1}\frac{z-2kH}{x} + \tan^{-1}\frac{z+2kH}{x}\right)\right] \tag{3.26}$$

Equation (3.26) represents the far field interpolating function to be used along the lines $z = 0$ and $z = -H$. However, it is somewhat awkward

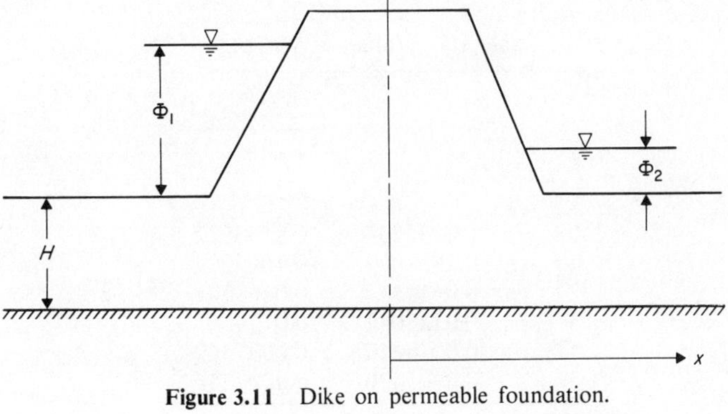


**Figure 3.11** Dike on permeable foundation.

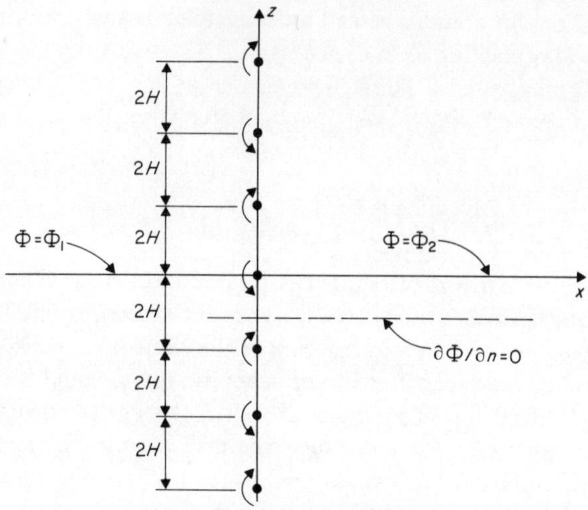


**Figure 3.12** Replacement of dike with a series of vortices.

because the infinite series must be evaluated at each quadrature point. An alternative is to terminate the foundation stratum some distance from the dam at $x = \pm nH$ $(n > 0)$. Then at some depth $z = -\alpha H$ $(1 \geqslant \alpha \geqslant 0)$

$$\Phi = \Phi_2 - \frac{\Phi_1 - \Phi_2}{\pi} R(n, \alpha) \tag{3.27}$$

in which

$$R(n, \alpha) = \tan^{-1}\frac{\alpha}{n} + \sum_{k=1}^{\infty} (-1)^k \left(\tan^{-1}\frac{\alpha - 2k}{n} + \tan^{-1}\frac{\alpha + 2k}{n}\right) \tag{3.28}$$

and a few values of $R$ are given in Table 3.3. Equations (3.27) and (3.28) are then used as boundary conditions along the lateral (artificial) boundaries of

**Table 3.3** Values of $R(n, \alpha)$.

| | α | |
|---|---|---|
| *n* | *0.5* | *1.0* |
| *(1)* | *(2)* | *(3)* |
| 0.5 | 0.6832254 | 0.8555622 |
| 1 | 0.2981078 | 0.4099209 |
| 2 | 0.0611517 | 0.0863741 |
| 3 | 0.0127047 | 0.0179661 |
| 4 | 0.0026410 | 0.0037349 |
| 5 | 0.0005490 | 0.0007764 |

the truncated region. (Often in truncating a region the condition imposed on such boundaries is either $\Phi = \text{constant}$ or $\partial\Phi/\partial n = 0$.) In this way the far field analytic solution is matched to the near field numerical solution. The pervious aquifer can be truncated quite close to the dam (say at $n = 2$ or double the thickness of the layer) with little error (Liu & Liggett 1979). The savings in computation are obvious.

## 3.6 Sources and sinks

Often pumping and recharge wells are represented by sources and sinks when the details of the flow in the vicinity of the well are unimportant. Such singularities usually appear in the interior of the solution region rather than on the boundaries and can cause considerable difficulty when using domain methods (finite differences or finite elements). In using the BIEM the calculation is easily done by the use of superposition

$$\Phi = \Phi_a + \Phi_b \tag{3.29}$$

in which $\Phi_a$ is the non-singular portion of the solution and $\Phi_b$ is given by

$$\Phi_b = \sum_{i=1}^{N_w} \frac{Q_i}{2\pi} \ln r_i \tag{3.30}$$

where $N_w$ is the number of sources, $Q_i$ is the strength of source $i$, and $r_i$ is the distance from the source. The object is to solve for $\Phi_a$ using the BIEM and then add $\Phi_b$ for the complete solution. However, the boundary conditions must be formulated in terms of the complete solution.

Assume that on the boundaries

$$\Phi = \Phi_c \quad \text{and/or} \quad \partial\Phi/\partial n = \partial\Phi_c/\partial n \tag{3.31}$$

Then in terms of $\Phi_a$

$$\Phi_a = \Phi_c - \sum_{i=1}^{N_w} \frac{Q_i}{2\pi} \ln r_i \quad \text{and/or} \quad \frac{\partial\Phi_a}{\partial n} = \frac{\partial\Phi_c}{\partial n} - \sum_{i=1}^{N_w} \frac{Q_i}{2\pi r_i} \frac{\partial r_i}{\partial n} \tag{3.32}$$

Even if the source(s) appear on the boundary of the region, this technique is valid.

# 4 Zones

As formulated up to this point the boundary integral equation method applies to problems of flow in porous media only where the permeability can be considered constant. This limitation is serious as a great many problems have variable permeabilities. In that case the governing equation is (in two dimensions)

$$\frac{\partial}{\partial x}\left(K\frac{\partial \Phi}{\partial x}\right)+\frac{\partial}{\partial y}\left(K\frac{\partial \Phi}{\partial y}\right)=0 \tag{4.1}$$

which reduces to Laplace's equation only if $K$ is a constant. If $K$ is not constant,

$$K\frac{\partial^2 \Phi}{\partial x^2}+K\frac{\partial^2 \Phi}{\partial y^2}+\frac{\partial K}{\partial x}\frac{\partial \Phi}{\partial x}+\frac{\partial K}{\partial y}\frac{\partial \Phi}{\partial y}=0 \tag{4.2}$$

Thus

$$\nabla^2 \Phi=-\frac{1}{K}\left(\frac{\partial K}{\partial x}\frac{\partial \Phi}{\partial x}+\frac{\partial K}{\partial y}\frac{\partial \Phi}{\partial y}\right) \tag{4.3}$$

Using Green's second identity (see (2.4)) produces

$$\int_\Gamma\left(\frac{\partial \Phi}{\partial n}\ln r-\frac{\Phi}{r}\frac{\partial r}{\partial n}\right)\mathrm{d}s=-\int_D\frac{1}{K}\ln r\left(\frac{\partial K}{\partial x}\frac{\partial \Phi}{\partial x}+\frac{\partial K}{\partial y}\frac{\partial \Phi}{\partial y}\right)\mathrm{d}A \tag{4.4}$$

The volume integral remains in the problem, and to perform the integration the solution domain must, in general, be discretized. But the outstanding advantage of the method is that the need to discretize the solution domain is avoided. However, some problems in solid mechanics have been done using boundary integral techniques where it was necessary to evaluate integrals over the domain (e.g. Banerjee & Butterfield 1977, Kumar & Mukherjee

1977). If there is a significant portion of the solution domain where the permeability is constant, the boundary integral method can be coupled with the finite element method, the latter being applied to the zone where the permeability varies.

In groundwater problems, however, the permeability is often not so well known that the specification of it as a continuously varying function in space is meaningful. (We exclude problems in unsaturated flow where the 'effective permeability' depends upon the degree of saturation and may change several orders of magnitude.) Instead, it is often specified as constants over large regions or zones, but there may be several such zones in the domain of any specific problem. Man-made earth structures, especially dams, are usually zoned where the permeability in one zone may be two or three orders of magnitude different than in an adjacent zone.

It is the zone concept which we will use to solve problems with changing permeability. Also it is sometimes expedient to divide a problem into different zones even though the properties of the medium are constant.

## 4.1 The compatibility equations

Consider a domain divided into two or more zones such as shown in Figure 4.1. The potential is continuous across the dividing line

$$\Phi^{(1)} = \Phi^{(2)} \qquad \text{on } \Gamma_D \tag{4.5}$$

where superscripts denote the zone number and $\Gamma_D$ is the line dividing the zones. Across any short length, say $\Delta s$, the amount of fluid leaving zone 1 must equal the amount of fluid entering zone 2, or (Fig. 4.2)

$$v_n^{(1)}\,\Delta s = v_n^{(2)}\,\Delta s \tag{4.6}$$

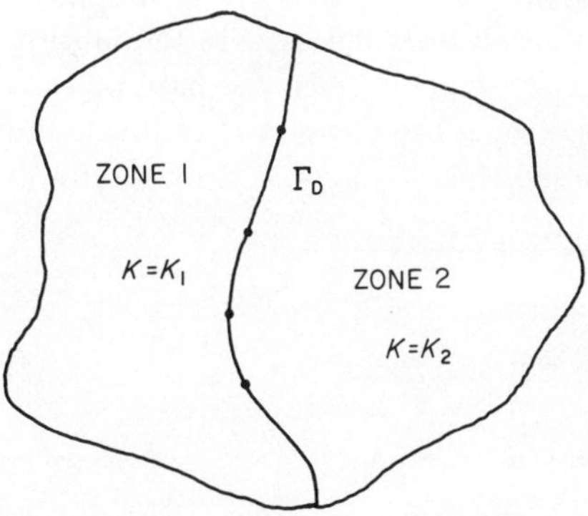


**Figure 4.1** A two-zone domain.

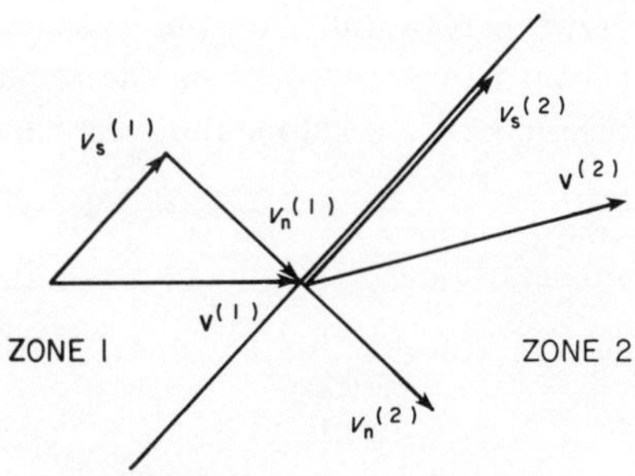


**Figure 4.2** Deflection of the velocity vector across the interzonal line.

In terms of the normal derivative

$$K_1 \frac{\partial \Phi^{(1)}}{\partial n} \Delta s = -K_2 \frac{\partial \Phi^{(2)}}{\partial n} \Delta s \tag{4.7}$$

in which $\mathbf{n}^{(1)} = -\mathbf{n}^{(2)}$ and the unit normal is still defined as pointing outward from the zone. Equations (4.5) and (4.7) form the compatibility equations or interzonal relationship between the potential and normal derivative for different zones.

These equations are used to supplement those formed by application of the BIEM. Suppose that zone 1 has $N_e^{(1)}$ exterior boundary points, zone 2 has $N_e^{(2)}$ exterior boundary points, and there are $M$ points on the interzonal boundary. There is one equation and one unknown for each exterior point; on the interzonal points there are two unknowns and (4.5) and (4.7) are available. Thus the entire system can be assembled into an $N_e^{(1)} + N_e^{(2)} + 2M$ set of equations and unknowns. The solution of these equations completes the specification of the boundary data, and both $\mathbf{\Phi}$ and $\partial\Phi/\partial n$ are then known along the exterior and interior boundaries. Subsequent solutions for interior points need only consider one zone at a time.

The same idea can be extended to any number of zones, in which case the total number of equations becomes $\Sigma N_e^{(i)} + 2\Sigma M^{(i)}$, where $N_e^{(i)}$ is the number of exterior points on the $i$th zone and $M^{(i)}$ is the number of points on the $i$th interior boundary. The coefficient matrix of these equations can be arranged in banded form where the band width may be as small as max $(N^{(i)} + 2M^{(i)})$, depending upon the configuration of the zones and the numbering of points and zones.

## 4.2 The sheetpile problem

When dealing with a cut-off wall (Fig. 4.3) the system of equations becomes singular if two separate boundary points occupy the same co-ordinates. Displacing the points along the cut-off wall does not help as the integral between two points on one side becomes singular when the base point is between those two but on the opposite side. The solution is to divide the region into zones along the sheetpile as shown in Figure 4.3. The cut-off wall

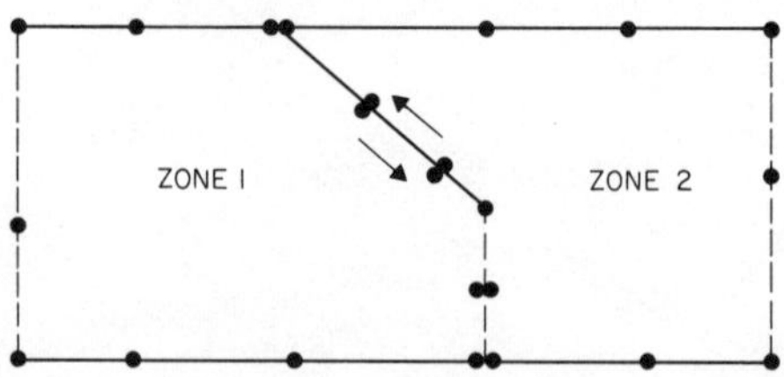


**Figure 4.3** Discretization of the cut-off wall problem.

is treated as an exterior boundary for each zone where the boundary conditions is $\partial\Phi/\partial n = 0$. The interzonal boundary conditions apply to the line dividing the zones.

The tip of the cut-off wall is, of course, a singularity where $\Phi \sim r^{\frac{1}{2}}$ and $\partial\Phi/\partial r \sim r^{-\frac{1}{2}}$ (except on wall where $\partial\Phi/\partial r = 0$), where $r$ is the distance from the tip. In order to get an accurate solution in the vicinity of the tip, special elements must be used (see Ch. 3). Any special element on the interzonal boundary for zone 1 must be compatible with a similar element for zone 2.

## 4.3 The interzonal singularity

In a general case several zones may intersect at a point, as shown in Figure 4.4. Such a point is a singularity (unless the permeability is the same in all zones). As an example consider the simpler case shown in Fig. 4.5, where the interzonal boundary has a corner. We take two points a distance ε on each side of the corner and apply the compatibility equations (4.5) and (4.7):

$$\Phi_1^{(1)} = \Phi_1^{(2)} \quad \Phi_2^{(1)} = \Phi_2^{(2)} \quad K_1 \frac{\partial \Phi_1^{(1)}}{\partial n} = -K_2 \frac{\partial \Phi_1^{(2)}}{\partial n}$$

$$K_1 \frac{\partial \Phi_2^{(1)}}{\partial n} = -K_2 \frac{\partial \Phi_2^{(2)}}{\partial n} \tag{4.8}$$

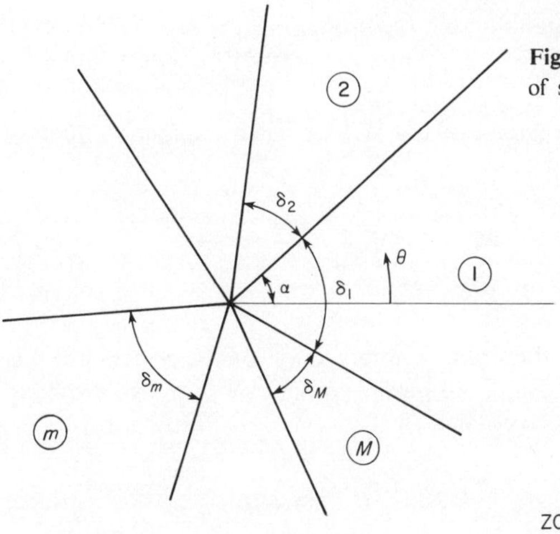


**Figure 4.4** The intersection of several zones.

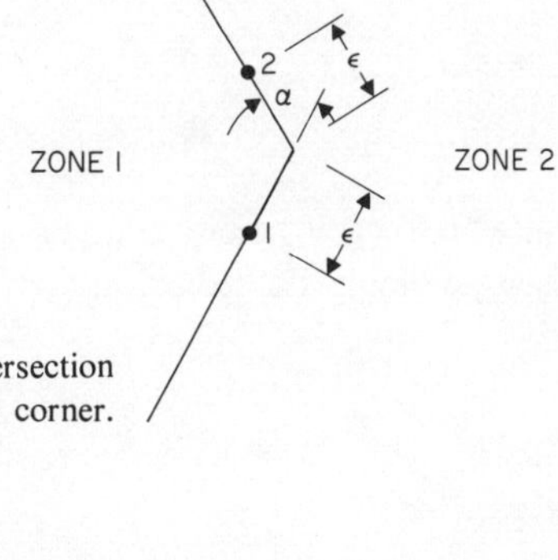


**Figure 4.5** The intersection of two zones at a corner.

Since the potential is continuous along the interzonal boundary and between zones, its derivative along the interzonal boundary is the same whether viewed from zone 1 or zone 2; that is, $\partial\Phi^{(1)}/\partial s = \partial\Phi^{(2)}/\partial s$. But, at the kink, $\partial\Phi/\partial s$ and $\partial\Phi/\partial n$ are not independent:

$$\left.\begin{aligned}
\frac{\partial\Phi_2}{\partial s} &= \frac{\partial\Phi_1}{\partial s}\cos\alpha - \frac{\partial\Phi_1^{(1)}}{\partial n}\sin\alpha \\
\frac{\partial\Phi_2^{(1)}}{\partial n} &= \frac{\partial\Phi_1}{\partial s}\sin\alpha + \frac{\partial\Phi_1^{(1)}}{\partial n}\cos\alpha \\
\frac{\partial\Phi_2}{\partial s} &= \frac{\partial\Phi_1}{\partial s}\cos\alpha + \frac{\partial\Phi_1^{(2)}}{\partial n}\sin\alpha \\
\frac{\partial\Phi_2^{(1)}}{\partial n} &= \frac{\partial\Phi_1}{\partial s}\sin\alpha - \frac{\partial\Phi_1^{(2)}}{\partial n}\cos\alpha
\end{aligned}\right\} \quad \varepsilon = 0 \tag{4.9}$$

Equating the first and third or second and fourth of (4.9) produces

$$\frac{\partial\Phi_1^{(1)}}{\partial n} = -\frac{\partial\Phi_1^{(2)}}{\partial n} \tag{4.10}$$

Obviously (4.10) is not compatible with (4.8) unless $K_1 = K_2$ (trivial case), $\partial\Phi_1^{(1)}/\partial n = \partial\Phi_1^{(2)}/\partial n = 0$ (stagnation point), or $\partial\Phi_1^{(1)}/\partial n = -\partial\Phi_1^{(2)}/\partial n = \infty$ (singular point).

In order to study the general case of Fig. 4.4 we write Laplace's equation in cylindrical co-ordinates:

$$\frac{\partial^2\Phi}{\partial r^2} + \frac{1}{r}\frac{\partial\Phi}{\partial r} + \frac{1}{r^2}\frac{\partial^2\Phi}{\partial\theta^2} = 0 \tag{4.11}$$

in which $r$ is measured from the point of intersection of the zones and $\theta$ is shown in Fig. 4.4. A general solution without vortices or sources in zone $m$ is

$$\Phi_m = r^\lambda(A_m \cos\lambda\theta + B_m \sin\lambda\theta) \tag{4.12}$$

in which $A_m$, $B_m$, and $\lambda$ are constants. The conditions to be satisfied on the lines dividing the zones are

$$\Phi_m = \Phi_{m+1} \tag{4.13}$$

$$K_m \frac{\partial \Phi_m}{\partial \theta} = K_{m+1} \frac{\partial \Phi_{m+1}}{\partial \theta} \tag{4.14}$$

Using (4.12) in (4.13) yields

$$A_1 \cos \lambda\alpha + B_1 \sin \lambda\alpha = A_2 \cos \lambda\alpha + B_2 \sin \lambda\alpha$$

$$A_2 \cos \lambda(\alpha + \delta_2) + B_2 \sin \lambda(\alpha + \delta_2) = A_3 \cos \lambda(\alpha + \delta_2) + B_3 \sin \lambda(\alpha + \delta_2)$$

$$\cdots$$

$$A_m \cos \lambda(\alpha + \delta_2 + \cdots + \delta_m) + B_m \sin \lambda(\alpha + \delta_2 + \cdots + \delta_m)$$

$$= A_{m+1} \cos (\alpha + \delta_2 + \cdots + \delta_m) + B_{m+1} \sin \lambda(\alpha + \delta_2 + \cdots + \delta_m)$$

$$\cdots$$

$$A_M \cos \lambda(\alpha + \delta_2 + \cdots + \delta_M) + B_M \sin \lambda(\alpha + \delta_2 + \cdots + \delta_M)$$

$$= A_1 \cos \lambda(\alpha - \delta_1) + B_1 \sin \lambda(\alpha - \delta_1) \tag{4.15}$$

Note that the angle in the last term is $\alpha - \delta_1$, not $\alpha + \delta_2 + \cdots + \delta_M$, since the range of $\theta$ in zone 1 is $\alpha - \delta_1 \leqslant \theta \leqslant \alpha$ and the angle $\alpha + \delta_2 + \cdots + \delta_M$ is outside that range. Using (4.12) in (4.14) gives a sequence of equations similar to (4.15). The general relationship is

$$\begin{aligned} K_m[-A_m \sin \lambda(\alpha + \delta_2 + \cdots + \delta_m) + B_m \cos \lambda(\alpha + \delta_2 + \cdots + \delta_m)] \\ = K_{m+1}[-A_{m+1} \sin \lambda(\alpha + \delta_2 + \cdots + \delta_m) \\ + B_{m+1} \cos \lambda(\alpha + \delta_2 + \cdots + \delta_m)] \end{aligned} \tag{4.16}$$

Equations (4.15) and (4.16) form $2M$ equations in the $2M$ unknowns $A_1, A_2, \ldots, A_M, B_1, B_2, \ldots, B_M$. Since these equations are homogeneous, the determinant of the coefficient matrix must be zero in order that the $A_m$ and $B_m$ are not all zero. This determinant forms an equation in the eigenvalues, $\lambda$. Only positive $\lambda$ is considered since a negative $\lambda$ would lead to a non-integrable singularity (an infinitely strong source). The lowest value of $\lambda$, say $\lambda_0$, determines the type of singularity at the intersection point. Since $\partial\Phi/\partial r \sim r^{\lambda-1}$, a value of $\lambda_0 < 1$ indicates a point of infinite velocity whereas a value of $\lambda_0 > 1$ indicates a stagnation point.

The equation for $\lambda$ which results from the coefficient matrix of (4.15) and (4.16) can be worked out for any number of zones which meet at a common point. For five zones

$$2\left(1-\prod_{i=1}^{5}\cos\lambda\delta_i\right)+\left[\prod_{i=1}^{5}\cos\lambda\delta_i\right]\left[\prod_{i=1}^{5}\sum_{\substack{j=1\\j\neq i}}^{5}\frac{K_i}{K_j}t_it_j\right.$$

$$\left.+\sum_{i=1}^{2}\sum_{\substack{j=2\\j\neq i}}^{3}\sum_{\substack{n=3\\n\neq j}}^{4}\sum_{\substack{l=4\\l\neq n}}^{5}\left(\frac{K_iK_n}{K_jK_l}+\frac{K_jK_l}{K_iK_n}\right)t_it_jt_nt_l\right]=0 \tag{4.17}$$

in which $t_i = \tan \lambda\delta_i$. The corresponding equation for four zones is obtained from (4.17) by setting $\delta_5 = 0$, for three zones by setting $\delta_5 = \delta_4 = 0$, and for two zones by setting $\delta_5 = \delta_4 = \delta_3 = 0$.

Consider the intersection of two zones. Then (4.17) reduces to

$$(K_1^2 + K_2^2)\sin\lambda\delta_1\sin\lambda\delta_2 - 2K_1K_2\cos\lambda\delta_2 + 2K_1K_2 = 0 \tag{4.18}$$

Using trigonometric identities (4.18) can be expressed as

$$\frac{\sin|(\lambda/2)(\delta_1+\delta_2)|}{\sin|(\lambda/2)(\delta_1-\delta_2)|} = \pm\frac{K_1-K_2}{K_1+K_2} \tag{4.19}$$

in which the $\pm$ sign results from taking a square root. To determine the proper sign consider the flow around a solid wedge where $K_1 = 0$ and the wedge angle is $\delta_1$. Then since $\delta_2 = 2\pi - \delta_1$

$$\frac{\sin\lambda\pi}{\sin\lambda(\pi-\delta_1)} = \pm(-1) \tag{4.20}$$

For flow around a sharp corner the potential behaves as

$$\Phi \sim r^\lambda \qquad \lambda = \pi/(2\pi - \delta) \tag{4.21}$$

Equation (4.20) fits the criterion of (4.21) if the minus sign is chosen. Then (4.19) can be written

$$K_1\tan(\lambda\delta_1/2) = -K_2\tan(\lambda\delta_2/2) \tag{4.22}$$

which is the equation to be used to find the eigenvalues $\lambda$.

Consider the case of flow through a channel with two zones separated by an interface at some angle $\alpha$ (Fig. 4.6). The solution in the neighborhood of point $B$ is a special case of the two zone problem with $\delta_2 = 2\alpha$ and $\delta_1 = 2\pi - 2\alpha$. Thus (4.22) becomes

$$K_1\tan\lambda(\pi-\alpha) = -K_2\tan\lambda\alpha \tag{4.23}$$

Figure 4.7 shows a plot of $\lambda_0$ *vs.* $\alpha$ for various values of $P = K_1/K_2$. From Fig. 4.7 it is evident that $1.5 > \lambda_0 > 0.5$. The minimum value of $\lambda$ occurs at $K_2/K_1 = \infty$ and is

$$\lambda = \pi/(2\alpha) \tag{4.24}$$

Thus the strongest possible singularity is $\lambda = 0.5$, which is the cut-off wall. Lafe *et al.* (1980) present experimental results of the type of interzonal flow considered above. Their results tend to confirm the analysis and calculations by the BIEM.

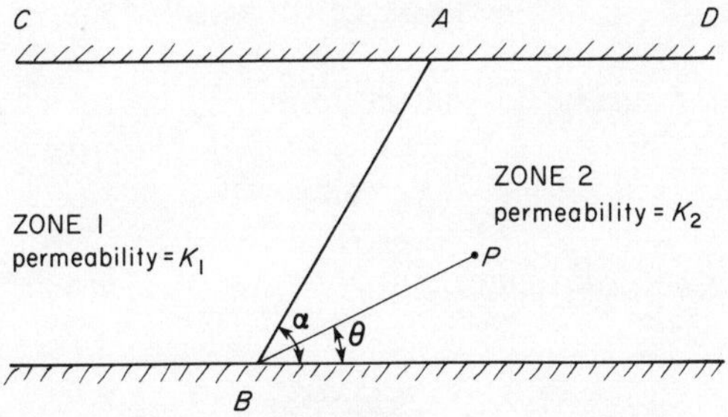


**Figure 4.6** Flow between zones of different permeability in a channel.

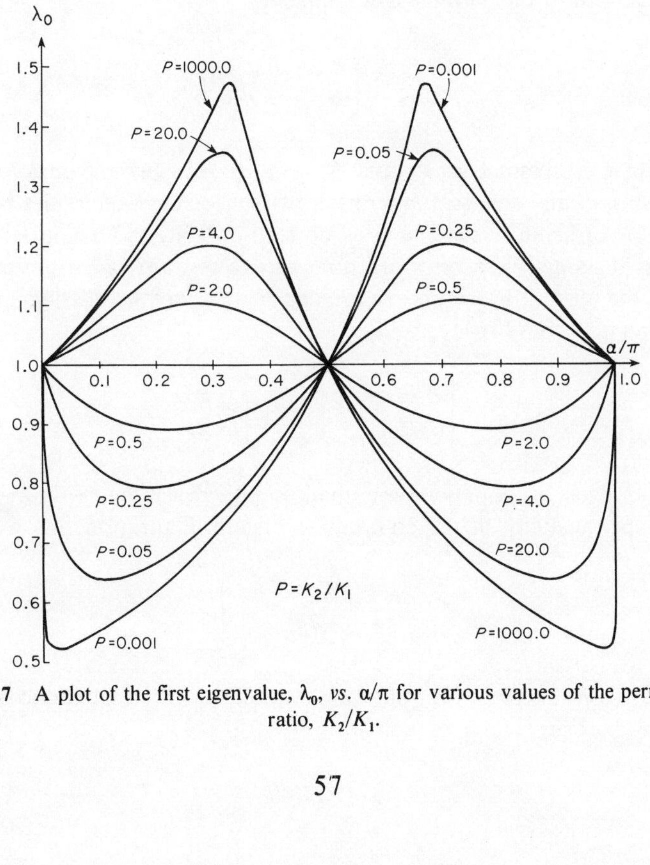


**Figure 4.7** A plot of the first eigenvalue, $\lambda_0$, *vs.* $\alpha/\pi$ for various values of the permeability ratio, $K_2/K_1$.

From a practical point of view it appears unnecessary to extend the above analysis to the more general case of Fig. 4.4. In most calculations the intersection point is simply avoided, but points are taken on each of the interzonal boundaries close to the intersection. The results may not be entirely accurate very close to the intersection but are usually more than accurate enough for practical purposes. The same point of view has been used when calculating with finite elements of finite differences, often without the analyst realizing that a singularity exists.

## 4.4 Anisotropic aquifers

Consider (1.9) for steady flow in the form

$$\nabla \cdot (\boldsymbol{K} \cdot \nabla\Phi) = 0 \tag{4.25}$$

in which $\boldsymbol{K}$ is the permeability tensor.

### *4.4.1 Principal axes*

In two dimensions $K$ has three independent components (see Section 1.5 for the parallel three-dimensional development)

$$\boldsymbol{K} = \begin{bmatrix} K_{xx} & K_{xy} \\ K_{yx} & K_{yy} \end{bmatrix} \tag{4.26}$$

in which $\boldsymbol{K}$ is symmetric, so that $K_{xy} = K_{yx}$. Any two-dimensional symmetric tensor can be written in terms of two independent quantities by rotating the co-ordinate axes. The axes of a co-ordinate system in which the tensor is diagonal, with only two non-zero values, are called the principal axes of the tensor. These two values can be obtained by solving for $\lambda$ the determinant (Long 1961)

$$\begin{vmatrix} K_{xx} - \lambda & K_{xy} \\ K_{yx} & K_{yy} - \lambda \end{vmatrix} = 0 \tag{4.27}$$

Since the algebraic equations are quadratic, there will be two values of $\lambda$, and the permeability in the co-ordinate system of the principal axes is

$$\boldsymbol{K}_* = \begin{bmatrix} K_{x_*} & 0 \\ 0 & K_{y_*} \end{bmatrix} \tag{4.28}$$

where $K_{x_*} = \lambda^{\mathrm{I}}$ (the first root of (4.27)) and $K_{y_*} = \lambda^{\mathrm{II}}$. The directions of the principal axes are found by solving

$$a_1(K_{xx} - \lambda) + a_2 K_{xy} = 0$$
$$a_1 K_{yx} + a_2(K_{yy} - \lambda) = 0 \tag{4.29}$$

for $a_1$ and $a_2$ with $\lambda = \lambda^{\text{I}}$. Then the direction of the principal $x$-axis is along the vector

$$\mathbf{a} = a_1 \boldsymbol{\delta}_{x_*} + a_2 \boldsymbol{\delta}_{y_*} \tag{4.30}$$

in which $\boldsymbol{\delta}_{x_*}$ and $\boldsymbol{\delta}_{y_*}$ are unit vectors in the $x_*$ and $y_*$ directions respectively. The other principal axis is found by a 90° rotation or by solving (4.29) with $\lambda = \lambda^{\text{II}}$. Note that the vector **a** of (4.30) cannot be found uniquely, but its direction is unique. We can find, for example,

$$(1/a_1)\mathbf{a} = \boldsymbol{\delta}_{x_*} + (a_2/a_1)\, \boldsymbol{\delta}_{y_*} \tag{4.31}$$

The angle the principal axes makes with the $x$–$y$ co-ordinate system is

$$\theta_* = \arctan \frac{K_{x_*} - K_{xx}}{K_{xy}} \tag{4.32}$$

Usually in groundwater basins the components of the permeability tensor are given as in (4.28) together with the direction of the principal axes instead of in the form of (4.26). Thus, there is no need to perform the preceding transformations.

### 4.4.2 *Stretched co-ordinates*

In the principal axes co-ordinate system (4.25) becomes

$$K_{x_*} \frac{\partial^2 \Phi}{\partial x_*^2} + K_{y_*} \frac{\partial^2 \Phi}{\partial y_*^2} = 0 \tag{4.33}$$

In order to reduce this equation to Laplace's equation, we define another co-ordinate system

$$x_{**} = x_* \qquad y_{**} = y_*(K_{x_*}/K_{y_*})^{\frac{1}{2}} \tag{4.34}$$

Then

$$\frac{\partial^2 \Phi}{\partial x_{**}^2} + \frac{\partial^2 \Phi}{\partial y_{**}^2} = 0 \tag{4.35}$$

and we have the simplicity of Laplace's equation.

### 4.4.3 Boundary conditions

In solving (4.35) it is necessary to take some care with the boundary conditions. For the Dirichlet problem ($\Phi$ given) there is no difficulty; the potential is given at each node as in any other problem. The locations of the nodal points are usually defined in a global co-ordinate system. If the problem contains zones, the permeabilities or transmissivities will be given in a zonal co-ordinate system $(x_*, y_*)$. If, in addition, the problem is anisotropic, the calculations will be done in a stretched co-ordinate system $(x_{**}, y_{**})$. In the latter case the boundary conditions are written in the global system and must be transformed into the stretched system.

In the case of Neumann boundary conditions the flux, $q_n$, normal to the boundary is given and that quantity must be written in terms of the normal derivative, $\partial\Phi/\partial n_{**}$, in the stretched co-ordinate system.

The flow normal to the boundary is

$$q_n = -\lfloor n_x \quad n_y \rfloor \begin{bmatrix} K_{xx} & K_{xy} \\ K_{yx} & K_{yy} \end{bmatrix} \begin{Bmatrix} \partial\Phi/\partial x \\ \partial\Phi/\partial y \end{Bmatrix} \tag{4.36}$$

in which $n_x$ and $n_y$ are the components of the unit normal vector in the $x$ and $y$ directions. In almost all problems the permeabilities are known in the co-ordinate system of the principal axes, so

$$q_n = q_{n_*} = -\lfloor n_{x_*} \quad n_{y_*} \rfloor \begin{bmatrix} K_{x_*} & 0 \\ 0 & K_{y_*} \end{bmatrix} \begin{Bmatrix} \partial\Phi/\partial x_* \\ \partial\Phi/\partial y_* \end{Bmatrix} \tag{4.37}$$

or

$$q_n = -n_{x_*} K_{x_*} \frac{\partial\Phi}{\partial x_*} - n_{y_*} K_{y_*} \frac{\partial\Phi}{\partial y_*} \tag{4.38}$$

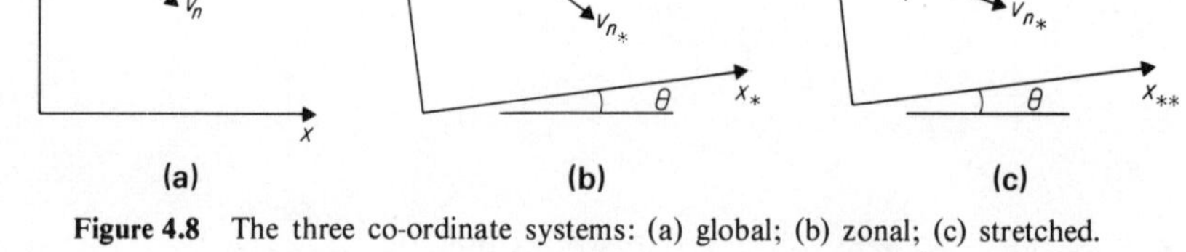


**Figure 4.8** The three co-ordinate systems: (a) global; (b) zonal; (c) stretched.

The three co-ordinate systems, global $(x, y)$, zonal $(x_*, y_*)$, and stretched $(x_{**}, y_{**})$ are shown in Figure 4.8. From the transformation of co-ordinates (4.34),

$$\frac{\partial\Phi}{\partial y_{**}} = \frac{\partial\Phi}{\partial y_*}\frac{dy_*}{dy_{**}} = \left(\frac{K_{y_*}}{K_{x_*}}\right)^{\frac{1}{2}}\frac{\partial\Phi}{\partial y_*} \tag{4.39}$$

Therefore†

$$q_{x_*} = q_{x_{**}} \qquad q_{y_*} = (K_{y_*}/K_{x_*})^{\frac{1}{2}}\, q_{y_{**}} \tag{4.40}$$

In the stretched co-ordinates

$$\begin{aligned} q_{x_{**}} &= q_{n_{**}} \sin\psi_{**} + q_{s_{**}} \cos\psi_{**} \\ q_{y_{**}} &= -q_{n_{**}} \cos\psi_{**} + q_{s_{**}} \sin\psi_{**} \end{aligned} \tag{4.41}$$

in which $q_{s_{**}}$ is the velocity component parallel to the boundary. In the zonal co-ordinates

$$q_{n_*} = q_{x_*} \sin\psi_* - q_{y_*} \cos\psi_* \tag{4.42}$$

Using (4.40) and (4.41) in (4.42) yields

$$\begin{aligned} q_{n_*} = {} & q_{n_{**}}(\sin\psi_{**} \sin\psi_* + (K_{y_*}/K_{x_*})^{\frac{1}{2}} \cos\psi_{**} \cos\psi_*) \\ & + q_{s_{**}}(\cos\psi_{**} \sin\psi_* - (K_{y_*}/K_{x_*})^{\frac{1}{2}} \sin\psi_{**} \cos\psi_*) \end{aligned} \tag{4.43}$$

The relationship between $\psi_*$ and $\psi_{**}$ is

$$\psi_{**} = \arctan\,[(K_{x_*}/K_{y_*})^{\frac{1}{2}}\psi_*] \tag{4.44}$$

Using (4.44) in (4.43) with the trigonometric identity

$$\cos\psi = \frac{1}{(1 + \tan^2\psi)^{\frac{1}{2}}} \tag{4.45}$$

yields

$$q_n = q_{n_*} = \left(\frac{K_{y_*}}{K_{x_*}}\right)^{\frac{1}{2}}\left[\frac{1 + \tan^2\psi_{**}}{1 + (K_{y_*}/K_{x_*})\tan^2\psi_{**}}\right]^{\frac{1}{2}} q_{n_{**}} \tag{4.46}$$

†Note that velocity $(q_y)$ transforms as a contravariant vector whereas the derivatives of the potential (e.g. $\partial\Phi/\partial y$) transform as a covariant vector (see, e.g., Lass 1950). In this case $q_{y_*} = q_{y_{**}}(dy_*/dy_{**})$.

Thus the normal velocity in zonal (and global) co-ordinates is written in terms of the normal velocity in stretched co-ordinates. If $q_n$ is given as a boundary condition, (4.46) can be solved for $q_{n_{**}}$ and that value used as $q_{n_{**}} = -\partial\Phi/\partial n_{**}$ to find the normal derivative in the stretched co-ordinate system. On the other hand if $q_{n_{**}}$ is a part of the solution on a Dirichlet boundary, (4.46) gives the real normal velocity, $q_n$.

On interior boundaries (between zones) the boundary conditions are given by (4.5) and (4.7). Equation (4.5) is applied in the usual manner. Equation (4.7) is used to find the relationship between $q_{n_{**}}^{(1)}$ and $q_{n_{**}}^{(2)}$. In the computer program, however, it is $\partial\Phi/\partial n_{**}$ which is computed on the boundary, not $q_{n_{**}}$ directly. Since $q_{x_*} = -K_{x_*}(\partial\Phi/\partial x_*)$ and $q_{y_*} = -K_{y_*}(\partial\Phi/\partial y_*)$, and since from (4.34)

$$\frac{\partial\Phi}{\partial x_{**}} = \frac{\partial\Phi}{\partial x_*} \quad \text{and} \quad \frac{\partial\Phi}{\partial y_{**}} = \left(\frac{K_{y_*}}{K_{x_*}}\right)^{\frac{1}{2}} \frac{\partial\Phi}{\partial y_*},$$

$$q_{x_{**}} = -K_{x_*}\frac{\partial\Phi}{\partial x_{**}} \qquad q_{y_{**}} = -K_{y_*}\frac{K_{x_*}}{K_{y_*}}\frac{\partial\Phi}{\partial y_{**}} \tag{4.47}$$

Thus the flow rates in the stretched co-ordinate system are a function of $K_{x_*}$ but not $K_{y_*}$. Using (4.46) and (4.47) in the second condition (4.7) gives

$$a^i\frac{\partial\Phi^i}{\partial n_{**}} = -a^k\frac{\partial\Phi^k}{\partial n_{**}} \tag{4.48}$$

in which the adjacent zones are numbered $i$ and $k$ and

$$a = \left[K_{x_*}K_{y_*}\frac{1 + \tan^2\psi}{1 + (K_{y_*}/K_{x_*})\tan^2\psi}\right]^{\frac{1}{2}} \tag{4.49}$$

# 5 Two-dimensional free surface flow problems

In this chapter we present a numerical scheme for implementing the free surface boundary conditions in the BIEM calculation. Although only two-dimensional problems are presented, the method is extended to axial symmetric and three-dimensional problems in later chapters.

A numerical example concerning the seepage flow in an earth dam is used to illustrate the assembly of matrix equations. Other examples include sudden drawdown in a zoned dam and recharge problems.

## 5.1 Free surface boundary conditions in a finite difference form

In order to solve the system of algebraic equations for $\Phi$ and $\partial\Phi/\partial n$ on the boundary, (2.34), the relation between $\Phi$ and $\partial\Phi/\partial n$, must be known on the free surface. For two-dimensional problems, the free surface boundary conditions can be reduced from (1.42a) and (1.42b) to be

$$\Phi = \eta(x, t) \qquad \text{on } z = \eta \tag{5.1}$$

$$\frac{\partial\eta}{\partial t} = -\left[1 + \left(\frac{\partial\eta}{\partial x}\right)^2\right]^{\frac{1}{2}} \frac{\partial\Phi}{\partial n} \qquad \text{on } z = \eta \tag{5.2}$$

If $\beta$ is defined as the angle between the free surface profile and the $x$-axis (see Fig. 5.1), then we have

$$\frac{\partial\eta}{\partial x} = -\tan\beta \tag{5.3}$$

Substitutions of (5.1) and (5.3) into (5.2) yield the following equation:

$$\left(\frac{\partial\Phi}{\partial t}\right)_x = -\frac{1}{\cos\beta}\frac{\partial\Phi}{\partial n} \qquad \text{on } z = \eta \tag{5.4}$$

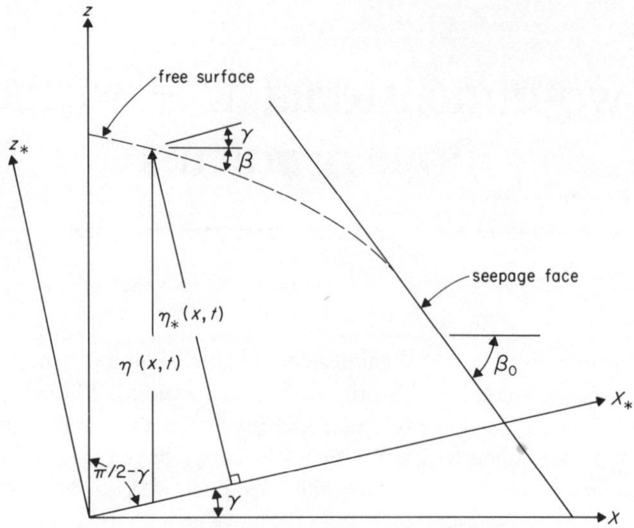


**Figure 5.1** The transformed co-ordinate system.

where the subscript $x$ indicates that (5.4) is the rate of change of the potential following the free surface in the vertical direction and holding the $x$ co-ordinate constant.

The free surface boundary condition (5.4) can be written in a finite difference form for a small time step, $\Delta t$, which gives the potential at time $k\,\Delta t$ as follows:

$$\Phi^{k+1} = \Phi^k - \frac{\Delta t}{\cos \beta^k}\left[\theta\left(\frac{\partial \Phi}{\partial n}\right)^{k+1} + (1-\theta)\left(\frac{\partial \Phi}{\partial n}\right)^k\right] \tag{5.5}$$

in which $\theta$ is a weighting factor that positions the derivative between time levels $k$ and $k+1$. Equation (5.5) establishes a relation between $\Phi$ and $\partial\Phi/\partial n$ at the time level $k+1$, and it can be used in (2.34) for solving unknown boundary variables. We reiterate that (5.5) is only applicable on the free surface with a fixed value of the $x$ co-ordinate. Moreover, since $\Phi$ and $\eta$ are exchangeable under the circumstance, (5.5) can also be used for computations of the new free surface elevation $\eta^{k+1}$ at time level $(k+1)\,\Delta t$ after the normal derivative of the potential $\partial\Phi/\partial n$ has been calculated. Finally, it should be pointed out that in (5.5) the angle $\beta$ is computed at the time level $k$ even though the equation is written for the level $k+1$. Although this problem can be avoided by iteration, a small time step, $\Delta t$, may provide sufficient accuracy.

For some problems it is convenient to be able to describe the change of the potential on the free surface along an arbitrary straight line. Thus a transformed co-ordinate system, $(x_*, z_*)$, is defined (Fig. 5.1) such that

$$
\begin{aligned}
x &= x_* \cos\gamma - z_* \sin\gamma \\
z &= x_* \sin\gamma + z_* \cos\gamma
\end{aligned}
\tag{5.6}
$$

in which $\gamma$ represents the angle of rotation. In terms of the new co-ordinate system the free surface elevation is described by $z_* = \eta_*(x_*, t)$. From (5.6) we have

$$\eta(x, t) = x_* \sin\gamma + \eta_*(x_*, t)\cos\gamma \qquad z_* = \eta_* \tag{5.7}$$

Substituting the boundary condition (5.1) into (5.7) yields

$$\Phi(x, t) = x_* \sin\gamma + \eta_*(x_*, t)\cos\gamma \qquad z_* = \eta_* \tag{5.8}$$

The change of the $\Phi$ on the free surface along $x_* =$ constant can be evaluated from (5.8) to be

$$(\Delta\Phi)_{x_*} = (\Delta\eta_*)_{x_*} \cos\gamma \qquad z_* = \eta_* \tag{5.9}$$

Referring to Fig. 5.1, the free surface slope can be redefined in terms of the transformed co-ordinates $(x_*, z_*)$ as follows:

$$\frac{\partial\eta_*}{\partial x_*} = -\tan(\gamma + \beta) \tag{5.10}$$

Similarly, the kinematic free surface boundary condition can be rewritten as

$$\frac{\partial\eta_*}{\partial t} = -\frac{1}{\cos(\gamma+\beta)}\frac{\partial\Phi}{\partial n} \qquad \text{on } z_* = \eta_* \tag{5.11}$$

Now we can substitute (5.9) into (5.11) to get the rate of change of the potential on the free surface in the $z_*$ direction,

$$\left(\frac{\partial\Phi}{\partial t}\right)_* = -\frac{\cos\gamma}{\cos(\gamma+\beta)}\frac{\partial\Phi}{\partial n} \qquad \text{on } z_* = \eta_* \tag{5.12}$$

Finally, the finite difference form of the above equation can be written as

$$\Phi^{k+1} = \Phi^k - \frac{(\Delta t)\cos\gamma}{\cos(\gamma+\beta^k)}\left[\theta\left(\frac{\partial\Phi}{\partial n}\right)^{k+1} + (1-\theta)\left(\frac{\partial\Phi}{\partial n}\right)^k\right] \tag{5.13}$$

## 5.2 Matrix formulation and assembly of equations

The seepage flow through an earth dam is used here as an example to illustrate the computational procedure for unsteady free surface flow problems. As shown in Fig. 5.2, a rectangular earth dam is fully saturated up to a constant level, $z = h_1$, initially. At $t = 0$ the water level in the downstream reservoir is suddenly dropped to a lower elevation $z = h_2 < h_1$, and it remains there for all time. The objective is to find the transient free

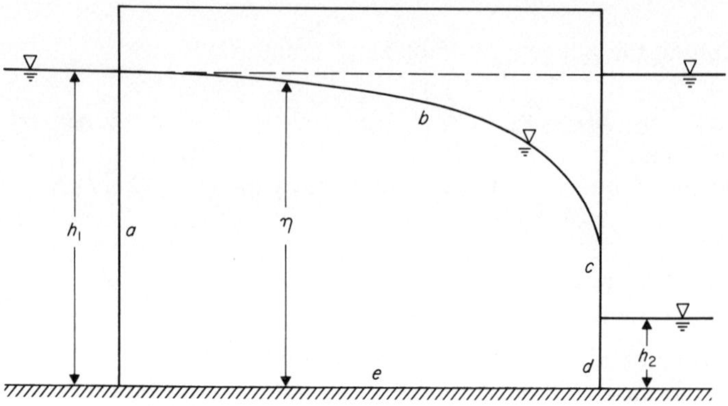


**Figure 5.2** The definition sketch of a sudden drawdown problem.

surface movement. If we assign $a$ as the upstream face, $b$ the free surface, $c$ the seepage surface, $d$ the downstream face, and $e$ the impervious bottom, the boundary conditions can be summarized as follows:

$$\Phi_a = h_1 \qquad \Phi_c = z \qquad \Phi_d = h_2 \qquad \frac{\partial \Phi_e}{\partial n} = 0 \tag{5.14}$$

$$\Phi_b^{k+1} = \Phi_b^k - \frac{\Delta t}{\cos \beta^k}\left[\theta\left(\frac{\partial \Phi_b}{\partial n}\right)^{k+1} + (1-\theta)\left(\frac{\partial \Phi_b}{\partial n}\right)^k\right] \qquad \eta^{k+1} = \Phi_b^{k+1}$$

in which $\eta$ measures the free surface elevation, and the superscript $k$ denotes the time level. The initial conditions are given as follows:

$$\Phi_a = \Phi_b^0 = \Phi_c = \Phi_d = \Phi_e = h_1 \qquad \eta^0 = h_1 \qquad \beta^0 = 0 \tag{5.15}$$

Assuming that the solution along the boundary has already been obtained at time $k\,\Delta t$, we shall try to find the solution at the next time step, $(k+1)\,\Delta t$, from (2.34). We can rewrite the equation in the following form

$$|R|\{\Phi\} = |L|\{\partial\Phi/\partial n\} \tag{5.16}$$

these matrices are partitioned into parts representing five segments of the boundary:

$$[-L_a \quad -L_b \quad -L_c \quad -L_d \quad R_e] \begin{Bmatrix} (\partial \Phi_a^{k+1}/\partial n) \\ (\partial \Phi_b^{k+1}/\partial n) \\ (\partial \Phi_c^{k+1}/\partial n) \\ (\partial \Phi_d^{k+1}/\partial n) \\ \Phi_e^{k+1} \end{Bmatrix}$$

$$= [-R_a \quad -R_b \quad -R_c \quad -R_d \quad L_e] \begin{Bmatrix} \Phi_a^{k+1} \\ \Phi_b^{k+1} \\ \Phi_c^{k+1} \\ \Phi_d^{k+1} \\ (\partial \Phi_e^{k+1}/\partial n) \end{Bmatrix} \tag{5.17}$$

Substitution of $\Phi_b^{k+1}$, from (5.14), into the right-hand side of the above equation yields

$$\left[-L_a \quad -L_b - R_b \frac{\Delta t \theta}{\cos \beta^k} \quad -L_c \quad -L_d \quad R_e\right] \begin{Bmatrix} (\partial \Phi_a^{k+1}/\partial n) \\ (\partial \Phi_b^{k+1}/\partial n) \\ (\partial \Phi_c^{k+1}/\partial n) \\ (\partial \Phi_d^{k+1}/\partial n) \\ \Phi_e^{k+1} \end{Bmatrix}$$

$$= [-R_a \quad -R_b \quad -R_c \quad -R_d \quad L_e] \begin{Bmatrix} \Phi_a^{k+1} \\ \Phi_b^k - \dfrac{\Delta t}{\cos \beta^k} (1 - \theta) \left( \dfrac{\partial \Phi_b^k}{\partial n} \right) \\ \Phi_c^{k+1} \\ \Phi_d^{k+1} \\ (\partial \Phi_e^{k+1}/\partial n) \end{Bmatrix} \tag{5.18}$$

in which all of the unknowns at time level $k + 1$ are written on the left-hand side. Boundary conditions and the solutions obtained from the previous time level are used in the right-hand side of (5.18). The $[R]$ and $[L]$ matrices and the surface slope $\beta$ are dependent upon the geometry of the boundary; therefore, they change according to the position of the free surface. In (5.18)

$|R|$, $|L|$, and $\beta$ are computed for time level $k$ even though the equation is written for time level $k+1$. This problem can be avoided by iteration. However, a small time step usually provides sufficient accuracy.

A numerical example of Figure 5.3 was computed assuming sudden drawdown from height $h_1 = 10$ to height $h_2 = 3$. The boundary was divided into 24 segments. Linear elements were used except the last free surface segment to the seepage face, where a quadratic element was installed. The reason for using this special element was to ensure that the free surface is always tangential to the seepage face. Consequently, the calculation was not begun at time $t = 0$, which gave a horizontal free surface and violated the tangency condition, but instead used the time series solution of Liggett and Hadjitheodorou (1965) to obtain the free surface at $t = 0.6$. In Figure 5.4

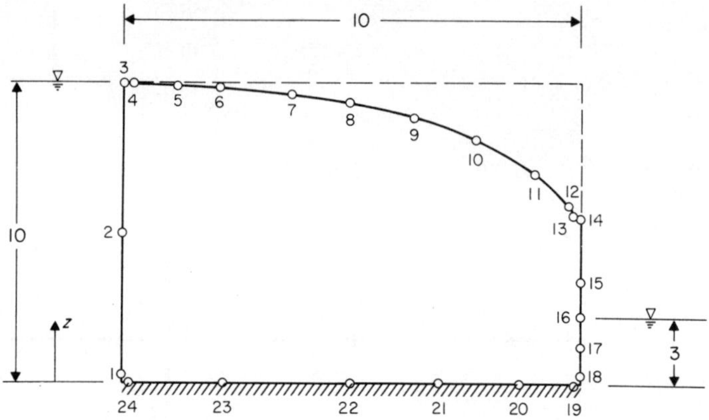


**Figure 5.3** Sample problem with a free surface and seepage face.

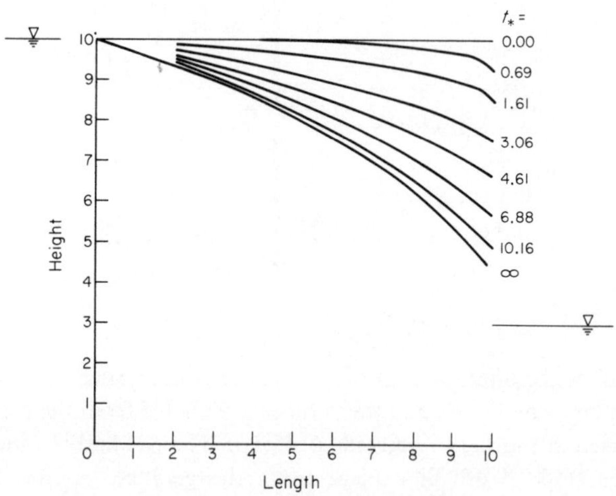


**Figure 5.4** Successive surface profiles for the sudden drawdown problem.

the successive free surface profiles are shown; the steady state was reached at $t = 30$.

Several remarks concerning the problem should be made here: (1) There is a singularity at the intersection of the seepage surface and the tail water. The special treatment of the singularity is presented in Chapter 3. In the computations shown in Figure 5.3, the singularity has been ignored, and it does not seem to affect the overall solutions. (2) The installation of the quadratic element on the free surface at its intersection with the seepage surface is not absolutely necessary. The replacement of the quadratic element by a linear element would violate the tangency condition; however, it has little effect on the accuracy of the entire solution. (3) The positions of nodal points on the seepage surface are changed in time such that they would not overlap with the free surface point.

An additional example of unsteady flow through a dam where the height of the water is increased on the upstream face is given in Section 11.3. The results of that computation are shown in Figure 11.5.

## 5.3 Free surface flow through a zoned dam

The assembly of matrix equation (5.16) is quite different for different boundary value problems. Sometimes it becomes a major effort as far as the computer programming is concerned. The numerical computations for $[R]$ and $[L]$ are basically the same for all problems. To demonstrate the involvement of the assembly of equations a problem concerning the seepage flow through a zoned dam is studied herein.

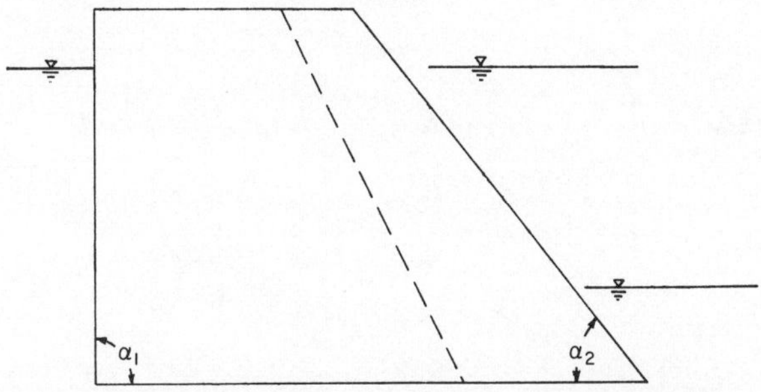


**Figure 5.5** The definition sketch of a zoned dam problem.

Shown in Figure 5.5 is an earth dam which consists of two zones of different permeabilities and is subject to sudden drawdown on the downstream sloping surface. We seek the position of the free surface as a function of time from $t = 0$ until it reaches steady state. To solve this problem by the BIEM, we divide the external and internal boundaries into small linear segments (Fig. 5.6). The external boundary conditions and the internal compatibility conditions can be summarized as follows:

Zone 1:
- pts 1 to $i_1$; $\Phi_1 = H_1$
- pts $i_1 + 1$ to $i_2$; $\Phi_1 = \eta_1$ and (5.13)
- pts $i_2$ to $i_4$; $K_1(\partial\Phi_1/\partial n) = -K_2(\partial\Phi_2/\partial n)$ and $\Phi_1 = \Phi_2$
- pts $i_4 + 1$ to $n_1$; $\partial\Phi_1/\partial n = 0$

Zone 2:
- pts 1 to $i_1 + 1$; $K_1(\partial\Phi_1/\partial n) = -K_2(\partial\Phi_2/\partial n)$ and $\Phi_1 = \Phi_2$
- pts $i_1 + 1$ to $i_2$; $\Phi_2 = \eta_2$ and (5.13)
- pt $i_2 + 1$; $\Phi_2 = \eta_2$ and $\partial\Phi_2/\partial n = 0$
- pts $i_2 + 2$ to $i_3$; $\Phi_2 = z$
- pts $i_3$ to $i_4$; $\Phi_2 = H_2$
- pts $i_4$ to $n_2$; $\partial\Phi_2/\partial n = 0$

The respective dimensionless coefficients of permeability $K_1$ and $K_2$ should be inserted in the second term of the right-hand side of (5.13) for each zone.

Using the boundary conditions in (5.16), we can assemble the matrix equation into the following form

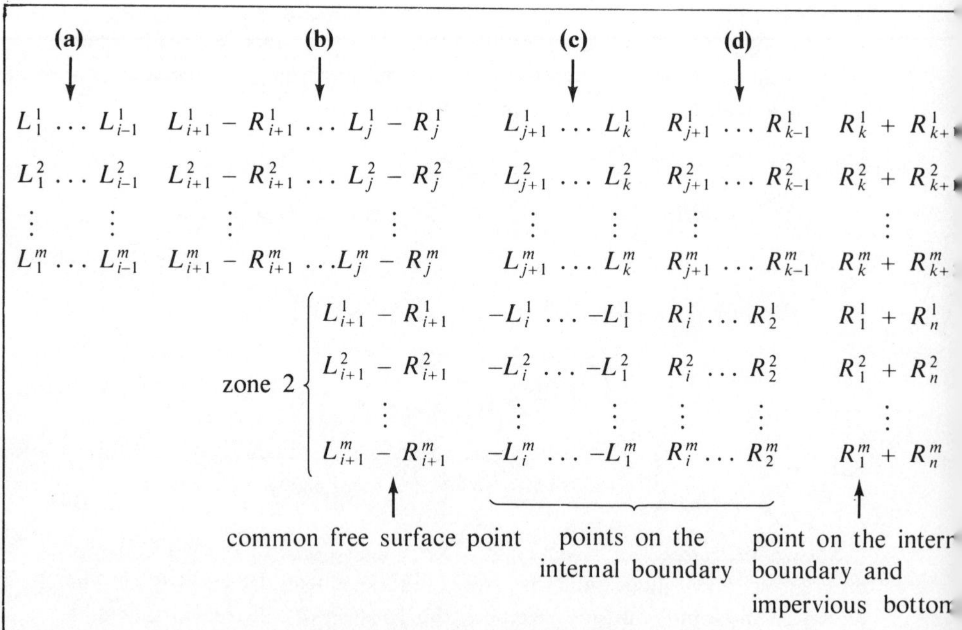


**Figure 5.7** The assembled coefficient matrix $[H]$.

$$[H]\{u\} = \{r\} \tag{5.19}$$

where $[H]$ is a coefficient matrix, $\{u\}$ is a column vector containing all unknowns and $\{r\}$ is a known column vector. The details of the coefficient matrix $[H]$ are shown in Figure 5.7 and are explained as follows:

(a) The first numbers are the coefficients of $\partial\Phi/\partial n$ along the upstream

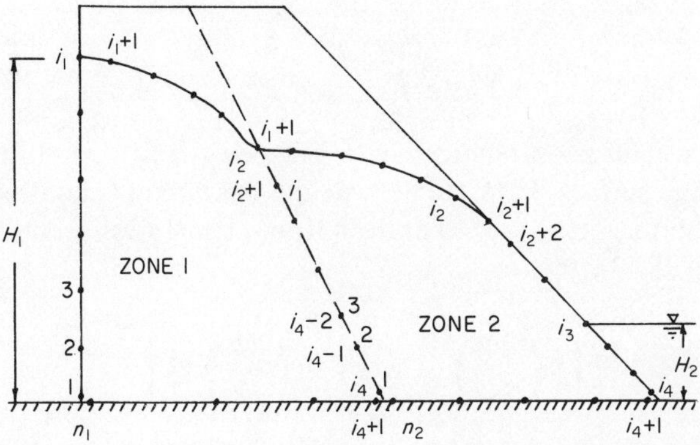


**Figure 5.6** The linear discretizations of boundaries of a zoned dam problem.

(e) ↓ *Note:* $i = i_1,\ j = i_2,\ k = i_4$

$\begin{matrix} {}^{1}_{k+2} & \cdots & R^{1}_{n-1} \\ {}^{2}_{k+2} & \cdots & R^{2}_{n-1} \\ \vdots & & \vdots \\ {}^{m}_{k+2} & \cdots & R^{m}_{n-1} \end{matrix}$ } zone 1

seepage surface-free surface point ↓

$$\begin{matrix} 0 & \cdots & 0 & L^1_{i+2} - \delta R^1_{i+2} & \cdots & L^1_j - \delta R^1_j & R^1_{j+1} & L^1_{j+2} & \cdots & L^1_k & R^1_{m+1} & \cdots & R^1_{n-1} \\ 0 & \cdots & 0 & L^2_{i+2} - \delta R^1_{i+2} & \cdots & L^2_j - \delta R^2_j & \mathrm{R}^2_{j+1} & L^2_{j+2} & \cdots & L^2_k & R^2_{m+1} & \cdots & R^2_{n-1} \\ \vdots & & \vdots & \vdots & & \vdots & \vdots & \vdots & & \vdots & \vdots & & \vdots \\ 0 & \cdots & 0 & L^m_{i+2} - \delta R^m_{i+2} & \cdots & L^m_j - \delta R^m_j & R^m_{j+1} & L^m_{j+2} & \cdots & L^m_k & R^m_{m+1} & \cdots & R^m_{n-1} \end{matrix}$$

(f)

face. Since $\Phi$ is known along this portion of the boundary, the corresponding right-hand side is

$$\sum_{j=1}^{i_1} R_{i,j}\Phi_j + L_{i,i_1}\,\partial\Phi_{i_1}/\partial n$$

Note that $L_{i,i_1}$ appears on the right-hand side and not in the coefficient matrix, since $\partial\Phi_{i_1}/\partial n$ is known:

$$\frac{\partial\Phi_{i_1}}{\partial n} = -\sin\left(\frac{\pi}{2} - \alpha_1\right), \quad \alpha_1 < \frac{\pi}{2}; \qquad \frac{\partial\Phi_{i_1}}{\partial n} = 0, \quad \alpha_1 \geqslant \frac{\pi}{2}$$

(b) The second set of numbers represents the coefficients of $\partial\Phi/\partial n$ along the free surface. In (5.16) both $\Phi$ and $\partial\Phi/\partial n$ are unknowns along the free surface. However, $\Phi$ at the advanced time step can be replaced by using (5.13), thus

$$[R]\left\{\Phi^k - \delta\left[\theta\left(\frac{\partial\Phi}{\partial n}\right)^{k+1} + (1-\theta)\left(\frac{\partial\Phi}{\partial n}\right)^{k}\right]\right\} = [L]\left\{\left(\frac{\partial\Phi}{\partial n}\right)^{k+1}\right\} \tag{5.20}$$

where

$$\delta = \frac{K(\Delta t)\cos\gamma}{\cos(\beta^k + \gamma)} \tag{5.21}$$

Rearranging so that the unknowns appear on the left produces

$$[L - \delta\theta R]\left\{\left(\frac{\partial\Phi}{\partial n}\right)^{k+1}\right\} = -[R]\left\{\Phi^k - \delta(1-\theta)\left(\frac{\partial\Phi}{\partial n}\right)^{k}\right\} \tag{5.22}$$

In the actual program these coefficients are divided by $\delta$ and the solution is for $\delta(\partial\Phi/\partial n)^{k+1}$ instead of the derivative alone. The advantage is that this quantity is the same for each zone at the point where the free surface crosses the dividing line between zones (i.e. points $i_2$ and $i_1 + 1$, Fig. 5.6). In zone 2 the coefficient of the first free surface point is placed in the same column as for the corresponding point of zone 1.

(c) On the line dividing the zones both $\Phi$ and $\partial\Phi/\partial n$ are unknown but related by the compatibility conditions. The numbers in this portion of the matrix represent the coefficients of $K(\partial\Phi/\partial n)$.

(d) The numbers in this portion of the matrix represent the coefficients of $\Phi$ along the dividing line. Points $i_4$ and $i_4 + 1$ of zone 1 are the same (or close together) as are points 1 and $n_2$ of zone 2. Thus the coefficients of these points are added in order to shorten the matrix and solve for a common $\Phi$.

(e) These numbers are the coefficients of $\Phi$ along the impervious bottom. The coefficient of $\Phi_{n_1}$ of zone 1 does not appear because the value of $\Phi$ is known at that point.

The right-hand (known) side of (5.19) follows obviously from the coefficient matrix. For each unknown there is a corresponding known potential or its normal derivative. Equation (5.19) can be solved by the direct Gaussian elimination method for all unknowns.

The conditions on the slope of the free surface as it crosses between zones are not as straightforward as presented here (Liggett & Liu 1979a). However, it appears that this feature can be mistreated with very little effect on the solution, and it is probably not wise to attempt to reproduce numerically the exact conditions at this point.

To illustrate the solution the case of sudden drawdown on the embankment of Figure 5.8 was computed and compared with data from a Hele–Shaw model. The triangular shaped embankment has two inclined surfaces where $\alpha_1 = \alpha_2 = 26.57°$. It is divided into two zones with the permeability ratio $K_1/K_2 = 0.25$. The dividing line has an angle $\gamma = 0.4007$ (22.95°). The embankment was first filled with fluid up to the elevation $z = 159$ mm, and a hinged gate was installed on the downstream face as a dam. At $t = 0$ the gate was suddenly opened so that the entire downstream face became a seepage surface. In the numerical computations the following data are used: zone 1; $i_1 = 4, i_2 = 8, i_4 = 12, K_1 = 1$: zone 2; $i_1 = 4, i_2 = 8, i_3 = 14, K_2 = 4$. Successive free surface locations are shown in Figure

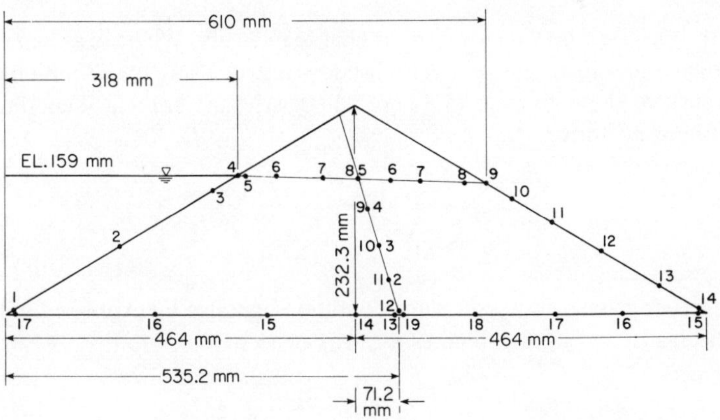


**Figure 5.8** Geometry and grid system of the numerical and experimental example.

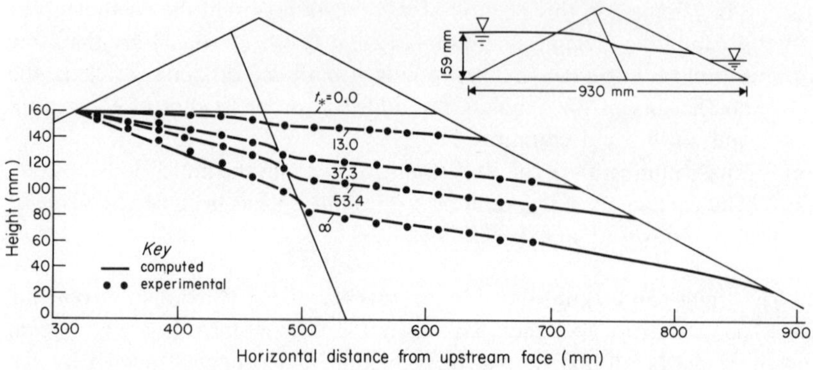


**Figure 5.9** Computed and experimental results of the sudden drawdown.

5.9. The times shown are to be considered dimensionless based on the permeability of zone 1 and the length scale. That is, $t_* = K_1 t/H_1$ ($K_1 = 0.7587$ mm/s and $H_1 = 159$ mm). The experimental data are also plotted in the same figure. Excellent agreement is observed.

## 5.4 Recharge problems

For two-dimensional groundwater recharge problems, the free surface boundary conditions should be modified to be

$$\Phi = \eta(x, t) \qquad \text{on } z = \eta \tag{5.23a}$$

$$\frac{\partial \eta}{\partial t} = -\left[1 + \left(\frac{\partial \eta}{\partial x}\right)^2\right]^{\frac{1}{2}} \frac{\partial \Phi}{\partial n} + W(x, t) \qquad \text{on } z = \eta \tag{5.23b}$$

where $W(x, t)$ is the dimensionless recharge intensity. We have assumed that the recharge velocity moves vertically downward. Using the definition of the free surface slope given in (5.3), we can rewrite the boundary condition in the following form:

$$\frac{\partial \eta}{\partial t} = -\frac{1}{\cos \beta} \frac{\partial \Phi}{\partial n} + W \qquad \text{on } z = \eta \tag{5.24}$$

On the other hand, this boundary condition can also be expressed in terms of the transformed co-ordinates $(x_*, z_*)$ of (5.6),

$$\frac{\partial \eta_*}{\partial t} = -\frac{1}{\cos (\gamma + \beta)} \frac{\partial \Phi}{\partial n} + \frac{W}{\cos \gamma}, \qquad \text{on } z_* = \eta_* \tag{5.25}$$

Using (5.9) we obtain the rate of change of the potential on the free surface in the $z_*$ direction to be

$$\left(\frac{\partial \Phi}{\partial t}\right)_{x_*} = -\frac{\cos \gamma}{\cos(\gamma + \beta)} \frac{\partial \Phi}{\partial n} + W \qquad \text{on } z_* = \eta_* \tag{5.26}$$

The numerical procedure uses the finite difference analog of (5.26).

$$\Phi^{k+1} = \Phi^k - \frac{(\Delta t)\cos\gamma}{\cos(\gamma + \beta^k)}\left[\theta\left(\frac{\partial \Phi}{\partial n}\right)^{k+1} + (1-\theta)\left(\frac{\partial \Phi}{\partial n}\right)^{k}\right] + \Delta t[\theta W^{k+1} + (1-\theta)W^k] \tag{5.27}$$

For the cases that free surface points are moved in the vertical direction ($z$ direction), (5.26) and (5.27) can be simplified by setting $\gamma = 0$,

$$\left(\frac{\partial \Phi}{\partial t}\right)_{x} = -\frac{1}{\cos \beta} \frac{\partial \Phi}{\partial n} + W \qquad \text{on } z = \eta \tag{5.28}$$

$$\Phi^{k+1} = \Phi^k - \frac{\Delta t}{\cos \beta^k}\left[\theta\left(\frac{\partial \Phi}{\partial n}\right)^{k+1} + (1-\theta)\left(\frac{\partial \Phi}{\partial n}\right)^{k}\right] + \Delta t[\theta W^{k+1} + (1-\theta)W^k] \tag{5.29}$$

### 5.4.1 *Uniform recharge in a Hele–Shaw model*

The Hele–Shaw experiment of Marino (1967) is shown schematically in Figure 5.10. To apply the numerical scheme developed in this chapter the boundary is replaced by 32 straight line segments as indicated in the same figure. For convenience the free surface is designated $a$, while $b$ represents the solid walls and bottom. These latter surfaces are impermeable boundaries on which $\partial\Phi/\partial n = 0$. Initially the fluid was at rest with a level

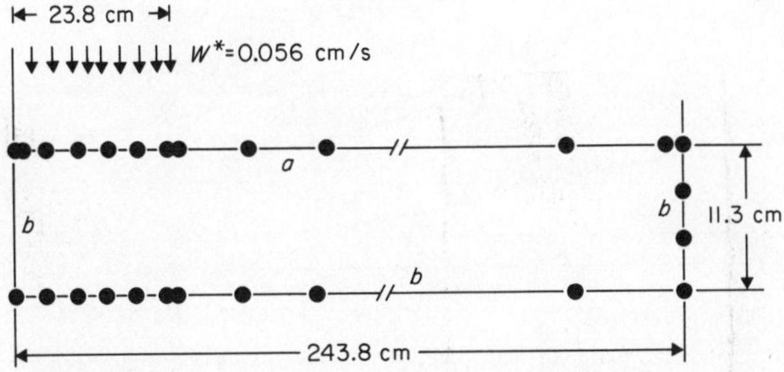


**Figure 5.10** Sketch of the Hele–Shaw model of Marino (1967).

free surface. At time $t = 0$ a uniform recharge rate of 0.056 cm/s was introduced in the leftmost 23.8 cm. The hydraulic conductivity of the model was $K = 0.42$ cm/s, and the effective porosity was $n = 1.0$. The dimensionless recharge intensity is $W = 0.056/0.42 = 0.133$.

The problem is solved by applying the boundary conditions, including the free surface conditions (5.26), to (5.16). The resulting matrix equations are

$$\left[ -L_a - R_a \frac{\Delta t\, \theta}{\cos \beta^k} \quad R_b \right] \left\{ \begin{matrix} (\partial \Phi_a/\partial n)^{k+1} \\ \Phi_b^{k+1} \end{matrix} \right\}$$

$$= |-R_a \quad L_b| \left\{ \begin{matrix} \Phi_a^k - \dfrac{\Delta t\,(1-\theta)}{\cos \beta^k} \left( \dfrac{\partial \Phi_a}{\partial n} \right)^{k+1} + \Delta t\, W^k \\ \dfrac{\partial \Phi_b}{\partial n} = 0 \end{matrix} \right\} \qquad (5.30)$$

in which $W^{k+1} = W^k$ has been employed. In (5.30) all unknowns at time level $(k + 1)\,\Delta t$ are placed on the left-hand side of the equation and can be solved successively in time steps with known initial conditions and boundary conditions.

In all numerical computations, the weighting factor $\theta$ was taken as 0.7, and the dimensionless time step was taken as 0.5. The numerical results for $t = 1, 5, 9$ min are plotted in Figure 5.11. The results of Neuman *et al.* (1976), who used a finite element method, are also shown by circles in the same figure. The present results compare favorably with those by Neuman *et al.* and also show good agreement with Marino's experimental data at

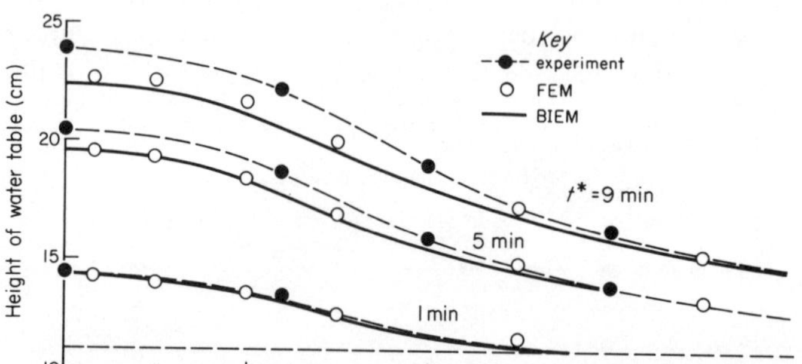


**Figure 5.11** Transient free surface profiles; comparisons between experimental data, numerical solutions by finite element method (FEM), and present results (BIEM).

early time ($t^* = 1$ min). At a later time, the agreement is somewhat less satisfactory near the crest of the water table. However, the experimental curves ($t^* = 5$ and 9 min) indicate a gain in mass (fluid in the model in excess of the original fluid plus accumulated recharge) of about 9–10%. The mass is always conserved in the present numerical results.

### 5.4.2 *Recharge to semi-infinite aquifer*

Figure 5.12 indicates a problem similar to the last one except that the aquifer extends to infinity in the lower half plane. The recharge occurs in the region $-1 < x < 1$ and the water is at rest at $t < 0$.

The integral equation (2.15) is to be solved around the entire boundary, including the infinite boundary. However, the first term on the right-hand side of (2.15) is zero on the infinite boundary due to the fact that it contains $1/r$ as $r \to \infty$. The second term is also zero since $\partial\Phi/\partial n$ is zero on the infinite boundary. There remains the integral on the free surface,

$$\alpha_i \Phi_p = \int_{-\infty}^{\infty} \left[ \Phi \frac{\partial}{\partial n} \ln r - \ln r \frac{\partial \Phi}{\partial n} \right] \mathrm{d}s \tag{5.31}$$

where the line integration is performed along the entire free surface, and the singular points $P_i$ are located on the free surface only.

The numerical integration over an infinite interval as required in equation (5.31) can be avoided by utilizing the analytical solutions at far field. As shown in Section 3.5.2, if the linearized solution forms are used in the far field, $|x| > |x_n|$, (5.31) can be approximated by

$$\alpha_i \Phi_p = \int_{-xn}^{xn} \left[ \Phi \frac{\partial}{\partial n} \ln r - \ln r \frac{\partial \Phi}{\partial n} \right] \mathrm{d}s + I_2 + I_3 \tag{5.32}$$

where

$$I_2 = \left( \frac{\partial \Phi}{\partial n} \right)_{xn} \left[ \frac{4t^2 + (1 - x_n^2 - t^2)^2}{1 - x_n^2 + t^2} \int_{xn}^{\infty} \ln|x - x_i| \frac{1 - x^2 + t^2}{4t^2 + (1 - x^2 - t^2)^2} \mathrm{d}x \right] \tag{5.33}$$

$$I_3 = \left( \frac{\partial \Phi}{\partial n} \right)_{-xn} \left[ \frac{4t^2 + (1 - x_n^2 - t^2)^2}{1 - x_n^2 + t^2} \int_{xn}^{\infty} \ln|x + x_i| \frac{1 - x^2 + t^2}{4t^2 + (1 - x^2 - t^2)^2} \mathrm{d}x \right] \tag{5.34}$$

The integrals in (5.33) and (5.34) can be integrated numerically. We remark here that the singular points are now located on the free surface in the interval $-x_n \leqslant x \leqslant x_n$.

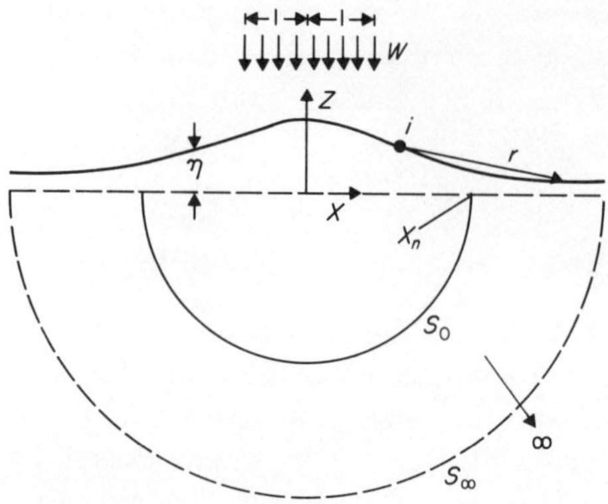


**Figure 5.12** Recharge to semi-infinite aquifer.

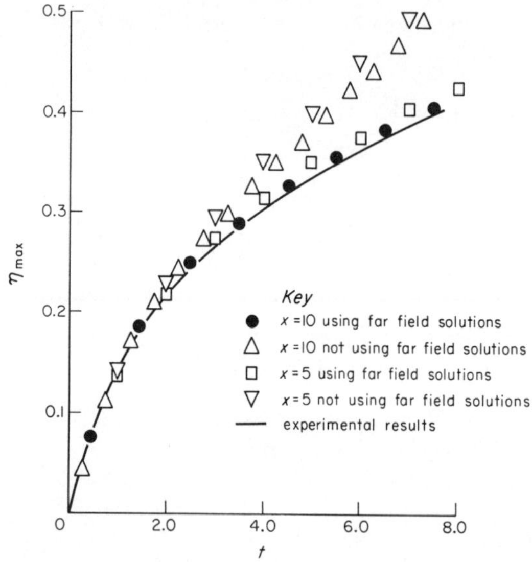


**Figure 5.13** Maximum height of water table (at $x = 0$) with $W = 0.2$ showing experimental data and numerical solutions with and without the use of the far field solutions.

The transient water table movement due to two different recharge rates, $W = 0.2$ and 1.0, was computed. In the case of $W = 0.2$ the maximum heights of the water table at $x = 0$ with $x_n = 5$ and 10 are plotted in Figure 5.13. Excellent agreement with experimental data (Bouwer 1962) is obtained for the case $x_n = 10$ by using the far field analytical solution. When $x_n = 5$ (which is not a large value) is chosen, the flow domain is reduced by half, and the agreement with experimental data is still good. In Figure 5.14, the transient free surface profiles are presented.

In the numerical computations, eight nodal points are always concentrated over the recharge area ($-1 \leqslant x \leqslant 1$). The symmetry about $z = 0$

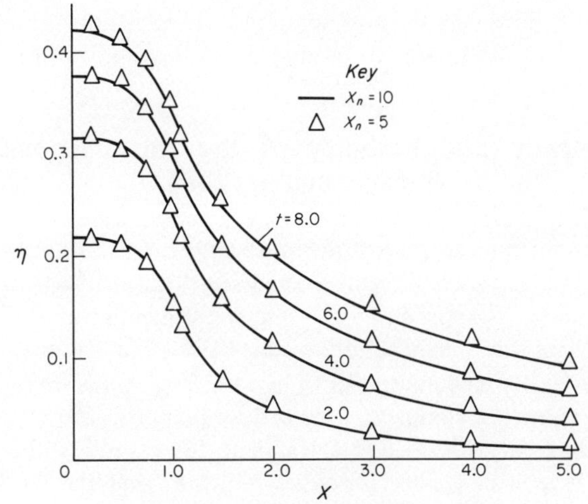


**Figure 5.14** Transient free surface profiles.

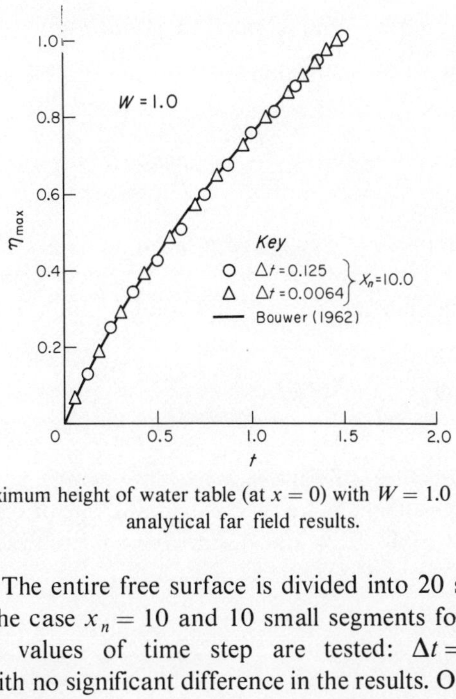


**Figure 5.15** Maximum height of water table (at $x = 0$) with $W = 1.0$ and $x_n = 10$ using the analytical far field results.

was not used. The entire free surface is divided into 20 small straight line segments for the case $x_n = 10$ and 10 small segments for the case $x_n = 5$. Two different values of time step are tested: $\Delta t = 0.125$ and 1.0, respectively, with no significant difference in the results. Obviously, by using the smaller time step a more detailed time history of the water table movement is obtained.

In the second case, the recharge rate $W$ is 1.0. The maximum height of the free surface at $x = 0$ for various times is shown in Figure 5.15. Excellent agreement with Bouwer's data is observed. In the computations $x_n = 10$ is chosen, and the far field analytical solution is adopted. The results of using two different time steps, $\Delta t = 0.064$ and 0.125, are shown in Figure 5.15.

## 5.5 Accuracy and stability of the time dependent free surface calculations

In this section the convergence of the finite difference approximation to the free surface boundary conditions is analyzed. The analysis assumes that Laplace's equation is satisfied by the numerical method used to discretize the space variables in our examples of the BIEM, but the analysis applies equally well to other techniques. As with most problems of this type, we are unable to analyze the complete, non-linear equations. Therefore, we consider the linearized problem and extrapolate the results to the non-linear case. Experience with similar problems indicates that the linear analysis gives a good guide to the non-linear results (e.g. Liggett & Cunge 1975).

### *5.5.1 The linearized conditions*

The linearized free surface equation (5.4) in dimensionless form is

$$\frac{\partial \Phi}{\partial t} = -\frac{\partial \Phi}{\partial z} \qquad \text{on } z = 0 \tag{5.35}$$

in which the origin of the $z$ axis has been taken on an average free surface, i.e. not only has (5.4) been linearized but the boundary conditions are now applied at the equilibrium surface, $z = 0$. A solution to Laplace's equation is

$$\Phi = \sum_{j=1}^{\infty} A_j \sin k_j x \frac{\cosh k_j(z + h)}{\cosh k_j h} e^{-\sigma_j t} \tag{5.36}$$

in which $A_j$ is a Fourier coefficient, $k_j$ is the wave number of the $j$th mode, $h$ is the depth of the medium, and $\sigma_j$ is the damping rate of the $j$th mode. For the purpose of this analysis we will consider only one representative term in the series and write

$$\Phi = A \sin kx \frac{\cosh k(z + h)}{\cosh kh} e^{-\sigma t} \tag{5.37}$$

Substitution of (5.37) into (5.35) gives a relationship between $\sigma$ and $k$ which is

$$\sigma = k \tanh kh \tag{5.38}$$

Since both $k$ and $h$ are real constants, the preceding equation implies that $\sigma$ is always a positive real constant and represents the damping rate of the physical system. The corresponding exact solution to the linearized problem is

$$\eta = A \sin kx \exp(-kt \tanh kh) \tag{5.39}$$

$$\Phi = A \sin kx \frac{\cosh k(z+h)}{\cosh kh} \exp(-kt \tanh kh) \tag{5.40}$$

In order to derive the numerical solution to the linearized equations, we begin with (5.5) with $\cos \beta^k = 1$. Assuming that the numerical representation of Laplace's equation is accurate, substitution of (5.37) into (5.5) gives

$$e^{-\sigma_N \Delta t} = 1 - \Delta t\, k(\tanh kh)[\theta e^{-\sigma_N \Delta t} + (1-\theta)] \tag{5.41}$$

in which $\sigma_N$ is the numerical damping rate. Solving for $\sigma_N$ yields

$$\sigma_N = -\frac{1}{\Delta t} \ln \left[ \frac{1-(1-\theta)k\,\Delta t \tanh kh}{1+\theta k\,\Delta t \tanh kh} \right] \tag{5.42}$$

which is to be compared to (5.38). If the numerical values of $\sigma$ in (5.38) and $\sigma_N$ in (5.42) differ, the numerica' approximation will add either damping or amplification to the real solution.

Let

$$\varepsilon = -k \tanh kh - \frac{1}{\Delta t} \ln \left[ \frac{1-(1-\theta)k\,\Delta t \tanh kh}{1+\theta k\,\Delta t \tanh kh} \right] \tag{5.43}$$

which is the departure of the numerical damping coefficient from the physical (linearized) damping coefficient given in (5.38) ($\varepsilon = \sigma_N - \sigma$). Thus a negative $\varepsilon$ indicates numerical amplification whereas a positive value indicates numerical damping. In order to achieve the best numerical approximation, $\varepsilon$ should be as near to zero as possible.

We define the parameter

$$q = k\,\Delta t \tanh kh \tag{5.44}$$

and substitute into (5.43) to obtain

$$\varepsilon\,\Delta t = -q - \ln \left[ \frac{1-(1-\theta)q}{1+\theta q} \right] \tag{5.45}$$

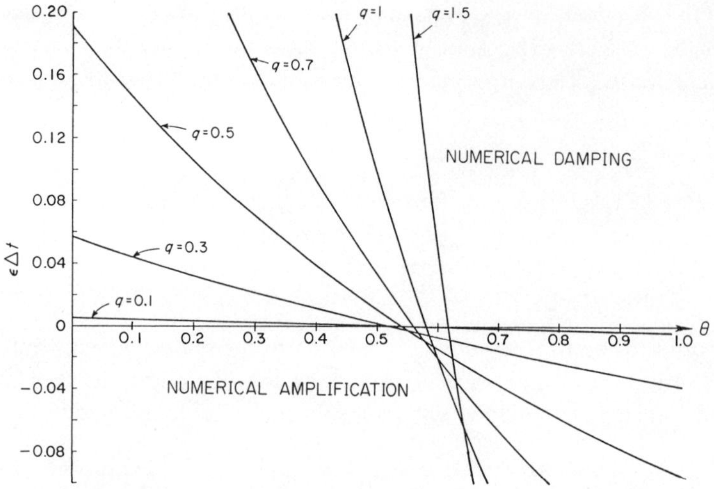


**Figure 5.16** A plot of θ *vs.* $\varepsilon\,\Delta t$ for various values of $q$.

Figure 5.16 shows a plot of $\varepsilon\,\Delta t$ *vs.* θ with $q$ as a parameter. From (5.45) and Figure 5.16 we note the following:

(a) The value of $q$ is limited by $q < (1-\theta)^{-1}$ in order that ε remain real. Conversely, $\theta > 1 - q^{-1}$. As θ approaches this limiting value the numerical damping becomes infinite and thus the solution is destroyed.

(b) Small values of $\Delta t$ result in less numerical distortion (either amplification or damping).

(c) Small values of $k$ (lower modes or leading terms in the Fourier series) are less distorted than the higher values.

(d) Optimum values of θ range upward from $\theta = 0.5$ for small values of $q$ to unity for extremely large $q$, but with a practical range of $0.5 < \theta < 0.6$.

(e) For a given value of $q$ there is a value of θ, $\theta = \theta_c$, where $\varepsilon = 0$ and there is no numerical distortion. Figure 5.17 shows a plot of $\theta_c$ *vs.* $q$ (the line labeled $\varepsilon = 0$).

(f) If $\theta < 1 - q^{-1}$, ε becomes complex and (5.45) can be written

$$\varepsilon\,\Delta t = -q - i\pi - \ln\frac{(1-\theta)q - 1}{1 + q\theta} = \Delta t\,(\varepsilon^{R} + i\varepsilon^{I}) \qquad (5.46)$$

where $i = \sqrt{(-1)}$, and the superscripts indicate the real and imaginary parts of ε. Thus the numerical damping coefficient becomes complex, indicating that the numerical solution will contain oscillations. These oscillations are damped if the real part of the numerical damping

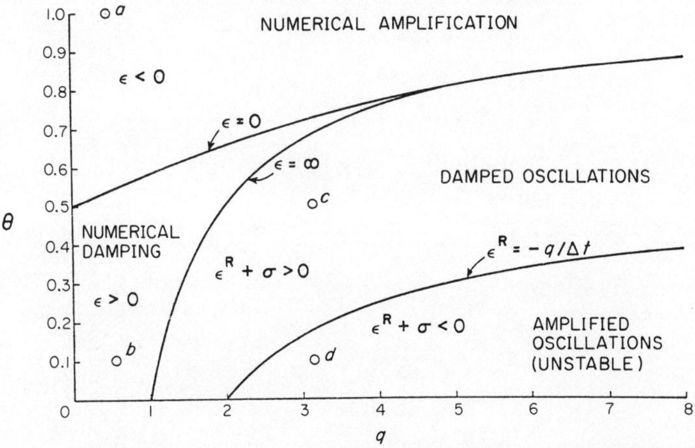


**Figure 5.17** Zones of amplification, damping and oscillation as a function of $\theta$ and $q$.

coefficient is positive ($\varepsilon^R + \sigma > 0$), but they grow in an unstable manner if the numerical damping coefficient is negative ($\varepsilon^R + \sigma < 0$) as indicated in Figure 5.17.

In any practical problem there is a range of the wave numbers $k$. It is therefore impossible to choose a single $\Delta t$ and/or $\theta$ which represents no numerical damping or amplification. The best alternative is to choose a sufficiently low value of $\Delta t$ for the important (lower) modes not to be seriously amplified and a value of $\theta$ such that little damping occurs at the higher modes. If the horizontal co-ordinate has a discretization of $\Delta x$, wave numbers higher than $k = \pi/\Delta x$ cannot be calculated in any case. Thus for a given $\Delta x$, Figure 5.16 can serve as a guide for the choice of $\Delta t$ and $\theta$.

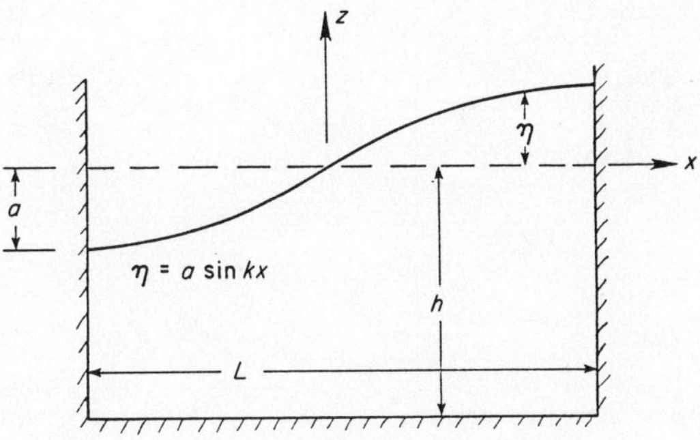


**Figure 5.18** The sine wave in a box as a test problem with a single wave number.

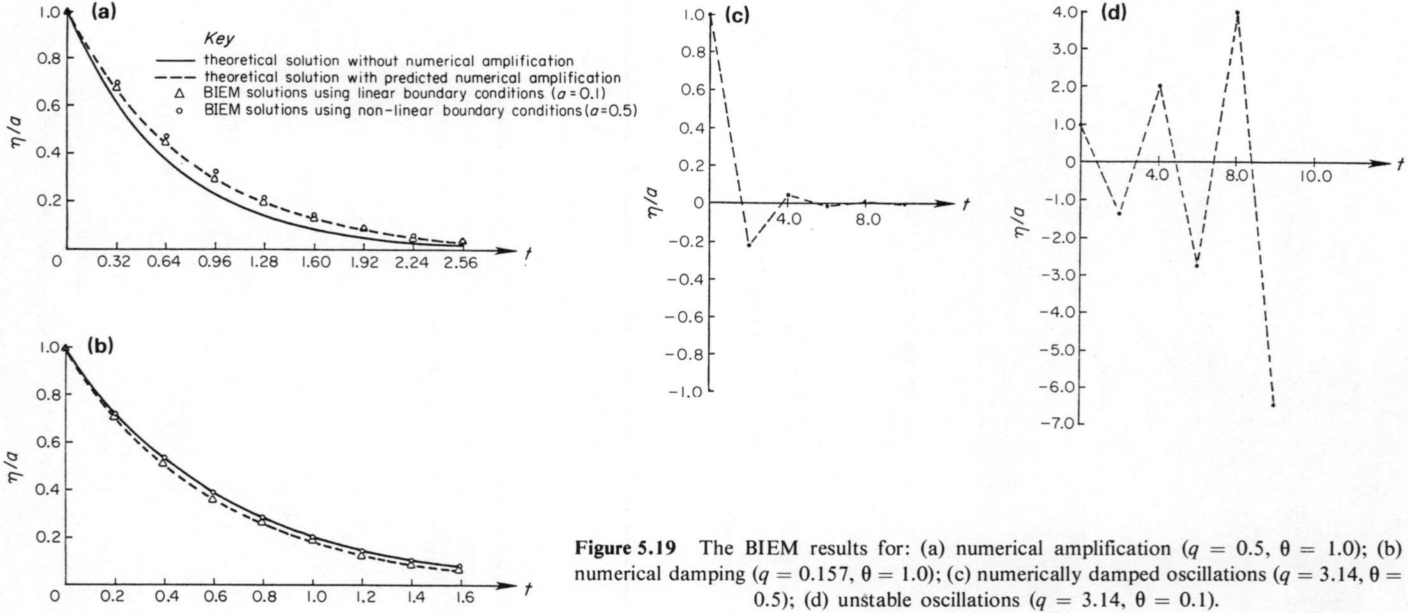


**Figure 5.19** The BIEM results for: (a) numerical amplification ($q = 0.5$, $\theta = 1.0$); (b) numerical damping ($q = 0.157$, $\theta = 1.0$); (c) numerically damped oscillations ($q = 3.14$, $\theta = 0.5$); (d) unstable oscillations ($q = 3.14$, $\theta = 0.1$).

### 5.5.2 Numerical verification

In order to verify (5.43) several numerical examples were run using the linearized boundary conditions as expressed by (5.35) with the BIEM used for the spatial integration. The initial free surface is in the shape of a sine wave and the wave is allowed to approach the equilibrium surface, $z = 0$, as time progresses (Figure 5.18). Since the free surface is a sine wave initially, it remains a sine wave with wave number $k = \pi/L$. In the numerical computations, the water depth $h$ is chosen to be the same as the width of the tank, i.e. $h = L = 2.0$. Therefore, the damping rate of the physical system can be computed from (5.38) and is equal to $\sigma = (\pi/2) \tanh \pi \approx \pi/2$.

The results of the numerical verification are shown in Figures 5.19a–d. In these four figures, the amplitude variations at $x = L/2$ are plotted. Figure 5.19a shows a typical case of numerical amplification. The values of $q$ and $\theta$ are indicated by point $a$ on Figure 5.17. The (linear) BIEM results are obtained by dividing the boundary into 40 segments. The amplitude of the sine wave is 0.1 for the linear problem. The numerical calculations depart from the exact results by the amount predicted in the above theory. The non-linear BIEM calculation (using (5.5) with boundary conditions evaluated at the free surface instead of the equilibrium surface and with $\cos \beta \neq 1$) is also shown on the figure. Of course the departure from the linearized solution is a function of amplitude. The contrasting case of numerical damping is shown in Figure 5.19b with values of $q$ and $\theta$ at point $b$ on Figure 5.17.

The linear BIEM oscillating solutions are shown in Figures 5.19c and 5.19d with corresponding points $c$ and $d$ on Figure 5.17, respectively. In Figure 5.19c the oscillations are damped, but the numerical solution is unacceptably far from the true solution. The oscillations shown in Figure 5.19d grow unstably and destroy the solution completely.

We should point out here that when using any discretization technique to represent the Laplace equation, additional numerical distortion is likely to be created, but in a spacewise rather than timewise sense. Although the present theory is derived for the linear problem, we believe that it also serves as a good guide to the choice of $\theta$ and $\Delta t$ for non-linear problems.

# 6 Axisymmetric flow problems

In this chapter, the BIEM is formulated for axial symmetric flow problems. Introducing the 'axisymmetric Green function', we reduce the general integral equation (2.21) to a line integral. Although the computational procedure for the resulting integral equation is quite similar to that of two-dimensional problems, the kernel functions are much more complicated and are discussed in Section 6.2.

The numerical examples given in this chapter include circular recharge problems and pumping well problems. In the pumping well problems, an iterative procedure is developed to compute the rate of change of the water level in a well and the volume flux rate withdrawn from an aquifer, which account for the total pumping rate prescribed at a given time.

## 6.1 Integral equation for axisymmetric problems

The integral equation for three-dimensional problems is derived in Chapter 2 and can be summarized in the following form:

$$-\alpha\Phi(x_i, y_i, z_i, t) = \oint_\Omega \left(\Phi\frac{\partial g}{\partial n} - g\frac{\partial \Phi}{\partial n}\right) \mathrm{d}A \tag{6.1}$$

where

$$\alpha = \begin{cases} 4\pi & \text{if } (x_i, y_i, z_i) \in D \quad \text{but } (x_i, y_i, z_i) \notin \Omega \\ \lim\limits_{R_0 \to 0} \mathrm{A_s}/\mathrm{R}_0^2 & \text{if } (x_i, y_i, z_i) \in \Omega \end{cases} \tag{6.2}$$

and $g$ is the three-dimensional free space Green function,

$$g = \frac{1}{R} \qquad R = [(x-x_i)^2 + (y-y_i)^2 + (z-z_i)^2]^{\frac{1}{2}} \tag{6.3}$$

In the case of axisymmetric problems all the physical variables and the geometry of the domain are independent of the co-ordinate θ in a cylindrical

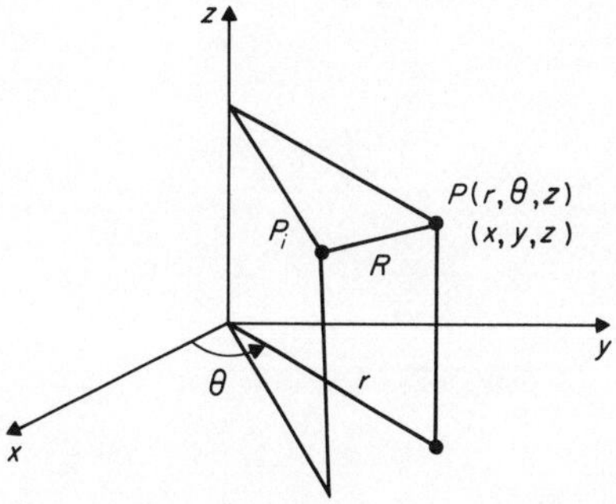


**Figure 6.1** Co-ordinate systems.

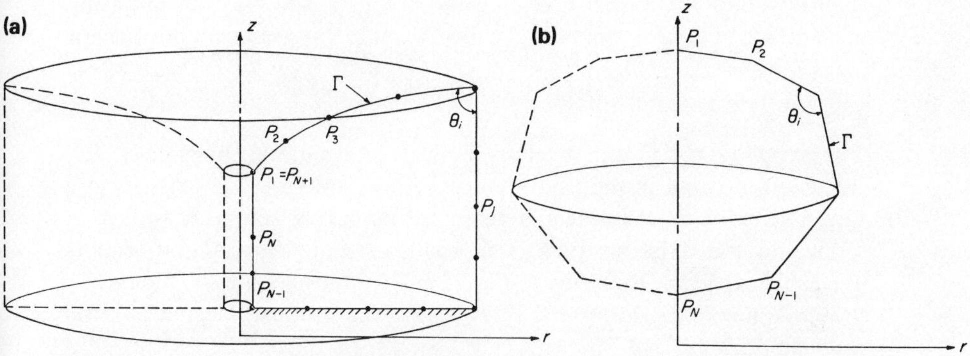


**Figure 6.2** Straight line segments approximation of the boundary contour Γ: (a) closed contour; (b) open contour.

co-ordinate system (Figure 6.1). However, the free space Green function, $g$, is a function of θ. Thus, we may rewrite (6.1) to be

$$-\alpha\Phi(r_i, z_i, t) = \oint_\Gamma \left[ \Phi \int_0^{2\pi} \frac{\partial g}{\partial n}\, d\theta - \frac{\partial \Phi}{\partial n} \int_0^{2\pi} g\, d\theta \right] r\, d\xi \tag{6.4}$$

in which the differential surface element, $dA$, has been expressed as $dA = r d\theta\, d\xi$ and Γ is the generating boundary contour which is the projection of Ω in the $r$–$z$ plane (Fig. 6.2). We define an 'axisymmetric Green function' as

$$G(r, z; r_i, z_i) = \int_0^{2\pi} g\, d\theta \tag{6.5}$$

and substitute it into (6.4). Thus

$$-\alpha\Phi(r_i, z_i, t) = \oint_\Gamma \left[ \Phi \frac{\partial G}{\partial n} - G \frac{\partial \Phi}{\partial n} \right] r\, d\xi \tag{6.6}$$

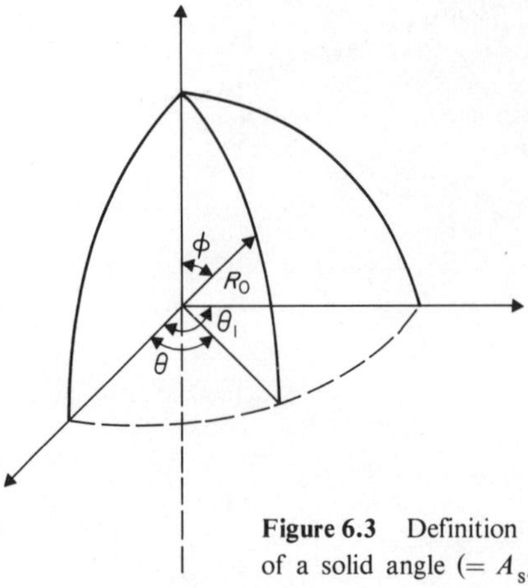


**Figure 6.3** Definition of the measure of a solid angle ($= A_s/R_0^2$).

The axisymmetric Green function $G$ must be evaluated numerically. It is discussed in detail in the next section. We can interpret the axisymmetric Green function as the potential generated by a ring source at $(r_i, z_i)$.

Due to the axial symmetry the solid angle defined in (6.2) can be evaluated explicitly. Referring to Figure 6.3, we consider the situation where $\theta_i$ is the interior angle formed by two adjacent boundary segments at the singular point $(r_i, z_i)$. The surface $A_s$ becomes

$$A_s = \int_0^{\theta_i} \int_0^{\pi} R_0^2 \sin \phi \, d\phi \, d\theta = 2R_0^2 \theta_i \tag{6.7}$$

for small radius $R_0$. Finally the solid angle $\alpha$ becomes

$$\alpha = \lim_{R_0 \to 0} \frac{A_s}{R_0^2} = 2\theta_i \tag{6.8}$$

For axisymmetric problems the integral equation is reduced to a line integral along the generating boundary contours. Thus the numerical procedures for the axisymmetric problems are similar to those for the two-dimensional problems. However, due to the complexity of the Green function, which cannot be expressed in terms of elementary functions, the details of the numerical integrations are somewhat more involved.

## 6.2 Computation of the Green function

To integrate the integral equation numerically, we must accurately compute the Green function, as defined by (6.5), and its normal derivative. Equation

(6.5) can be rewritten in terms of the cylindrical co-ordinates $(r, z)$ in the following form:

$$G(r, z; r_i, z_i) = \int_0^{2\pi} \frac{d\theta}{(a - b\cos\theta)^{\frac{1}{2}}} = \frac{4}{(a+b)^{\frac{1}{2}}} K(m) \tag{6.9}$$

where

$$a = r_i^2 + r^2 + z_d^2 \qquad b = 2rr_i \quad z_d = z - z_i \tag{6.10}$$

and $K(m)$ is the complete elliptic integral of the first kind,

$$K(m) = \int_0^{\pi/2} \frac{d\theta}{(1 - m\sin^2\theta)^{\frac{1}{2}}} \qquad 0 \leqslant m < 1 \tag{6.11}$$

The modulus, $m$, of the elliptic integral and its complementary modulus, $m_1 = 1 - m$, can be expressed specifically as follows:

$$m = \frac{4rr_i}{(r + r_i)^2 + z_d^2} \qquad m_1 = \frac{(r - r_i)^2 + z_d^2}{(r + r_i)^2 + z_d^2} \tag{6.12}$$

The complete elliptic integral of the first kind has a logarithmic singularity as $m \to 1$ $(m_1 \to 0)$, which corresponds to $r \to r_i$ and $z \to z_i$.

For the convenience of numerical computations, the elliptic integral may be approximated by (Abramowitz & Stegun 1974)

$$K(m) = [a_0 + a_1 m_1 + \cdots + a_4 m_1^4] + [b_0 + b_1 m_1 + \cdots + b_4 m_1^4] \ln(1/m_1) + \varepsilon(m),$$

$$|\varepsilon(m)| \leqslant 2 \times 10^{-8} \tag{6.13}$$

where

$$\begin{aligned} a_0 &= 1.38629436112 & b_0 &= 0.5 \\ a_1 &= 0.09666344259 & b_1 &= 0.12498593597 \\ a_2 &= 0.03590092383 & b_2 &= 0.06880248576 \\ a_3 &= 0.03742563713 & b_3 &= 0.03328355346 \\ a_4 &= 0.01451196212 & b_4 &= 0.00441787012 \end{aligned} \tag{6.14}$$

and $\varepsilon(m)$ indicates the upper bound of the error for this approximation. For our purposes, the fourth order of polynomial approximation is sufficiently accurate. However, Hart *et al.* (1968) have presented higher order polynomial approximations (up to the 10th order) which of course give smaller errors, $\varepsilon(m)$.

The normal derivatives of the Green function, $G$, along the boundary contour, $\Gamma$, can be written as

$$\frac{\partial G}{\partial n} = \frac{\partial G}{\partial r} n_r + \frac{\partial G}{\partial z} n_z$$
$$= \frac{2n_r}{r(a+b)^{\frac{1}{2}}} [E(m) - K(m)] - 4 \frac{[n_r(r - r_i) + n_z z_d]}{(a-b)(a+b)^{\frac{1}{2}}} E(m) \quad (6.15)$$

where $(n_r, n_z)$ are direction cosines, and $E(m)$ is the complete elliptic integral of the second kind. As for (6.13), the polynomial approximation is also used for $E(m)$,

$$E(m) = [1 + c_1 m_1 + \cdots + c_4 m_1^4] + [d_1 m_1 + \cdots + d_4 m_1^4] \ln(1/m_1) + \varepsilon(m),$$
$$|\varepsilon(m)| < 2 \times 10^{-8} \quad (6.16)$$

in which

$$\begin{aligned} c_1 &= 0.44325141463 & d_1 &= 0.24998368310 \\ c_2 &= 0.06260601220 & d_2 &= 0.09200180037 \\ c_3 &= 0.04757483546 & d_3 &= 0.04069697526 \\ c_4 &= 0.01746596451 & d_4 &= 0.00426449639 \end{aligned} \quad (6.17)$$

We now can use (6.9) and (6.15) in (6.6) for the numerical integrations with the help of the polynomial approximations for the elliptic integrals.

## 6.3 Numerical integration of the integral equation

The integral equation, (6.6), may be solved by numerical integration along the boundary $\Gamma$ on the $r$–$z$ plane. Since the form of the integral equation for an axisymmetric problem is the same as that of a two-dimensional problem, the method of discretization of boundary developed in Section 2.2 for two-dimensional problems can be readily applied here for axisymmetric problems. Using the linear interpolation function for $\Phi$ and $\partial\Phi/\partial n$ between a pair of consecutive points $P_j$ and $P_{j+1}$ along the boundary as shown in (2.24) and (2.25), we can express the integral on the right-hand side of (6.6) over a linear segment as follows:

$$I_e = \int_{\xi_j}^{\xi_{j+1}} r\left(\Phi \frac{\partial G}{\partial n} - G\frac{\partial \Phi}{\partial n}\right) d\xi = [K_1^e] \begin{Bmatrix} \Phi_j \\ \Phi_{j+1} \end{Bmatrix} - [K_2^e] \begin{Bmatrix} (\partial\Phi/\partial n)_j \\ (\partial\Phi/\partial n)_{j+1} \end{Bmatrix} \quad (6.18)$$

where $\xi$ denotes the distance along the segment measured from an origin with reference to the singular point $P_i$. In (6.18) the row matrices $\lfloor K_1^e \rfloor$ and $\lfloor K_2^e \rfloor$ take the same forms as defined in (2.27). The elements in the matrices, $I_{ij}$ $(i, j = 1, 2)$, are given as follows:

$$
\begin{aligned}
I_{11} &= \frac{1}{\xi_{j+1} - \xi_j} \int_{\xi_j}^{\xi_{j+1}} \left( \frac{\partial G}{\partial n} \right) r\xi \, d\xi \\
I_{12} &= \frac{1}{\xi_{j+1} - \xi_j} \int_{\xi_j}^{\xi_{j+1}} \left( \frac{\partial G}{\partial n} \right) r \, d\xi \\
I_{21} &= \frac{1}{\xi_{j+1} - \xi_j} \int_{\xi_j}^{\xi_{j+1}} G r\xi \, d\xi \\
I_{22} &= \frac{1}{\xi_{j+1} - \xi_j} \int_{\xi_j}^{\xi_{j+1}} G r \, d\xi
\end{aligned}
\tag{6.19}
$$

The analytical results of above integrals are not obtainable due to the complexity of the integrands, namely $G$ and $\partial G/\partial n$. These integrals can be evaluated by Gaussian quadrature. Two different situations are considered separately.

When the base point $P_i$ does not coincide with either $P_j$ or $P_{j+1}$, the elliptic integrals have regular behavior (see (6.13) and (6.16)). Therefore, the integrands of the integrals, $I_{ij}$, are regular functions over the linear segment connecting $P_j$ and $P_{j+1}$. Equations (6.19) can be evaluated by the regular Gaussian quadrature

$$
\begin{aligned}
I_{ij} &= \frac{1}{\xi_{j+1} - \xi_j} \int_{\xi_j}^{\xi_{j+1}} f\left( G \text{ or } \frac{\partial G}{\partial n}, r, \xi \right) d\xi \\
&\approx \frac{1}{2} \sum_{k=1}^{M} \omega_k f\left[ G(\xi_k) \text{ or } \frac{\partial G(\xi_k)}{\partial n}, r(\xi_k), \xi_k \right]
\end{aligned}
\tag{6.20}
$$

with

$$
\xi_k = [(\xi_{j+1} - \xi_j)\,\chi_k + \xi_{j+1} + \xi_j]/2 \tag{6.21}
$$

where the Gaussian quadrature weights, $\omega_k$, and the positions of Gaussian quadrature points, $\chi_k$, are available in Abramowitz and Stegun (1974) for different values of the order $M$. A more detailed discussion on the Gaussian quadrature is in Appendix A.5.

For cases that the singular point $P_i$ is located at either $P_j$ or $P_{j+1}$, the integrals $I_{ij}$ have an integrable singularity at $\xi \to \xi_j$ (or $\xi_{j+1}$) due to the fact that the complete elliptic integral of the first kind, $K(m)$, has a logarithmic singularity as $m \to 1$ (or $m_1 \to 0$). Thus, the regular Gaussian does not give accurate results in this circumstance. We can, however, remedy the situation by rearranging the polynomial approximation for $K(m)$ as follows:

$$K(m) = K^*(m) - b_0 \ln m_1 \tag{6.22}$$

with

$$K^*(m) = a_0 + \sum_{k=1}^{4} [a_k m_1^k + b_k m_1^k \ln (1/m_1)] \tag{6.23}$$

where $K^*(m)$ is a regular function along the segment $\xi_j \leqslant \xi \leqslant \xi_{j+1}$. Using (6.22) and (6.23) in the integrals $I_{ij}$, we observe that there are two types of integrals: one with a regular function as its integrand and one with an integrand which has a logarithmic singularity. The latter integral can be accurately evaluated by the special Gaussian quadrature (see App. A.5), i.e.,

$$\int_0^1 f(x) \ln x \, dx = \sum_{k=1}^{M} \omega_k^s f(\chi_k^s) \tag{6.24}$$

where the $\omega_k^s$ and $\chi_k^s$ are the special Gaussian quadrature weights and locations, respectively. Abramowitz and Stegun (1974) provide values of $\omega_k^s$ and $\chi_k^s$ for $M = 2$, 3 and 4. Stroud and Secrest (1966) give values of $\omega_k^s$ and $\chi_k^s$ for $M = 2$–16.

Once the integrals $I_{ij}$ are computed for all segments along the boundary, the global equations for $\Phi_j$ and $(\partial\Phi/\partial n)_j$ can be obtained by summing (6.18) from all segments. Thus

$$R_{i,j}\Phi_j = L_{i,j}\left(\frac{\partial\Phi}{\partial n}\right)_j \qquad i,j = 1, 2, 3, \ldots, N \tag{6.25}$$

where

$$R_{i,j} = \Sigma[(k_1^e)_{i,j} + 2\delta_{i,j}\theta_i] \qquad \delta_{i,j} = \begin{cases} 1 & i = j \\ 0 & i \neq j \end{cases} \tag{6.26}$$

$$L_{i,j} = \Sigma(k_2^e)_{i,j} \tag{6.27}$$

For a given problem either $\Phi$ or $\partial\Phi/\partial n$ is known along each segment of the boundary, and (6.25) is used to solve for the corresponding unknowns.

### 6.3.1 Numerical example – uniform flow in a circular pipe

The uniform potential flow in a circular pipe is used here as an illustrative example. Referring to Figure 6.4, the diameter of the pipe is 20 units and the length is 10 units. The boundary conditions are given as follows: the potential on the top surface of the pipe is kept at $\Phi = 10$, while the potential on the bottom surface of the pipe is zero. The side boundary is assumed to be a solid wall, so that the radial velocity must vanish. The analytical solution of this problem is

$$\Phi = z \tag{6.28}$$

where $z$ is the vertical axis as shown in Figure 6.4a.

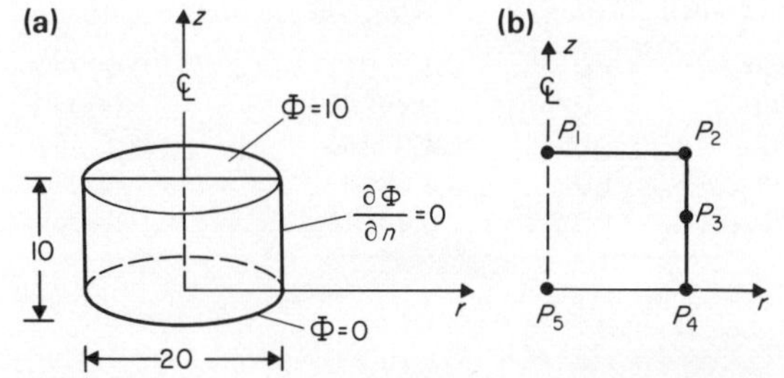


**Figure 6.4** (a) Geometry and boundary conditions of the numerical example. (b) The five-points grid system used for numerical computations.

To solve this problem numerically by using the scheme developed in the previous section, we first define the boundary contour on the $r$–$z$ plane as shown in Figure 6.4b. The boundary contour is not a closed one; the segment between $P_1$ and $P_5$ in Figure 6.4b is not part of the exterior boundary. This boundary is divided into four linear segments with five nodal points ($N = 5$). The co-ordinates of nodal points and the corresponding boundary conditions are summarized in Table 6.1. The normal derivative of the potential, $\partial\Phi/\partial n$, is only piecewise continuous along the boundary and discontinuous at the two corner points, $P_2$ and $P_4$.

**Table 6.1** Input data.

| *Point* | $r$ | $z$ | $\theta_i$ | $\Phi$ | $\partial\Phi/\partial n$ |
|---|---|---|---|---|---|
| 1 | 0.0 | 10.0 | $\pi$ | 10.0 | – |
| 2 | 10.0 | 10.0 | $\pi/2$ | 10.0 | –*/0.0† |
| 3 | 10.0 | 5.0 | $\pi$ | – | 0.0 |
| 4 | 10.0 | 0.0 | $\pi/2$ | 0.0 | 0.0*/–† |
| 5 | 0.0 | 0.0 | $\pi$ | 0.0 | – |

*Normal derivative prior to the point.
†Normal derivative after the point.
– Unknown to be solved.

Based on the computational procedures discussed in this chapter, two computer runs were made, in which four and eight Gaussian quadrature points were used to evaluate the integrals numerically. In both runs fourth-order polynomial approximations were adopted for the complete elliptic integrals. In Table 6.2, numerical results are shown to be in excellent agreement with the analytical solutions given by (6.28). The case using eight Gaussian quadrature points shows more accurate results.

**Table 6.2** Comparison between numerical results and the analytical solution.

| *Point* | *Unknown* | *Analytical solution* | *Numerical results* | |
|---|---|---|---|---|
| | | | *Four Gauss points* | *Eight Gauss points* |
| 1 | $\partial\Phi/\partial z$ | 1.0 | 1.00011 | 0.99999 |
| 2 | $\partial\Phi/\partial z$ | 1.0 | 0.99999 | 1.00000 |
| 3 | $\Phi$ | 5.0 | 4.99836 | 4.99998 |
| 4 | $\partial\Phi/\partial z$ | 1.0 | 1.00000 | 1.00000 |
| 5 | $\partial\Phi/\partial z$ | 1.0 | 0.99993 | 0.99999 |

## 6.4 Axisymmetric free surface flow problems

For axisymmetric free surface flow problems, the free surface boundary conditions can be deduced from (1.42a) and (1.43),

$$\Phi = \eta \qquad \text{on } z = \eta(r, t) \tag{6.29}$$

$$\frac{\partial \eta}{\partial t} = -\left[1 + \left(\frac{\partial \eta}{\partial r}\right)^2\right]^{\frac{1}{2}} \frac{\partial \Phi}{\partial n} + W \qquad \text{on } z = \eta(r, t) \tag{6.30}$$

where $W(r, t)$ is the dimensionless recharge intensity. Due to the property of axial symmetry, the normal derivative of the potential, $\partial\Phi/\partial n$, on the free surface is independent of the $\theta$ co-ordinate. From (6.29) and (6.30) the rate of change of the potential of a point on the free surface with a fixed $r$ co-ordinate becomes

$$\left(\frac{\partial \Phi}{\partial t}\right)_r = -\frac{1}{\cos \beta} \frac{\partial \Phi}{\partial n} + W \tag{6.31}$$

where $\tan \beta$ is defined as the free surface slope; i.e. $\partial\eta/\partial r = -\tan \beta$. The finite difference analog of (6.31) is

$$\Phi^{k+1} = \Phi^k - \frac{\Delta t}{\cos \beta^k} \left[ \theta \left( \frac{\partial \Phi}{\partial n} \right)^{k+1} + (1-\theta) \left( \frac{\partial \Phi}{\partial n} \right)^k \right]$$

$$+ \Delta t \left[ \theta W^{k+1} + (1-\theta) W^k \right] \tag{6.32}$$

in which $k$ denotes the $k\,\Delta t$ time level and $\theta$ is a weighting factor. Equation (6.32) represents a relation between $\Phi$ and $\partial\Phi/\partial n$ at the $(k+1)\,\Delta t$ time level on the free surface and can be used in (6.25) to replace each $\Phi$ by a $(\partial\Phi/\partial n)$ on the free surface.

### 6.4.1 Recharge flows

In this section we consider the response of a free surface, initially at rest, due to a uniform circular recharge. The boundary value problem can be summarized as follows:

$$\nabla^2 \Phi = 0 \text{ in the flow region;} \qquad \nabla^2 = \frac{1}{r}\frac{\partial}{\partial r}\left( r \frac{\partial}{\partial r} \right) + \frac{\partial^2}{\partial z^2} \tag{6.33}$$

$$\frac{\partial \Phi}{\partial z} = 0 \qquad z = -h \tag{6.34}$$

$$\Phi = \eta \qquad z = \eta \tag{6.35}$$

$$\frac{\partial \eta}{\partial t} = -\left[ 1 + \left( \frac{\partial \eta}{\partial r} \right)^2 \right]^{\frac{1}{2}} \frac{\partial \Phi}{\partial n} + W \qquad z = \eta \tag{6.36}$$

$$\Phi \to \text{constant}, \; r \to \infty \tag{6.37}$$

The unconfined aquifer has a constant depth, $h$, and infinite lateral extent. The constant recharge $W$ is limited to an area $r \leqslant 1$ (Fig. 6.5).

The numerical scheme developed in previous sections is first applied to a problem for an infinitely deep aquifer (i.e. $h \to \infty$). Equation (6.25) is to be solved around the entire boundary, including the infinite boundary. However, the left-hand term of that equation is zero on the infinite boundary, $\Gamma_\infty$ (Fig. 6.6), due to the fact that it contains $1/R_s$ and $R_s \to \infty$. The right-hand term is also zero since $\partial\Phi/\partial n$ is zero on the infinite boundary. There remains the computation of unknown quantities on the free surface, $0 \leqslant r < \infty$. The truncation method is used herein to reduce the problem to a finite domain. The free surface is assumed to be undisturbed at $r \geqslant r_n$; i.e.

$$\eta = \Phi = 0, \; r_n < r < \infty.$$

Figure 6.5 A circular recharge area.

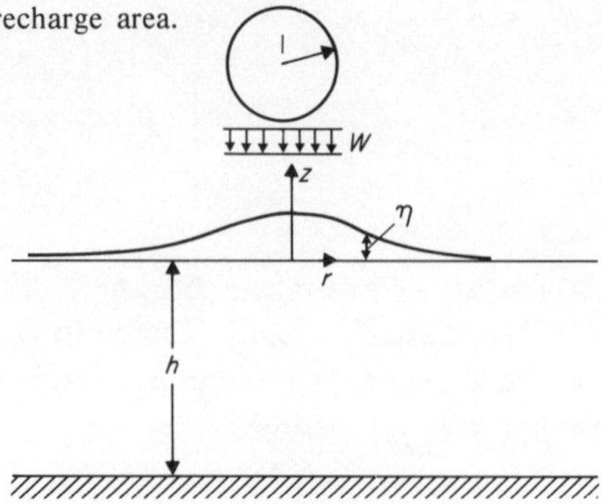


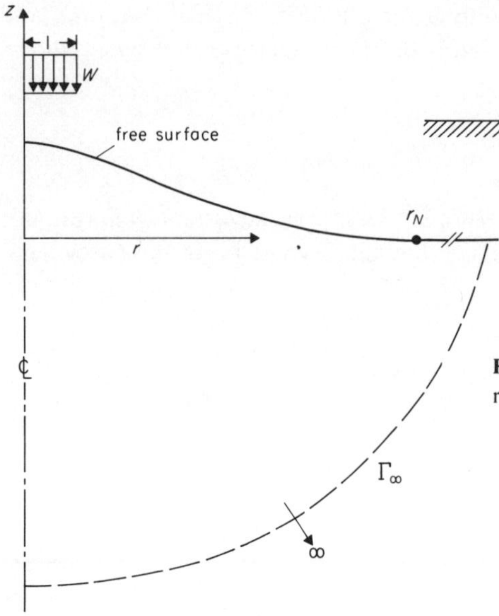


Figure 6.6 Bounding surfaces for recharge to an infinitely deep aquifer.

To ensure that the truncation of the free surface boundary has no effect on the solution (i.e. on significant free surface elevation change) in the region of interest, $r_n$ was continually moved outward until no change in the solution was detected.

An alternate method of modeling the infinite boundary is to use a linearized solution in the far field to approximate $\Phi$ and $\partial\Phi/\partial n$ at these distant boundaries in a manner analogous to the two-dimensional case discussed in Section 3.5.1. However, because the variable changes in axisymmetric problems are confined to a quite limited radial domain, this increased far field accuracy was not required.

In Figure 6.7 the BIEM results for the free surface profiles for the recharge intensity $W = 0.02$ are compared with the linearized theory of Dagan (1967a). Excellent agreement is achieved. In the BIEM computations the 22 nodal point grid was chosen between $r = 0$ and $r = r_n$ to ensure that a relatively smooth curve was obtained. Different grids were also chosen by using 11–30 nodal points. The 22 nodal point grid seems to be an efficient and accurate grid.

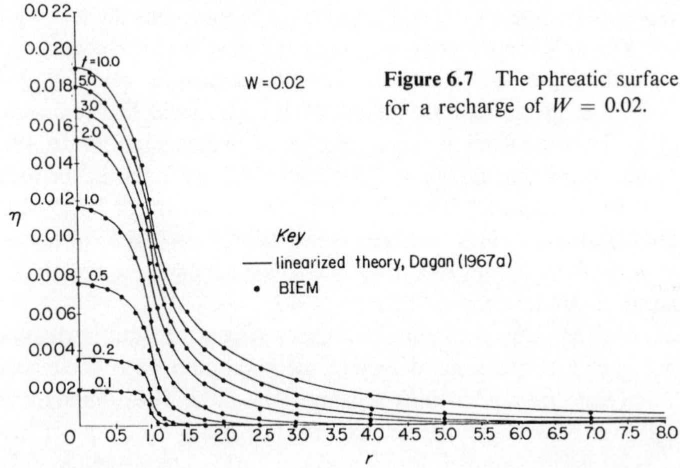


**Figure 6.7** The phreatic surface for a recharge of $W = 0.02$.

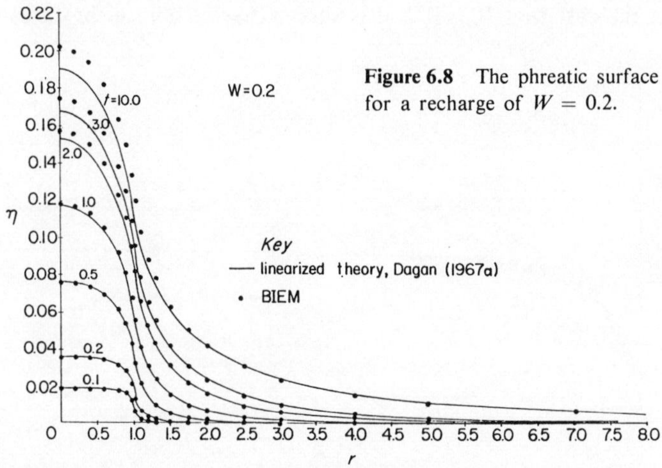


**Figure 6.8** The phreatic surface for a recharge of $W = 0.2$.

The linearized theory of Dagan (1967a) can be written as

$$\eta = \frac{W}{2\pi} \int_0^1 \int_0^{2\pi} \{[r^2 + \rho^2 - 2r\rho \cos \theta]^{-\frac{1}{2}}$$

$$- [t^2 + r^2 + \rho^2 - 2r\rho \cos \theta]^{-\frac{1}{2}}\} \rho \, d\rho \, d\theta + O(W^2) \quad (6.38)$$

The curves in Figure 6.7 were produced by integrating (6.38). The integration was performed by employing elliptic integrals and Gaussian quadrature in ρ.

The 22 nodal point grid was used for the case using $W = 0.2$. The comparisons between the BIEM results and the linearized theory developed by Dagan (1967a) are shown in Figure 6.8. It is seen that excellent agreement is obtained for early times, $t < 1.0$. However, at larger times the leading order solution of Dagan's theory underestimates the peak free surface at $r = 0$ by roughly 5%. In Figure 6.9 the maximum free surface elevation, $\eta_{max}/W$, is plotted versus time for various values of $W$. The solid line represents all values of $W$ from the leading order solution of Dagan's theory, (6.38). The BIEM results show that higher $\eta_{max}/W$ are obtained for a higher recharge intensity. It is obvious that Dagan's theory breaks down for large values of $W$. Figure 6.9 also indicates that the perturbation theory becomes invalid at large time even for a small value of $W$ due to the accumulation of water near the recharge mound.

Calculations are also performed for cases where the aquifer thickness is finite. In Figure 6.10 the time history of the maximum water table is shown for various values of aquifer thickness and $W = 0.02$. The numerical results agree quite well with the linearized theory. For $h > 10$ the aquifer behaves virtually as an infinite aquifer. Similar sets of curves are presented in Figure 6.11 for the case that $W = 0.2$. It is shown that except for the cases of very

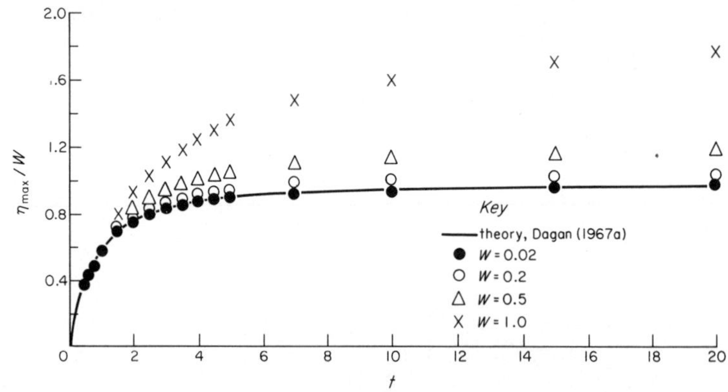


**Figure 6.9** The normalized maximum free surface rise for various values of the recharge (infinite aquifer thickness).

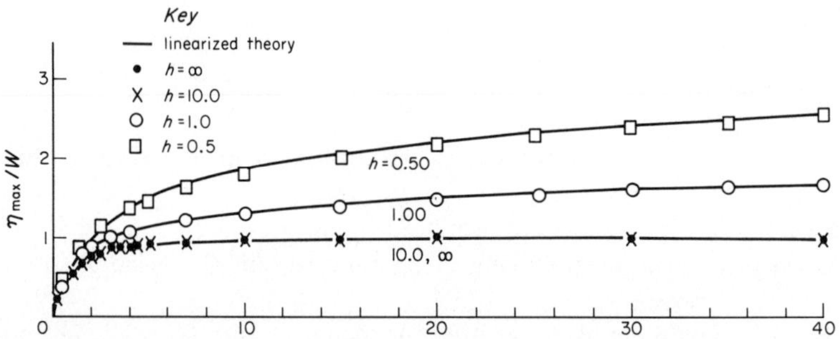


**Figure 6.10** The normalized maximum free surface rise for various values of aquifer thickness ($W = 0.02$).

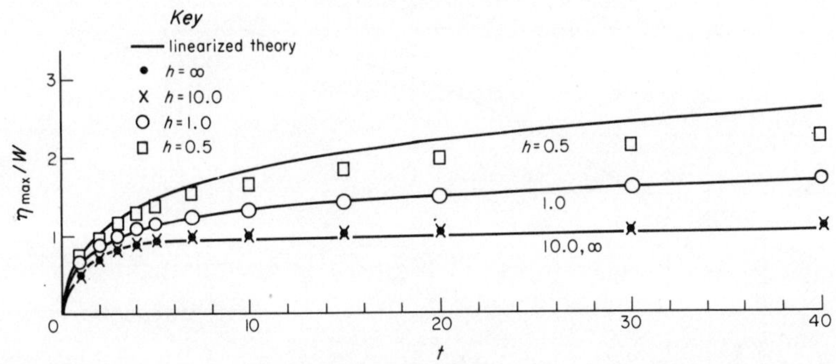


**Figure 6.11** The normalized maximum free surface rise for various values of aquifer thickness ($W = 0.2$).

shallow aquifers the BIEM results agree with Dagan's theory. However, for a shallow aquifer, e.g. $h = 0.5$, the linearized theory seems to overpredict the maximum water table.

### 6.4.2 *Well problems*

In this section radial flow towards an axisymmetric opening (i.e. a well) is considered. The main differences between the well problem and the recharge problem discussed in the previous section are the presence of a seepage face and a vertical boundary very close to the axis of symmetry. Also two types of far field horizontal boundary conditions are investigated:

(1) The phreatic surface is maintained at a constant level a finite distance away from the well at $r = r_e$, which is the case of a well in a circular island.

(2) The well is placed in an aquifer of infinite horizontal extent.

The boundary value problem for free surface well flows can be described as follows (see Fig. 6.12):

$$\nabla^2\Phi = 0 \qquad \text{in the flow region} \tag{6.39}$$

$$\Phi = h_w \qquad r = r_w,\ 0 \leqslant z \leqslant h_w \tag{6.40}$$

$$\Phi = z \qquad r = r_w,\ h_w \leqslant z \leqslant h_s \tag{6.41}$$

$$\left.\begin{aligned} &\Phi = z && (6.42a)\\ &\frac{\partial \eta}{\partial t} = -\left[1 + \left(\frac{\partial \eta}{\partial r}\right)^2\right]^{\frac{1}{2}} \frac{\partial \Phi}{\partial n} && (6.42b) \end{aligned}\right\} r_w \leqslant r \leqslant r_e,\ z = \eta$$

$$\frac{\partial \Phi}{\partial z} = \frac{\partial \Phi}{\partial n} = 0 \qquad r_w \leqslant r \leqslant r_e,\ z = 0 \tag{6.43}$$

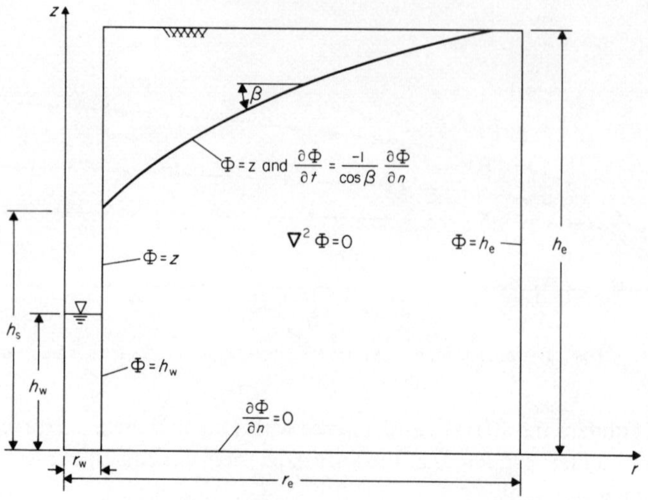


**Figure 6.12** Definition sketch of the well problem.

All variables appearing in (6.39)–(6.43) are dimensionless. The well radius is designed as $r_w$. For the case of a well in a circular island the external horizontal boundary condition requires

$$\Phi = h_e \qquad r = r_e \qquad 0 \leqslant z \leqslant h_e \tag{6.44}$$

In the case of an infinite aquifer, $r_e \to \infty$, the boundary condition given in (6.44) is still valid. However, all relevant integrals, (6.19), are zero on the boundary $r = r_e$, $0 \leqslant z \leqslant h_e$ as $r_e \to \infty$.

The initial conditions are

$$\eta(r, t = 0) = h_e \qquad h_w(t = 0) = h_s(t = 0) = h_e \tag{6.45}$$

in which $h_e$ is fixed as a constant for all time. Two types of flow problems are considered: a quick drawdown and a pumping well.

For quick drawdown problems, the water level inside the well is dropped from $h_e$ to a constant value $h_w$ at $t = 0$ and stays at that level for all time. For a pumping well problem, the rate of withdrawal from the well, $Q_w(t)$, is prescribed. The principle of conservation of mass requires that $h_w(t)$ satisfies the following dimensionless differential equation (Gambolati 1976):

$$Q_w(t) = Q_I(t) - \frac{\pi r_w^2}{n}\frac{dh_w}{dt} \qquad \frac{dh_w}{dt} < 0 \tag{6.46}$$

in which $Q_I(t)$ is the volume rate of flow into the well from the aquifer,

$$Q_{\mathrm{I}}(t) = -2\pi r_{\mathrm{w}} \int_0^{h_s} \left(\frac{\partial \Phi}{\partial n}\right)_{r=r_{\mathrm{w}}} \mathrm{d}z \tag{6.47}$$

It is recalled that $(\partial\Phi/\partial n)_{r=r_{\mathrm{w}}} = (-\partial\Phi/\partial r)_{r=r_{\mathrm{w}}}$. The second term on the right-hand side of (6.46) represents the well storage or rate of change of the water volume inside the well. In actual wells the well storage is modified to include the effects of a production pipe, i.e. $\pi r_{\mathrm{w}}^2$ should be replaced by $\pi(r_{\mathrm{w}}^2 - r_{\mathrm{p}}^2)$ where $r_{\mathrm{p}}$ is the radius of the production pipe (Gambolati 1976).

The solution algorithm which solves the matrix equation (6.25)

$$[R]\{\Phi\} = [L]\{\partial\Phi/\partial n\} \tag{6.48}$$

follows in a manner analogous to the two-dimensional problems. Using the matrix partitioning concept, we may rewrite equation (6.48) as

$$[R_{\mathrm{a}} \quad R_{\mathrm{sp}} \quad R_{\mathrm{b}} \quad R_{\mathrm{c}} \quad R_{\mathrm{d}}]\begin{Bmatrix} \Phi_{\mathrm{s}} \\ \Phi_{\mathrm{sp}} \\ \Phi_{\mathrm{b}} \\ \Phi_{\mathrm{c}} \\ \Phi_{\mathrm{d}} \end{Bmatrix}$$

$$= [L_{\mathrm{a}} \quad L_{\mathrm{sp}} \quad L_{\mathrm{b}} \quad L_{\mathrm{c}} \quad L_{\mathrm{d}}]\begin{Bmatrix} (\partial\Phi/\partial n)_{\mathrm{a}} \\ (\partial\Phi/\partial n)_{\mathrm{sp}} \\ (\partial\Phi/\partial n)_{\mathrm{b}} \\ (\partial\Phi/\partial n)_{\mathrm{c}} \\ (\partial\Phi/\partial n)_{\mathrm{d}} \end{Bmatrix} \tag{6.49}$$

where the subscripts refer to the following portions of the boundary:

- a — well boundary ((6.40) and (6.41))
- sp — single point where the seepage face intersects the well boundary
- b — free surface boundary (6.42)
- c — external horizontal boundary ((6.44) for finite $r_{\mathrm{e}}$; excluded for infinite $r_{\mathrm{e}}$)
- d — aquifer basement (6.43)

On a free surface the relation between $\Phi$ and $\partial\Phi/\partial n$ at time level $k$ is given in (6.32) with $W = 0$.

Rearrangement of the known and unknown values of $\Phi$ and $\partial\Phi/\partial n$ on the boundary results in the following set of simultaneous equations:

$$[H]\{u\} = \{r\} \tag{6.50}$$

where $[H]$ is the coefficient matrix, $\{u\}$ is the vector of unknowns, and $\{r\}$ is a known right-hand side, and where

$$[H] = \left[ -L_{\mathrm{a}} \quad R_{\mathrm{sp}} \quad -L_{\mathrm{b}} - R_{\mathrm{b}} \frac{\Delta t\,\theta}{\cos\beta^{k}} \quad -L_{\mathrm{c}} \quad R_{\mathrm{d}} \right]$$

$$\{u\}^{\mathrm{T}} = \lfloor (\partial\Phi/\partial n)_{\mathrm{a}} \quad \Phi_{\mathrm{sp}} \quad (\partial\Phi/\partial n)_{\mathrm{b}} \quad (\partial\Phi/\partial n)_{\mathrm{c}} \quad \Phi_{\mathrm{d}} \rfloor$$

$$\{r\} = [-R_{\mathrm{a}} \quad L_{\mathrm{sp}} \quad -R_{\mathrm{b}} \quad -R_{\mathrm{c}} \quad L_{\mathrm{d}}]\{C\} \tag{6.51}$$

$$\{C\}^{\mathrm{T}} = \left\lfloor \Phi_{\mathrm{a}} \quad (\partial\Phi/\partial n)_{\mathrm{sp}} \quad \Phi_{b}^{k} - \frac{\Delta t(1-\theta)}{\cos\beta^{k}} (\partial\Phi/\partial n)_{b}^{k} \quad \Phi_{\mathrm{c}} \quad (\partial\Phi/\partial n)_{\mathrm{d}} \right\rfloor$$

Known quantities on the $k$th time level are denoted by a $k$ superscript; all quantities without a superscript are assumed to be values on the $(k + 1)$th time level. All unknowns are in $\{u\}$ and all knowns are in $\{C\}$. It should also be noted that the final system of simultaneous equations may be formed directly without the use of $[R]$ and $[L]$, thus increasing the efficiency of the algorithm and reducing the required storage. Furthermore, a significant portion of the matrices remain unchanged and need not be re-evaluated at each time level.

Three cases are considered here: quick drawdown with finite boundaries, and pumping wells with both finite and infinite boundaries. In all computations the well radius $r_{\mathrm{w}}$ has been used as the characteristic length scale. Therefore, $r = r_{\mathrm{w}}$ describes the well surface.

*Case 1: quick drawdown with finite horizontal boundary*

The solution for the free surface with $r_{\mathrm{e}} = 51$ is shown in Figure 6.13 at selected times for

$$h_{\mathrm{e}} = h_{\mathrm{w}}(t = 0) = 10 \qquad h_{\mathrm{w}}(t > 0) = 3 \tag{6.52}$$

The solution is essentially steady for $t > 250$. The Dupuit solution is also shown in Figure 6.13 (Bear 1972). The Dupuit theory ignores the seepage face, but 'far' from the well the Dupuit free surface is very close to the exact solution. Thus for radial distances greater than $r = r_{\mathrm{w}} + fh_{\mathrm{e}}$, where $f$ is *very* roughly equal to one (Bear 1972), these solutions should match. The Dupuit solution for the non-dimensional free surface, $h_{\mathrm{D}}$, is given by

$$h_{\mathrm{D}} = \left\{ \frac{Q_{\mathrm{D}}}{\pi} \ln (r/r_{\mathrm{e}}) + h_{\mathrm{e}}^{2} \right\}^{\frac{1}{2}} \tag{6.53}$$

where $Q_{\mathrm{D}}$ is the Dupuit flow rate, known to be the exact value (ignoring capillarity effects), and is given by

$$Q_D = \pi \frac{h_e^2 - h_w^2}{\ln(r_e/r_w)} \tag{6.54}$$

where $Q_D = Q_D^*/K^*(r_w^*)^2$. The error in the steady state solutions,

$$\varepsilon = \frac{Q_D - Q(r_w, t \to \infty)}{Q_D} \tag{6.55}$$

for the case shown in Figure 6.13 is less than 0.03 (see Table 6.3, runs 3D & 3E).

To verify the present numerical scheme with available experimental data and other numerical solutions, a quick drawdown problem with $r_e = 16$, $h_w(t > 0) = 2.5$ is presented. In Figure 6.14, the BIEM steady state results are plotted with Hall's (1955) experimental curve from a sector sand model. There is a small but very noticeable difference between the solutions, mainly due to the unsaturated flow and capillary effects not taken into account with the BIEM. Also plotted in the same figure is the steady state finite element solution (which also ignores the unsaturated effects) obtained by Neuman and Witherspoon (1971). The agreement between the BIEM and finite element solution is excellent. For real field situations, the relative discrepancy due to capillary effects will, in general, be less than indicated in Figure 6.14. The Dupuit solution is also shown in Figure 6.14 (Bear 1972). The error in flow rate defined in (6.55) for this case is very small (see Table 6.3, run 2).

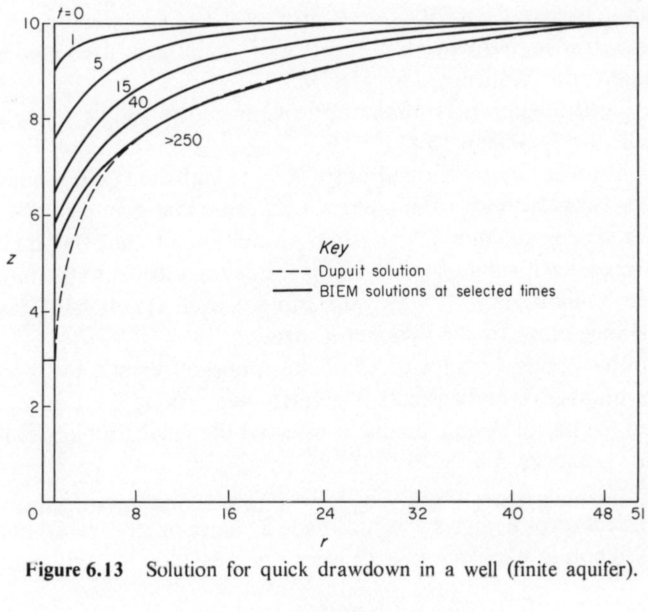


**Figure 6.13** Solution for quick drawdown in a well (finite aquifer).

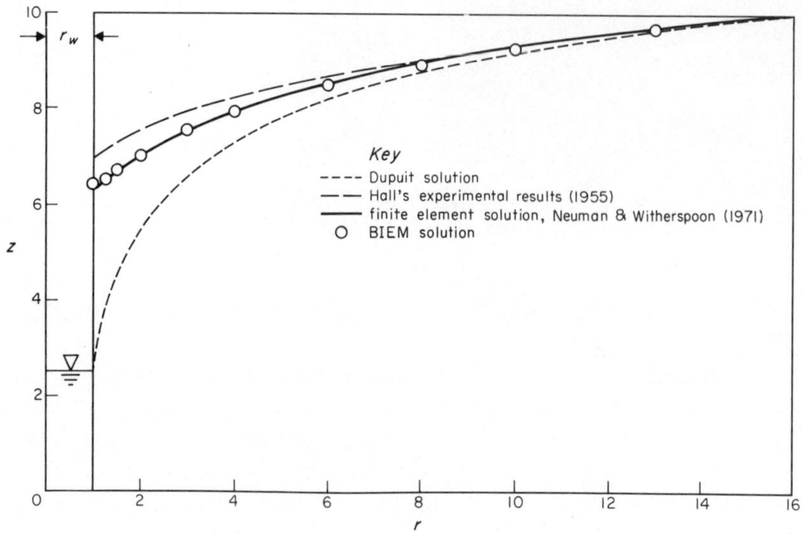


**Figure 6.14** Comparison of steady state results (finite aquifer).

Numerical experiments were carried out for the quick drawdown problem by using different numbers of nodal points as well as Gauss points in computing the integrals shown in (6.19). Some of the steady state results are summarized in Table 6.3. The following observations can be made based on the data presented in Table 6.3:

(1) The error $\varepsilon$ increases as $r_e$ increases (run sequence 1, 2, 3). Note that run 1 used a 21-point grid; a finer grid should yield $\varepsilon \leqslant O(10^{-5})$.

(2) The error, $\varepsilon$, is reduced as more nodal points are used to define the region (run sequence 3A, 3D, 3E).

(3) A fourth degree polynomial approximates the elliptic integrals very well ((6.13) and (6.16)).

(4) Eight point Gaussian quadrature is quite sufficient (run sequence 3A, 3B). Near the well, each interval between nodal points can be broken into several subintervals and four to eight point Gaussian quadrature used on each subinterval to increase accuracy because the integrands (see Voinov *et al.* 1974) vary rapidly as a function of boundary distance close to the symmetric axis.

(5) Double precision helps to eliminate round-off errors, but only when the number of nodal points is greater than around 50. For a 23 point grid the use of double precision affected the results by less than 0.1% (run sequence 3A, 3C).

A transient error analysis was not made because other detailed numerical solutions ignoring capillary effects were not available for this case.

**Table 6.3** Runs for case 1: quick drawdown, finite aquifer as $t \to \infty$. Unless otherwise noted, 8 Gauss points are used for each interval, and computations performed using single precision (IBM 370/168).

| *Run* | *Comments* | *Nodal points* | $r_w$ | $r_e - r_w$ | $h_w$ | $h_e$ | $Q_D$ | *Q at* $r_w$ *(BIEM)* | *% error × 100* |
|---|---|---|---|---|---|---|---|---|---|
| 1 | $r_e$ smaller than Hall (1955) | 21 | 1 | 10 | 3 | 10 | 119.2 | 118.7 | + 0.42 |
| 2 | Hall's (1955), $r_w^* = 4.8$ inches | 31 | 1 | 15 | 2.5 | 10 | 106.23 | 106.23 | $O(10^{-3})$ |
| 3A | $r_e$ greater than Hall (1955) | 23 | 1 | 50 | 3 | 10 | 72.71 | 62.76 | +13.7 |
| 3B | run 3A, 16 Gauss points | 23 | 1 | 50 | 3 | 10 | 72.71 | 62.74 | +13.7 |
| 3C | run 3A, double precision | 23 | 1 | 50 | 3 | 10 | 72.71 | 62.76 | +13.7 |
| 3D | run 3A, more nodes | 27 | 1 | 50 | 3 | 10 | 72.71 | 70.64 | + 2.8 |
| 3E | run 3D, more nodes | 33 | 1 | 50 | 3 | 10 | 72.71 | 70.94 | + 2.4 |

*Case 2: pumping well with finite horizontal boundary*

The major difference between a pumping well and the quick drawdown problem is that rather than prescribing a value of water level in the well, conservation of mass provides an additional relationship, (6.46). For a period of time $\Delta t = T_2 - T_1$, (6.46) can be integrated to yield

$$\int_{T_1}^{T_2} Q_{\mathrm{w}}(t)\,\mathrm{d}t = \int_{T_1}^{T_2} Q_{\mathrm{I}}(t)\,\mathrm{d}t - \int_{T_1}^{T_2} \left(\frac{\pi r_{\mathrm{w}}^2}{n}\right) \frac{\mathrm{d}h_{\mathrm{w}}}{\mathrm{d}t}\,\mathrm{d}t \qquad (6.56)$$

where $Q_{\mathrm{w}}$ is a specified function of time. The mean

$$\bar{Q}_{\mathrm{w}} = \frac{1}{T_2 - T_1} \int_{T_1}^{T_2} Q_{\mathrm{w}}(t)\,\mathrm{d}t$$

is known, and if $Q_{\mathrm{I}}(t)$ is assumed to vary linearly between $T_1$ and $T_2$, then (6.56) becomes

$$\bar{Q}_{\mathrm{w}}\,\Delta t = \left(\frac{Q_{\mathrm{I}}^{k} + Q_{\mathrm{I}}^{k+1}}{2}\right) \Delta t - \frac{\pi r_{\mathrm{w}}^2}{n} (h_{\mathrm{w}}^{k+1} - h_{\mathrm{w}}^{k}) \qquad (6.57)$$

where the superscript $k$ refers to quantities at time $T_1$, and the superscript $(k + 1)$ refers to unknown quantities at time $T_2$. Thus (6.57) contains two unknowns, $h_{\mathrm{w}}^{k+1}$ and $Q_{\mathrm{I}}^{k+1}$. An iterative procedure is developed to solve (6.57) and to compute the free surface profile using the BIEM at time $T_2$. The iteration procedure is described as follows: an estimated value of $Q_{\mathrm{I}}^{k+1}$ is first used in (6.57) to solve for $h_{\mathrm{w}}^{k+1}$ ($= h_{i=0}$, $i$ being the iteration number). The flow problem may be viewed as a portion of a quick drawdown problem between times $T_1$ and $T_2$; the water level in the well drops from $h_{\mathrm{w}}^{k}$ at $T_1$ to $h_{i=0}$ and stays there until time $T_2$. The BIEM is used to compute the free surface profile and $Q_{\mathrm{I}}^{k+1}$. The new value of $Q_{\mathrm{I}}^{k+1}$ is then used to calculate $h_{\mathrm{w}}^{k+1} = h_{i+1}$. After $(i + 1)$ iterations, the absolute error

$$\varepsilon_{\mathrm{h}} = \left| \frac{h_{i+1} - h_i}{h_{\mathrm{e}}} \right| \qquad (6.58)$$

is calculated. This completes a cycle. The new $h_{\mathrm{w}}^{k+1}$ is used to calculate $Q_{\mathrm{I}}^{k+1}$ for another cycle. The iteration procedure is stopped and the solution is accepted when at last $\varepsilon_{\mathrm{h}}$ falls below a prescribed small number. The majority of the runs discussed herein use $\varepsilon_{\mathrm{h}} = 0.002$. To begin the next iteration, (6.57) is rewritten as

$$\bar{Q}_{\mathrm{w}}\,\Delta t = \left(\frac{Q_{\mathrm{I}}^{k+1} + Q_{\mathrm{I}}^{k+2}}{2}\right) \Delta t - \frac{\pi r_{\mathrm{w}}^2}{n} (h_{\mathrm{w}}^{k+2} - h_{\mathrm{w}}^{k+1}) \qquad (6.59)$$

The converged value of $h_w^{k+1}$ is used in (6.59) to solve for $h_w^{k+2}$ and the iteration cycle begins anew. In order to aid in convergence the actual guess may be taken as

$$h_{i+1}(\text{used}) = h_i + \alpha(h_{i+1}(\text{predicted}) - h_i) \tag{6.60}$$

where $\alpha = 1$ corresponds to using $h_w^{k+1}$ (predicted), and $\alpha$ is usually taken between 0.75 and 0.95. An alternative to the iterative procedure is to use the two-step scheme developed by Neuman and Witherspoon (1971).

Figure 6.15 shows the transient BIEM solutions for the same case as that analyzed in Figure 6.14, assuming a well is being pumped at a constant rate. In order to reach a steady state with $h_w(t \to \infty) = 2.5$, the withdrawal rate, $Q_w$, must approach the Dupuit flow rate, 106.23 (see run 2, Table 6.3). Also plotted in Figure 6.15 are the solutions obtained by Neuman and Witherspoon (1971). The finite element solution for the free surface is seen to be only slightly higher than the BIEM solution for a corresponding nondimensional time. However, significant differences appear in the solutions for $h_w(t)$. In Figure 6.16 the time history of the water level in the well is presented. Two different time step sizes are used in the BIEM computations: 20 and 40 time steps between the initial time, $t = 0$, and $t \approx 60$, which is roughly the time required to obtain the steady state solution. It is seen that the solutions converge as the smaller time steps are used. Results obtained by Neuman and Witherspoon are also plotted in Figure 6.16. Their time steps are much larger than those employed in the example, using about 10 time steps to reach steady state.

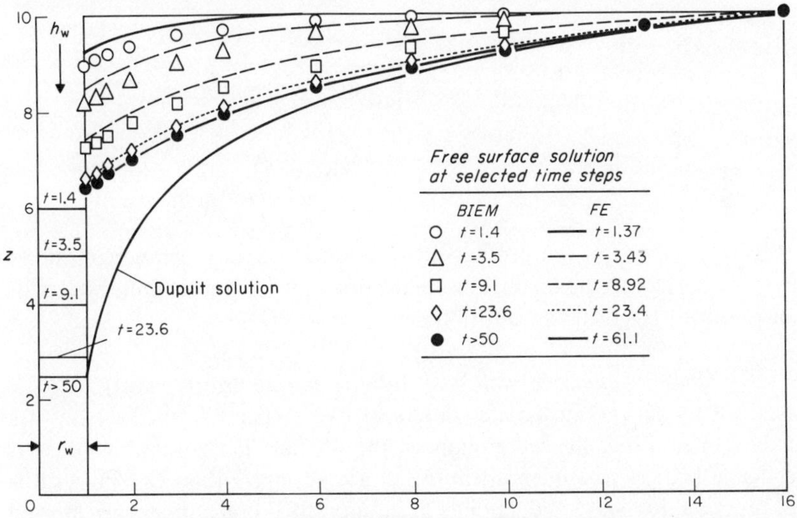


**Figure 6.15** Comparison of BIEM and finite element method (Neuman & Witherspoon 1971) results for the solution of a pumping well (finite aquifer).

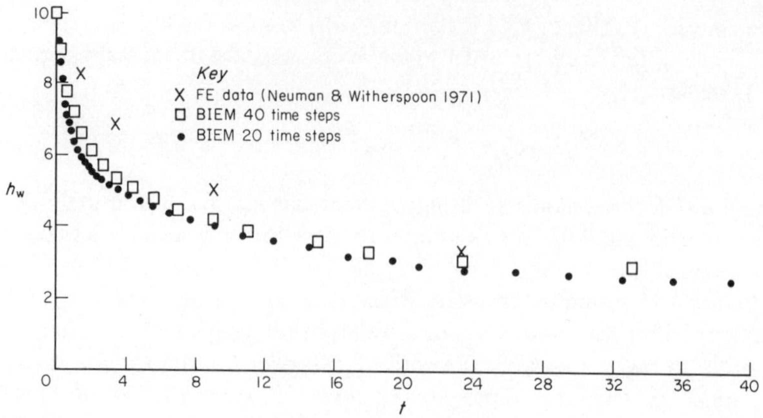


**Figure 6.16** Comparison of BIEM and finite element results for the water level in the well.

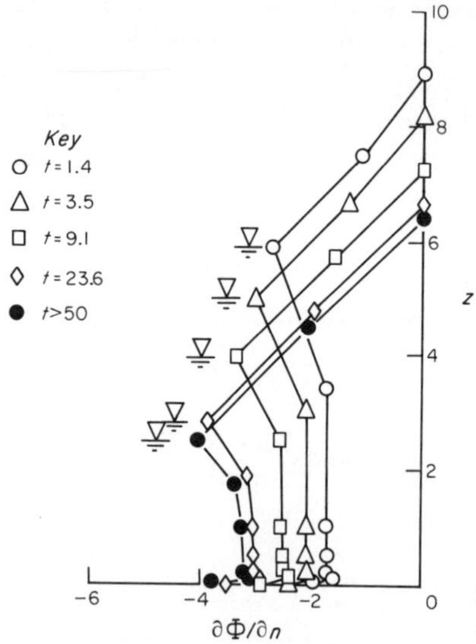


**Figure 6.17** BIEM calculations of the inflow into the well.

Figure 6.17 shows the profiles of the normal velocity component along the well face at various times. The steady state profile agrees quite well with that obtained by Neuman and Witherspoon (1971).

*Case 3: pumping well with infinite horizontal boundary*

The case for an infinite aquifer differs in two respects from the case of a finite aquifer. First, the integration on the far field horizontal boundary is negligible because the integrals in (6.19) are no larger than $O(1/r)$. As it is not practical to locate nodal points at infinity, final nodal points are located at $r = r_{I3}$ (Fig. 6.18) on the free surface and $r = r_{I4}$ on the aquifer basement. The integrals $r_w \leqslant r \leqslant r_{I3}$ on the free surface and $r_w \leqslant r \leqslant r_{I4}$ on the aquifer

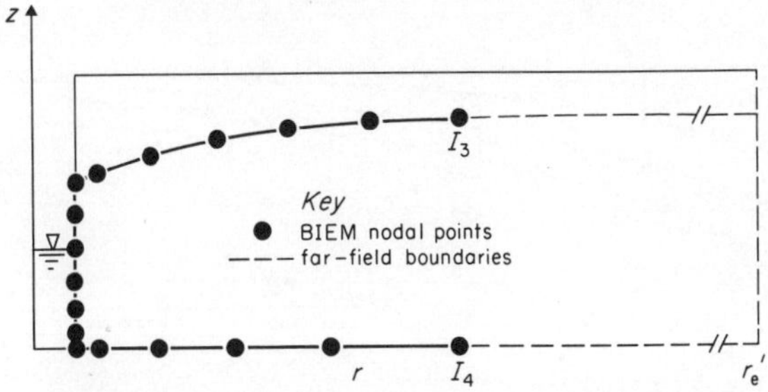


**Figure 6.18** The well problem for an infinite aquifer.

basement are performed as usual. Beyond these points ($r > r_{I_3}$ on the free surface, $r > r_{I_4}$ on the basement) a different shape function (special element, see Ch. 3) than linear is used for $\Phi$ and $\partial\Phi/\partial n$. The one chosen is based upon the steady state Dupuit solution,

$$\Phi(r, t) = \begin{cases} \Phi(r_I, t)\dfrac{N(r)}{N(r_I)} & r_I < r < r'_e \\ 0 & r'_e < r < \infty \end{cases} \tag{6.61}$$

where $r'_e$ is an extremely large number $O(10^3\text{–}10^4)$, the subscript $I$ is either $I = I_3$ (free surface) or $I = I_4$ (aquifer basement), and $N(r)$ is chosen as

$$N(r) = \left\{\frac{Q_f}{\pi}\ln\left(\frac{r}{r_I}\right) + h_I^2\right\}^{\frac{1}{2}} \tag{6.62}$$

with

$$Q_f = \frac{\pi(h_e^2 - h_I^2)}{\ln(r'_e/r_I)} \tag{6.63}$$

In an analogous manner

$$\frac{\partial\Phi}{\partial n}(r, t) = \begin{cases} \dfrac{\partial\Phi(r_I, t)}{\partial n}\dfrac{\partial N(r)}{\partial n}\Big/\left(\dfrac{\partial N(r)}{\partial n}\right)_{r=r_I} & r_I < r < r'_e \\ 0 & r'_e < r < \infty \end{cases} \tag{6.64}$$

where

$$\frac{\partial N(r)}{\partial n} = \frac{n_r(r)}{r}\left\{\frac{Q_f}{\pi}\ln(r/r_I) + h_I^2\right\}^{-\frac{1}{2}} \tag{6.65}$$

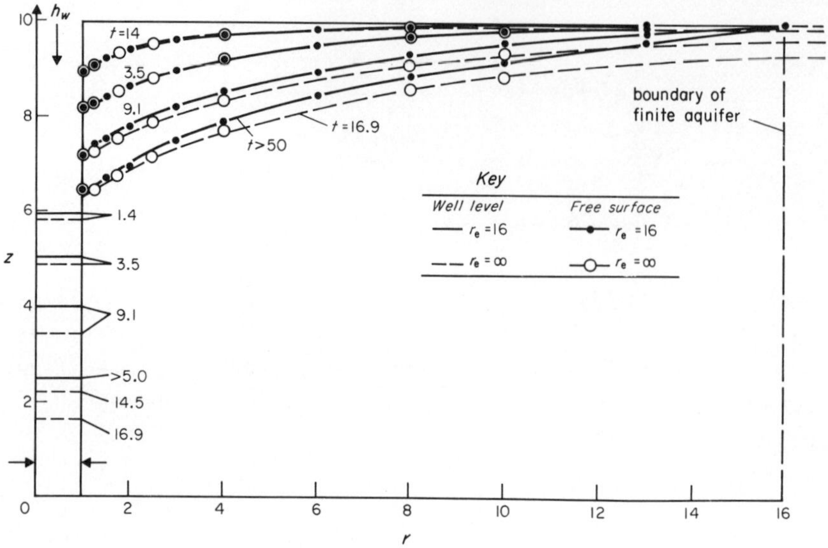


**Figure 6.19** Comparison of BIEM results for a pumping well with a finite aquifer and with an infinite aquifer.

and $n_r(r)$ is the $r$ component of the normal on the free surface. For ease in computation, the region $r_I < r < r'_e$ is broken into a number of straight line segments. Contributions on these segments to the coefficient matrices, $[R]$ and $[L]$, are accomplished by substitution of (6.61) and (6.64) into (6.6) and collecting coefficients of $\Phi(r_I)$ and $\partial\Phi(r_I)/\partial n$.

In order to compare the results for an infinite aquifer, the steady state pumping rate $Q_w = 106.23$ (BIEM solution run 2, Table 6.3) for the finite aquifer is used as the pumping rate. In Figure 6.19 solutions for $r_e = 16$ and $r_e = \infty$ are compared. For small times the two solutions virtually coincide. However, once the drawdown effects reach $r = 16$, the two solutions begin to diverge. Since there is no steady state solution for the infinite aquifer problem, the well will be pumped dry eventually.

In conclusion, it should be stated how easy it is to simulate the pumping of a well in an infinite aquifer, once a run has been made for a finite aquifer. The only changes required are to assign $(r, z)$ co-ordinates to a few extra points on the free surface and basement and change the program flag to indicate that an infinite horizontal aquifer is being modeled. These alterations can be done in a few minutes, as opposed to the substantial changes required to extend a finite element or finite difference grid to approximate an infinite boundary.

# 7 Three-dimensional problems

In this chapter, the BIEM is formulated for three-dimensional problems. The boundary surface is divided into finite elements. Linear triangular elements are used in the analysis presented herein. The numerical integrations are performed by using the method of Gaussian quadrature. The numerical results are shown for recharge problems, for seepage flow from a surface pond, and for well fields.

## 7.1 Discretization of the integral equation

For three-dimensional problems, the integral equation defined by (2.21) and (2.22) can be solved by choosing a finite number of singular points $P_i, i = 1, 2, 3, \ldots, N$, on the boundary $\Omega$ and performing the surface integration for each successive point. By applying a finite element discretization to the surface, the surface integrals reduce to a system of $N$ simultaneous equations in $N$ unknowns. The unknowns are either $\Phi$ or $\partial\Phi/\partial n$ at each boundary point $P_i$.

The surface, $\Omega$, is broken into subregions (finite elements). The actual surface is approximated in an isoparametric fashion, where the shape of the surface as well as the global co-ordinates are assumed to vary as a polynomial of the same order (linear, quadratic, cubic). The global co-ordinates $(x, y, z)$ can be written in terms of the local co-ordinates $(\xi, \eta)$ within an element as

$$\begin{aligned} x &= \{N(\xi, \eta)\}^{\mathrm{T}}\{x\}_e \\ y &= \{N(\xi, \eta)\}^{\mathrm{T}}\{y\}_e \\ z &= \{N(\xi, \eta)\}^{\mathrm{T}}\{z\}_e \end{aligned} \qquad (7.1)$$

where $\{N\}$ is a vector of $M$ components, one for each node in a given element. The expressions for $\{N\}$ for a variety of elements can be found in Zienkiewicz (1971). The global co-ordinates of the $j$th node of the element,

$(x_j, y_j, z_j)$, are the $j$th respective entries to the three vectors $\{x\}_e$, $\{y\}_e$ and $\{z\}_e$. A more detailed discussion describing the transformation to the $\xi$–$\eta$ co-ordinates can be found in Wu (1976) and Jeng and Wexler (1977).

For a typical element, the integrals in (2.21) have the form

$$\iint_{A_e} f(x, y, z; x_i, y_i, z_i)\, dA = \iint_{A_e} F(\xi, \eta; x_i, y_i, z_i) J(\xi, \eta)\, d\xi\, d\eta \tag{7.2}$$

where $A_e$ is the area of the $e$th element and $J$, the Jacobian of the co-ordinate transformation, is

$$J = \left| \frac{\partial \mathbf{R}}{\partial \xi} \times \frac{\partial \mathbf{R}}{\partial \eta} \right| \tag{7.3}$$

with

$$\mathbf{R} = x\mathbf{i} + y\mathbf{j} + z\mathbf{k} \tag{7.4}$$

We can also approximate $\Phi$ and $\partial\Phi/\partial n$ in an element using the same interpolation functions $\{N\}$ as follows:

$$\begin{aligned} \Phi &= \{N\}^{\mathrm{T}}\{\Phi\}_e \\ \frac{\partial \Phi}{\partial n} &= \{N\}^{\mathrm{T}}\left\{\frac{\partial \Phi}{\partial n}\right\}_e \end{aligned} \tag{7.5}$$

where $\{\Phi\}_e$ and $\{\partial\Phi/\partial n\}_e$ are column vectors in terms of the potentials and their normal derivatives, respectively, evaluated at the $M$ nodal points of the $e$th element.

Substitution of (7.1)–(7.5) into (2.21) yields

$$\alpha_i \Phi_i + \sum_{e=1}^{m} \{a\}_e^{\mathrm{T}}\{\Phi\}_e = \sum_{e=1}^{m} \{b\}_e^{\mathrm{T}} \left\{\frac{\partial \Phi}{\partial n}\right\}_e \tag{7.6}$$

with the point $P_i$ at $(x_i, y_i, z_i)$ and $\alpha_i$ being the solid angle at $P_i$. In (7.6), $m$ denotes the total number of elements on the boundary $\Omega$. For a typical $e$th element $\{a\}_e$ and $\{b\}_e$ are vectors depending only upon the local geometry of the boundary (element). In this chapter linear triangular elements are used and further derivations are particular for this type of element. The linear interpolation functions $\{N\}$ are given as follows:

$$\{N\} = \begin{Bmatrix} N_1 \\ N_2 \\ N_3 \end{Bmatrix} = \begin{Bmatrix} \xi \\ \eta \\ 1-\xi-\eta \end{Bmatrix} \tag{7.7}$$

Then $J = 2A_e$ and

$$\{a\}_e = \begin{Bmatrix} a_1 \\ a_2 \\ a_3 \end{Bmatrix}_e = 2A_e \int_{Ae} \begin{Bmatrix} N_1 \\ N_2 \\ N_3 \end{Bmatrix} \frac{\partial}{\partial n}\left(\frac{1}{R}\right) \mathrm{d}\xi\, \mathrm{d}\eta \tag{7.8}$$

$$\{b\}_e = \begin{Bmatrix} b_1 \\ b_2 \\ b_3 \end{Bmatrix}_e = 2A_e \int_{Ae} \begin{Bmatrix} N_1 \\ N_2 \\ N_3 \end{Bmatrix} \left(\frac{1}{R}\right) \mathrm{d}\xi\, \mathrm{d}\eta \tag{7.9}$$

The numerical evaluation of the integrals in (7.8) and (7.9) is performed by Gaussian quadrature (see App. A.6). Zienkiewicz (1971) presents a discussion on the use of Gaussian quadrature to evaluate integrals of the form appearing in (7.8) and (7.9) in which

$$\int_{Ae} f(\xi, \eta)\, \mathrm{d}\xi\, \mathrm{d}\eta = \sum_{j=1}^{M} \omega_j f(\xi_j, \eta_j) + E \tag{7.10}$$

where $(\xi_j, \eta_j)$ are the local co-ordinates for the $j$th quadrature point and $\omega_j$ is the corresponding weighting factor. The size of the error, $E$, depends on the behavior of the integrand $f(\xi, \eta)$ and the number of quadrature points, $M$, chosen in the approximation. Cowper (1973) provided some detailed discussions on the Gaussian quadrature formulas shown in (7.10).

When the point $P_i$ coincides with one of the three nodes of a triangle, the vector $\{a\}_e$ given by (7.8) is zero due to the fact that $\partial R/\partial n = 0$. One of the integrands for the vector $\{b\}_e$ in (7.9) contains an integrable singularity at $R = 0$. The singular integration is examined by assigning the point $P_i$ to node number 3 of the triangle without loss of generality. As the third node is approached, $R \to 0$, $N_1 \to 0$, $N_2 \to 0$, and $N_3 \to 1$. The integrands of $b_1$ and $b_2$ are regular functions and the corresponding integrals can be evaluated by the Gaussian quadrature shown in (7.10). The integral given by $b_3$ can be broken into two parts, one with a non-singular integrand,

$$b'_3 = 2A_e \iint_{Ae} (N_3 - 1)\frac{1}{R}\, \mathrm{d}\xi\, \mathrm{d}\eta \tag{7.11}$$

and one with an integrable singularity,

$$b_3^* = 2A_e \iint_{A_e} \frac{1}{R}\, d\xi\, d\eta \tag{7.12}$$

where $b_3 = b_3' + b_3^*$. Again, the usual quadrature is used for the evaluation of the integral in (7.11). Employing polar co-ordinates, (7.12) becomes

$$b_3^* = \frac{2A_e b}{(1 + m^2)^{\frac{1}{2}}} \ln \left| \frac{\tan\,[(\theta_1 + \alpha)/2]}{\tan\,(\alpha/2)} \right| \tag{7.13}$$

Figure 7.1 illustrates the graphical interpretation of $m$ and $b$, the slope and intercept of the line connecting nodes (1) and (2). The angle $\theta_1$ is formed by the triangle sides adjacent to node 3 and $\alpha$ is a constant given by $\cos\alpha = (1 + m^2)^{-\frac{1}{2}}$, $\sin\alpha = -m(1 + m^2)^{-\frac{1}{2}}$ (Dwight 1961). Equation (7.13) is a generalization of the special case that Wu (1976) presents for a right triangle.

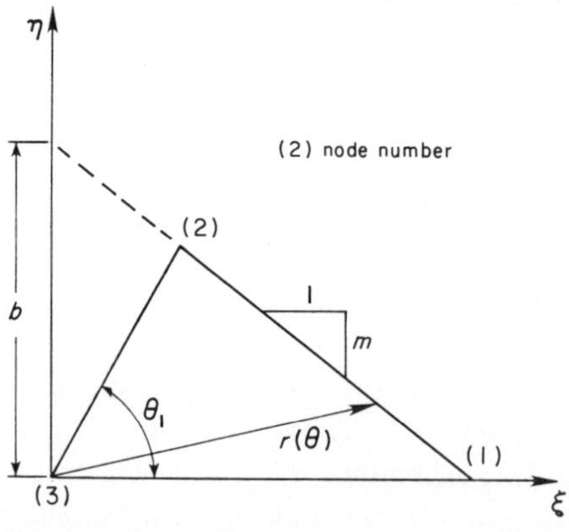


**Figure 7.1** Local co-ordinate system for a singular element.

To complete the information required in (7.6), the solid angle $\alpha_i$ (see (2.22) for definition) must be calculated at point $P_i$. The solid angle is formed by the intersection of $L$ flat triangular elements, where $L \geqslant 3$. There are also $L$ intersection lines formed where these triangular elements meet. The angle $\tau_j$ (interior to $D$) is determined by the angle of intersection of the two planes containing two adjacent triangles. Then (Durell & Arnold 1932)

$$\alpha_i = \sum_{j=1}^{L} \tau_j - (L - 2)\pi \tag{7.14}$$

For example, if all the triangular elements at $P_i$ lie in the same plane, all the

$\tau_j$ are equal to $\pi$ and $\alpha_i = 2\pi$. An alternate method of computing $\alpha_i$ is to apply (7.6) to a fictitious problem where $\partial\Phi/\partial n = 0$ and $\Phi = 1.0$ along the boundary. Thus

$$\alpha_i = -\sum_{e=1}^{m} \{a\}_e^{\mathrm{T}} \begin{Bmatrix} 1 \\ 1 \\ 1 \end{Bmatrix} \tag{7.15}$$

Since $\{a\}_e^{\mathrm{T}}$ must be computed for (7.6), the latter method for computing $\alpha_i$ does not require any additional computing steps. However, this method becomes inaccurate when the problem has an infinite domain.

After applying (7.6) at each of $N$ nodal points, a system of $N$ linear equations in $N$ unknowns is formed. By rearranging the equations and collecting all the unknowns together, the equations can be recast as

$$[H]\{u\} = \{r\} \tag{7.16}$$

where $[H]$ is a known coefficient matrix, $\{u\}$ is a vector of the single unknown (either $\Phi$ or $\partial\Phi/\partial n$) at each node, and $\{r\}$ is a vector of known values. The solution for $\{u\}$ can be obtained by direct Gaussian elimination or an iterative method.

## 7.2 Simple example

To illustrate the computational procedure developed in the previous section, we consider here a simple problem concerning a uniform potential flow in a pipe with a square cross section. As shown in Figure 7.2, the pipe has a dimension of $1 \times 1 \times 1$ units. The potential on the front face is kept at $\Phi = 1.0$, while on the back face the potential is $\Phi = 0$. On the four sides of the pipe the normal velocity must vanish, i.e. $\partial\Phi/\partial n = 0$. The analytical solution to this problem is

$$\Phi = 1 - x \quad 0 \leqslant x \leqslant 1 \tag{7.17}$$

The analytical solution indicates that both $\Phi$ and $\partial\Phi/\partial n$ are either linear or constant on each face of the pipe. Hence, the use of linear triangular elements and linear interpolation functions for $\Phi$ and $\partial\Phi/\partial n$ should not introduce any error. In other words, the errors in the numerical solutions should be the result of the quadrature used to evaluate $\{a\}_e$ and $\{b\}_e$ and the round-off errors resulting from the equation solving routine.

Figure 7.2 shows the grid used herein with 14 nodes and 24 linear triangular elements. Results were obtained using various numbers of Gauss points, $M$, to evaluate $\{a\}_e$ and $\{b\}_e$. Table 7.1 depicts the convergence of

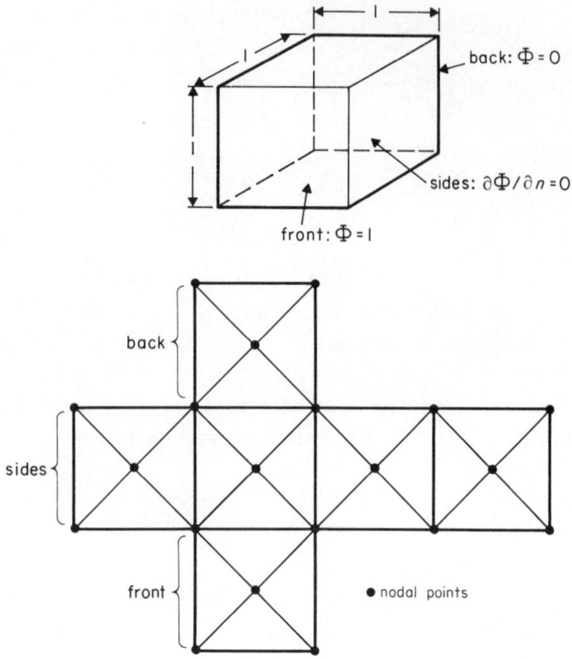


**Figure 7.2** Elements and boundary conditions for flow through a box.

the numerical solutions to the analytical result for different numbers of quadrature points used on non-singular elements. On the singular elements which border $P_i$, 64-point quadrature is always used to evaluate $\{b\}_e$.

Table 7.1 compares the CPU time required for using different quadrature points. The CPU time is based on the Cornell University IBM 370/168 and a FORTRAN G compiler. It is clear that although 64-point quadrature gives the best solution, the use of a lower value of $M$ can yield a sufficiently accurate result with substantial computational savings.

We should mention that Wu (1976) studied the same flow problem by assuming constant values of $\Phi$ and $\partial\Phi/\partial n$ on each linear element. Wu used 64-point quadrature on all elements and obtained a result with a 7.8% error. Our numerical results suggest that using a higher order interpolation func-

**Table 7.1** Error and computing times, numerical example.

| *M* | $E_\Phi \times 100\%$ | $E_{\partial\Phi/\partial n} \times 100\%$ | *CPU time (s)* |
|---|---|---|---|
| 4 | 2.3% | 13% | 0.56 |
| 7 | 0.62% | 1.6% | 0.61 |
| 13 | 0.04% | 0.26% | 0.71 |
| 25 | 0.0014% | 0.012% | 0.94 |
| 64 | 0.0004% | 0.0016% | 1.62 |

tion in the formulation one may obtain more accurate results with less computing time.

## 7.3 Free surface flow problems

For three-dimensional free surface flow problems, the free surface boundary conditions are given in (1.42) and (1.43). In terms of the Cartesian coordinates $(x, y, z)$, they may be rewritten as

$$\Phi = \eta \qquad \text{on } z = \eta \tag{7.18}$$

$$\frac{\partial \eta}{\partial t} = -\left[1 + \left(\frac{\partial \eta}{\partial x}\right)^2 + \left(\frac{\partial \eta}{\partial y}\right)^2\right]^{\frac{1}{2}} \frac{\partial \Phi}{\partial n} + W(x, y, t) \qquad \text{on } z = \eta \tag{7.19}$$

in which $W$ is the recharge intensity. As in the two-dimensional and axial symmetric problems, using finite differences, (7.18) and (7.19) become

$$\Phi^{k+1} = \Phi^k - (\Delta t)\gamma^k \left[\theta\left(\frac{\partial \Phi}{\partial n}\right)^{k+1} + (1-\theta)\left(\frac{\partial \Phi}{\partial n}\right)^{k}\right]$$
$$+ \Delta t \, [\theta W^{k+1} + (1 - \theta) W^k] \tag{7.20}$$

where

$$\gamma^k = \left\{1 + \left[\left(\frac{\partial \eta}{\partial x}\right)^k\right]^2 + \left[\left(\frac{\partial \eta}{\partial y}\right)^k\right]^2\right\}^{\frac{1}{2}} \tag{7.21}$$

and the superscript $k$ denotes the time level, $\Delta t$ is the difference in time between the $(k + 1)$th and $k$th time levels, and $\theta$ is a weighting function $(0 \leqslant \theta \leqslant 1)$ which positions the time derivative between the $k$th time level and the $(k + 1)$th time level. Equation (7.20) is used in the matrix equation, (7.6), such that values of $(\partial \Phi / \partial n)$ are the remaining unknowns in the matrix equation. Equation (7.20) is used to compute $\Phi$ on the free surface at time level $(k + 1)\,\Delta t$. The new location of the free surface can be determined from (7.18), i.e. $\eta^{k+1} = \Phi^{k+1}$.

Three numerical examples are presented herein: (1) uniform recharge problems, (2) pond recharge problems, and (3) well field problems.

### *7.3.1 Recharge problems*

We consider the free surface response due to a uniform recharge, $W$, over a rectangular area with dimensions $2 \times 2B$ (Fig. 7.3). The aquifer is either infinite or finite in depth.

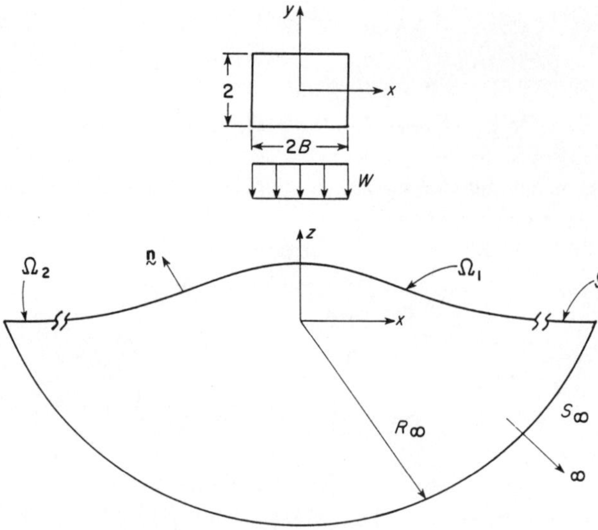


**Figure 7.3** A rectangular recharge area in an aquifer of infinite depth.

*Case 1: aquifer of infinite depth*

Far from the recharge area, $\eta \approx 0$, $\Phi \approx 0$, and $\partial\Phi/\partial n \approx 0$. Thus, as a first approximation the free surface is discretized only for the area which is 'significantly' close to the recharge area; the solution region is truncated at a finite distance where changes remain small. This method is chosen due to its simplicity as well as the fact that the analogous treatment of axisymmetric problems (Ch. 6) showed that excellent results are obtained. A more thorough treatment is to use a shape function based on a linearized theory for the free surface outside the truncated region in a manner similar to the two-dimensional case considered in Chapters 3 and 5. Solutions were obtained with the outer boundary of the grid located at various distances from the recharge area in order to determine where to truncate the grid without affecting the solution near the recharge area. For the boundary $S_\infty$, as $R_\infty \to \infty$,

$$\int_{S_\infty} \Phi \frac{\partial}{\partial n}\left(\frac{1}{R}\right) \mathrm{d}A \to 0 \qquad \int_{S_\infty} \frac{\partial \Phi}{\partial n}\frac{1}{R}\,\mathrm{d}A \to 0$$

$$\Phi \to 0 \quad \text{on } S_\infty \qquad\qquad \frac{\partial\Phi}{\partial n} \to 0 \quad \text{on } S_\infty \tag{7.22}$$

The solution is started from a flat surface, $\eta = 0$. Equation (7.6) is applied at the advanced time level $t^{k+1} = t^k + \Delta t$ for all $N$ points. The unknowns are all the $\{\Phi\}$ and $\{\partial\Phi/\partial n\}$ values at time $(k+1)$ and at location $\eta^{k+1} = \Phi^{k+1}$; thus, there is a total of $2N$ unknowns. However, by applying (7.20), all of the unknown values of $\Phi$ in (7.6), $\Phi^{k+1}$, are rewritten in terms of the unknown $(\partial\Phi/\partial n)^{k+1}$, leaving $N$ unknown quantities. Collecting coefficients of $(\partial\Phi/\partial n)^{k+1}$ yields the system of simultaneous equations (7.16) with $\{u\} = \{(\partial\Phi/\partial n)^{k+1}\}$. The known coefficient matrix is $[H]$ and $\{r\}$ is the known right-hand vector.

Recharge over a square area is discretized using the grid shown in Fig. 7.4 with $m = 312$ and $N = 165$. Figure 7.5 presents the solution for $\eta(x, 0, z)$ plotted against $x$ ($x \geqslant 0$) for selected time steps. The inset of Figure 7.5 shows a contour plot (plan view) of $\eta(x, y, \infty)$, i.e. the steady state solution. Also shown in Figure 7.5 is Dagan's (1967a) linearized solution which is valid for small recharge ($W \ll 1$).

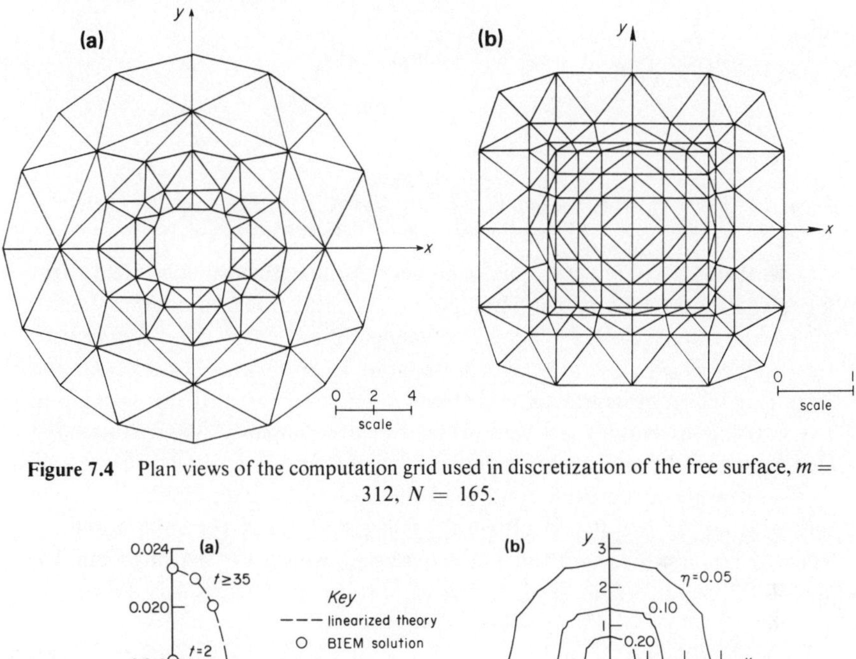


**Figure 7.4** Plan views of the computation grid used in discretization of the free surface, $m = 312$, $N = 165$.

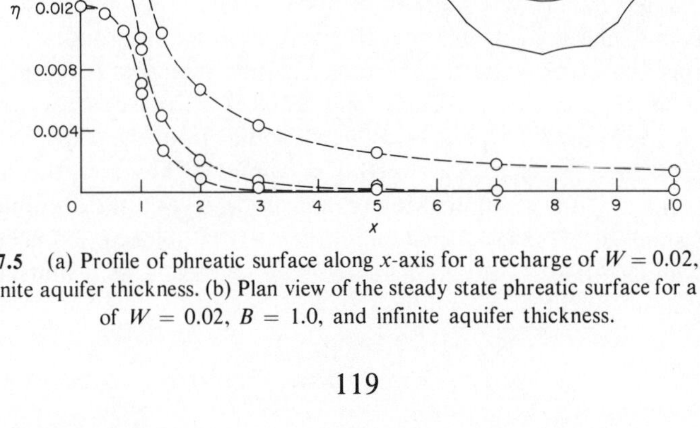


**Figure 7.5** (a) Profile of phreatic surface along $x$-axis for a recharge of $W = 0.02$, $B = 1.0$, and infinite aquifer thickness. (b) Plan view of the steady state phreatic surface for a recharge of $W = 0.02$, $B = 1.0$, and infinite aquifer thickness.

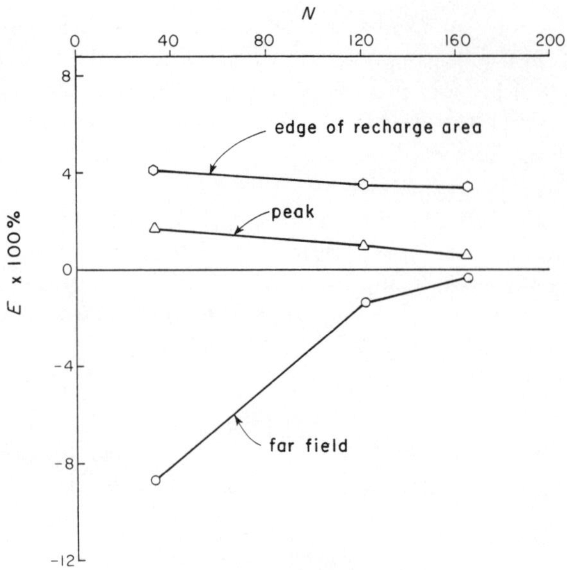


**Figure 7.6** Error in BIEM solution *vs.* mesh refinement for a recharge of $W = 0.02$, $B = 1.0$, and infinite aquifer thickness.

The BIEM results agree quite well with the linearized theory. The slight lag in the BIEM solution for small time is probably due to a combination of coarse time step and grid. That is, the value of $\partial\eta/\partial t$ at $t = 0$ is taken as zero (rest), whereas the actual temporal variation of $\partial\eta/\partial t$ reaches a peak soon after $t = 0$ and monotonically decreases thereafter. The BIEM solution at $t = 35.0$ has essentially arrived at steady state. Dagan (1967a) states that steady state is reached 'theoretically at $t = \infty$ and practically at $t \approx 25$'.

A convergence test was conducted for the steady state results to insure convergence of the BIEM. Figure 7.6 presents the error in $\eta$ for three selected nodal points and three selected grids, where it is assumed that for $W = 0.02$ Dagan's solution is sufficiently accurate. The error is defined as

$$E = (\eta_{\text{DAGAN}} - \eta_{\text{BIEM}})/\eta_{\text{DAGAN}} \tag{7.23}$$

and is plotted against the number of nodal points that were used. At the nodal points, excellent results are obtained, even for the crudest of grids. The rather flat error slope for the peak solution indicates that for a large increase in computational effort, only a small increase in accuracy is obtained. The error for a point at a large distance from the origin is greatly reduced as the number of nodal points is increased; however, the solution far from the origin is small in absolute magnitude, so that the absolute error is quite small in all cases. Furthermore, the BIEM solution for $\eta$ between nodal points (a linear variation) improves dramatically as a more refined mesh is used.

**Table 7.2** Computing time for various grid arrangements.

| *Grid no.* | *Symmetry used?* | *N* | *n** | *Approximate CPU time per timestep (s)* |
|---|---|---|---|---|
| 0 | yes | 33 | 9 | 0.62 |
| 0 | no | 33 | — | 2.4 |
| 1 | yes | 121 | 31 | 8.0 |
| 2 | yes | 165 | 42 | 13.9 |

Table 7.2 presents computation times for the square recharge region using a FORTRAN G compiler and Cornell University's IBM 370/168. Times are presented on a per-time step basis. Considerable savings are obtained by utilizing the double symmetry of the problem. Solutions can be performed over a single quadrant, reducing the work by a factor of roughly four, i.e. the number of unknowns are approximately $n^* = (N - 1)/4 + 1$. By changing a single program flag, the symmetry may be ignored, solving for the entire grid of $N$ points. Using the grid in Figure 7.4, $n^* = 42$ and thus Gaussian elimination was determined to be an adequate method for solving the system of $n^*$ equations. For larger systems, an iterative equation solver is suggested such as that used by Jeng and Wexler (1977). Because of the large number of quadrature points required to evaluate the integrals, a significant portion of the computational effort is devoted to calculating $\{a\}_e$ and $\{b\}_e$ for all elements. Thus, further efforts to improve the efficiency should be devoted to improving the efficiency of the numerical quadrature as well as the equation solving algorithm.

The times presented here are greatly affected by the value of $M$, the number of quadrature points used to evaluate $\{a\}_e$ and $\{b\}_e$. For the cases in Table 7.2, the average value of $M$ is approximately 10. The error in evaluating $\{a\}_e$ and $\{b\}_e$ for a given value of $M$ depends upon the smoothness of the integrands, which in turn depends upon the distance from the point $P_i$, and the shape, size, and orientation of the $e$th element. Five quadrature formulas were included in the program for $M = 4$, 7, 13 (Cowper 1973), $M = 25$ (Zienkiewicz 1971) and $M = 64$ (Stroud 1971). The lowest value of $M$ should be used which keeps the error in approximating $\{a\}_e$ and $\{b\}_e$ below a prescribed maximum error. Rizzo and Shippy (1977) present a brief discussion of these errors and cite references for further details. Unfortunately the formulas for $M = 25$ and 64 are not uniformly symmetric with respect to the three triangle vertices, having a concentration of quadrature points near one vertex. However, the results are superior to the highest symmetric quadrature rule available in Cowper (1973).

Dagan (1967a) presents only the first term in the perturbation series for infinitely deep aquifers and rectangular recharge rates. Because the theory becomes quite involved for other terms in the series, a considerable amount of numerical calculation is required to obtain the solution for $\eta(x, y, t)$ with large recharge rates. Thus the BIEM provides a powerful extension to this basic theory. Virtually any shape, intensity and distribution of recharge, with or without symmetry, can be approximated using the BIEM, including temporal variations of recharge. The case in Figure 7.5 is re-analyzed with $W = 0.2$ in order to demonstrate the non-linear behavior of the free surface for larger values of recharge (see Fig. 7.7).

A rectangular recharge region ($B = 1.5$) was also analyzed and compared to linear theory. The normalized peak, $\eta(0, 0, t)/W$, is shown in Figure 7.8 for various values of $W$. Figure 7.8 includes Dagan's (1967a) solution which collapses into one curve.

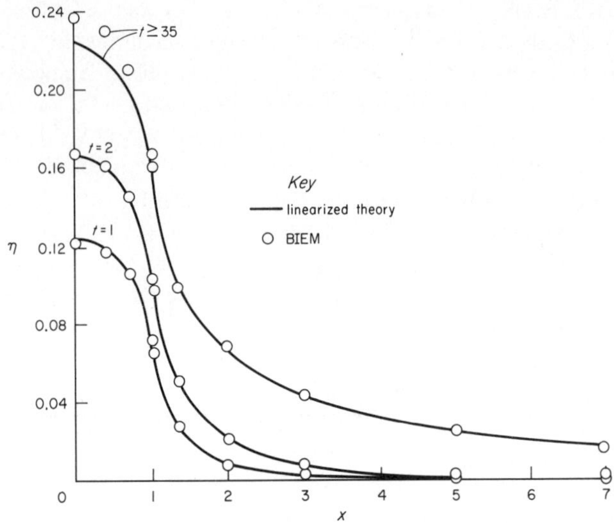


**Figure 7.7** Profile of phreatic surface along $x$-axis for a recharge of $W = 0.2$, $B = 1.0$, and infinite aquifer thickness.

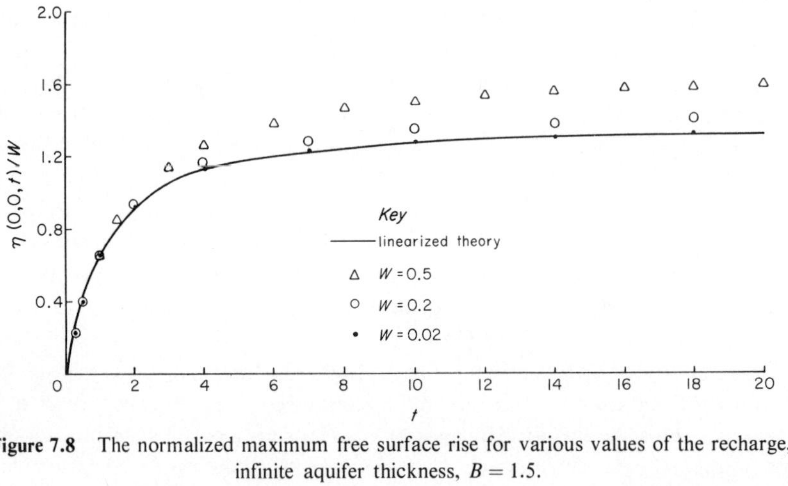


**Figure 7.8** The normalized maximum free surface rise for various values of the recharge, infinite aquifer thickness, $B = 1.5$.

For all the cases analyzed here, it is obvious that for small recharge rates, the agreement between the BIEM results and Dagan's (1967a) solution is excellent. For larger recharge rates, those for which the linearized theory is not valid, the BIEM illustrates the increasing importance of the non-linear behavior. The linearized theory underpredicts the free surface elevation in and near the recharge area.

*Case 2: aquifer of finite depth*

Figure 7.9 illustrates the case of recharge to an aquifer of finite depth. No linearized theory has been presented for rectangular recharge in aquifers of finite depth, so the results are compared to the axisymmetric case, which has both a linearized solution and numerical BIEM solutions.

For nodal points on the basement of an aquifer, $\partial\Phi/\partial n$ is the prescribed boundary condition, and $\Phi$ is the unknown variable. The free surface is again truncated about 10 units from the recharge area. By deforming the grid shown in Figure 7.4, a roughly circular recharge region is formed. The aquifer basement is discretized into a rather course grid of 64 elements within a radial distance of 10 units of the recharge area. Contributions beyond this limit are negligible for the solution near the recharge area. A finer basement discretization has a nominal effect on the solution for the free surface while the computational effort increases considerably. The vertical aquifer sides, located at infinity, are ignored.

Table 7.3 shows that for finite aquifer depths and $W = 0.2$, excellent results are obtained when compared to the axisymmetric BIEM results of Lennon *et al.* (1979a). A typical CPU execution time per time step on the Cornell University IBM 370/168 is about 16 s, compared to about 14 s for

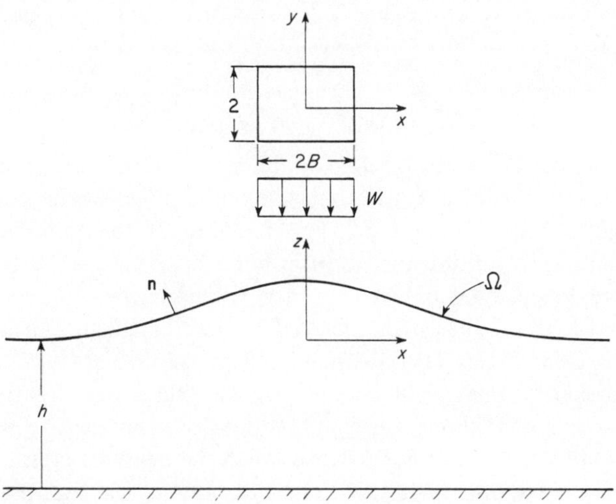


**Figure 7.9** A rectangular recharge area in an aquifer of finite depth.

**Table 7.3** Comparison of peak free surface for axisymmetric and three-dimensional BIEM, a circular recharge area, various aquifer depths and $W = 0.20$.

| | | *Time* | | | |
|---|---|---|---|---|---|
| *Depth* | *Case* | *4* | *10* | *20* | *30* |
| $\infty$ | axi. | 0.185 | 0.202 | 0.207 | 0.208 |
| | 3-D | 0.183 | 0.203 | 0.209 | 0.211 |
| 1 | axi. | 0.215 | 0.265 | 0.295 | 0.314 |
| | 3-D | 0.214 | 0.263 | 0.297 | 0.318 |
| 0.5 | axi. | 0.257 | 0.336 | 0.390 | 0.420 |
| | 3-D | 0.255 | 0.326 | 0.373 | 0.402 |

the infinite depth case. The increase is moderate because the basement grid is rather coarse. The choice of $W = 0.2$ provides a comparison for cases where the non-linear effects are important. Agreement with the axisymmetric BIEM solutions is better for smaller recharge ($W = 0.02, 0.002$) than for the recharge of $W = 0.2$. This better agreement is attributed to a smaller discretization error, because for small recharge, $W \leqslant 0.02$, the triangular surface elements are an excellent approximation to the virtually flat water table geometry.

Also the case of various recharge rates for a square recharge region, $B = 1.0$, is analyzed (Fig. 7.10). The linearized solution is available only for infinite depth aquifers, and thus no comparison was possible for rectangular recharge and finite depths.

The BIEM is not restricted to an aquifer of constant depth; virtually any shape of aquifer can be discretized and used as illustrated by the simple examples chosen here. Solutions at interior points may be obtained for $\Phi$ by using (2.23).

### 7.3.2 Pond seepage problem

The last problem considered herein is the steady state flow from a pond through a homogeneous porous material down to a water table located at a finite distance below the pond. The condition at the pond surface is $\Phi =$ constant, and the unknown on the pond boundary is $\partial\Phi/\partial n$.

The three-dimensional BIEM approach is employed in much the same manner as for the recharge problems. First, an initial guess for the free surface location is made. Then the pond and free surface are discretized into linear triangular elements. The plan view of the grid is somewhat similar to the recharge grid of Figure 7.4 except that the $z$ co-ordinates are assigned to form the pond and sloping free surface. Then the new free surface is computed at successive time levels until a steady state is reached; that is, until successive solutions for the free surface and $\partial\Phi/\partial n$ are smaller than some

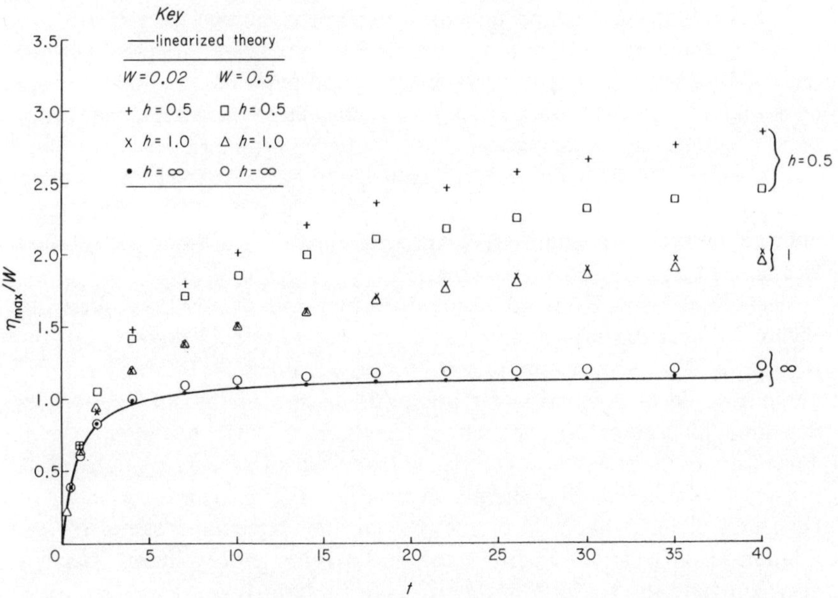


**Figure 7.10** The normalized maximum free surface rise for $B = 1.0$ and various values of aquifer thickness and recharge rates.

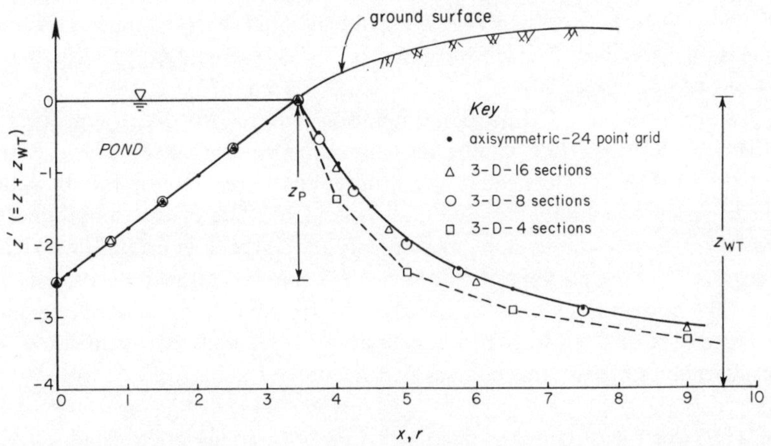


**Figure 7.11** The steady state flow from a pond down to a water table at $z_{WT} = 4$.

error criteria. The error criteria used here for solutions at successive time levels with $\Delta t = 5$ are: (1) the $z$ co-ordinate of points on the free surface, $z = \eta$, change less than 0.1% of $z_{WT}$, (2) the changes in $\partial\Phi/\partial n$ value on the boundary of the pond is less than 0.01% of any $\partial\Phi/\partial n$, and (3) the $\partial\Phi/\partial n$ on the free surface is less than 0.1% of a typical $\partial\Phi/\partial n$ through the pond boundary.

The three-dimensional steady state free surface solution for $y = 0$, $x > 0$ is shown in Figure 7.11 for a water table 7.4 units below the pond surface ($z_{WT} = 4$). The shape of the pond is an inverted pyramid of $n_s$ sides, where $n_s$ is taken to be 4, 8, and 16. As $n_s$ becomes large the three-dimensional pond solution should approach that of the axisymmetric case (a cone-shaped pond). Results for the axisymmetric pond (using the formulation described in Lennon *et al.* 1979a) are also shown in Figure 7.12. The results indicate that $n_s = 8$ is quite sufficient to represent the axisymmetric solution and is used in the following discussion.

The flux through the pond, represented by a plot of $\partial\Phi/\partial n$, is shown in Figure 7.13 for the axisymmetric solution with 12 and 24 nodal points, as well as the three-dimensional solution for $n_s = 8$ along $y = 0$ (an edge of the pyramid). The 12 point axisymmetric grid has the same nodal locations as the three-dimensional grid along the $x$-axis. It is obvious that the three-dimensional $\partial\Phi/\partial n$ solution compares quite well with the 12 point axisymmetric grid, although it is slightly higher than the axisymmetric solution. However, along a flat side of the pyramid the three-dimensional $\partial\Phi/\partial n$ solution is, in general, slightly below the axisymmetric solution, yielding approximately the same flow rate, $Q$ (see Table 7.4).

The singular behavior at the tip of the pond is better represented by concentrating more elements close to the pond tip. This effect is demonstrated in Figure 7.12 by comparing the sharper axisymmetric $\partial\Phi/\partial n$ profile with 24 points (with a point at $r = 0.05$) to the 12 point axisymmetric grid and three-dimensional grid which have points at $r = 0.75$ and $(x, y) = (0.75, 0)$, respectively.

The approximate CPU times per iteration required to solve the problems using a FORTRAN G compiler on Cornell University's IBM 370/168 are shown in Table 7.4. Between 4 and 8 iterations were required for the solution to reach essentially steady conditions, depending on the initial guess, size of grid, and time step size. As in the three-dimensional recharge problems $n^*$ is the number of simultaneous equations after the symmetry is used. The grid with $n_s = 1.6$ required about the same computational time as for the grid with $n_s = 8$ because a coarser grid (per section) was used, $n^*$ is considerably smaller, and coarser quadrature formulas were employed.

**Table 7.4** Computer times and flow for various grids, pond problem.

| *Grid* | *N* | *n** | *Q* | *CPU time per iteration (s)* |
|---|---|---|---|---|
| axisymmetric | 12 | — | 46.9 | 0.21 |
| axisymmetric | 24 | — | 47.1 | 0.84 |
| 3-D, 4 sections | 117 | 30 | 38.0 | 7.2 |
| 3-D, 8 sections | 209 | 27 | 44.4 | 8.9 |
| 3-D, 16 sections | 305 | 20 | 47.3 | 9.0 |

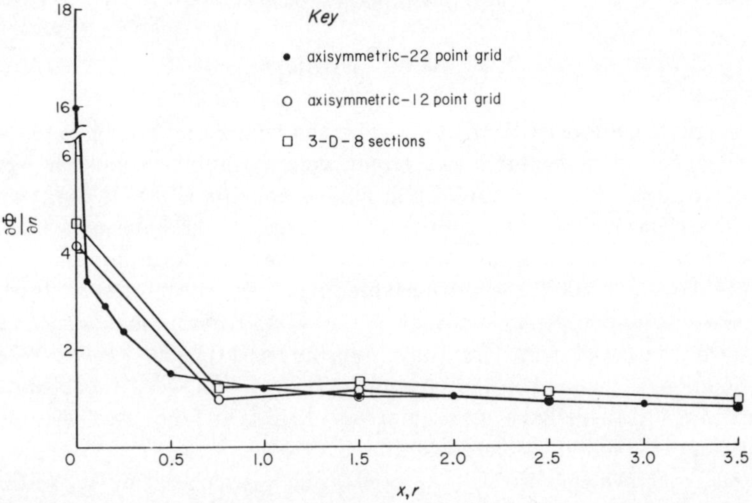


**Figure 7.12** The comparison of three-dimensional and axisymmetric BIEM solutions for the flux through a pond.

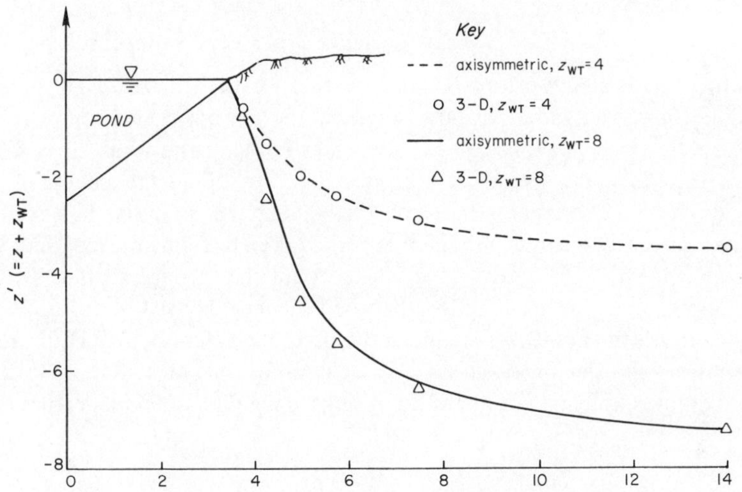


**Figure 7.13** The steady state flow from a pond down to a water table for $z_{WT} = 4$ and $z_{WT} = 8$.

Finally, Figure 7.13 presents solutions with the same pond, but a deeper water table ($z_{WT} = 8$). The 24-point axisymmetric grid is compared to the three dimensional grid with $n_s = 8$. The results for $z_{WT} = 4$ are also plotted in Figure 7.13 for these two grids. The results show excellent comparison for the free surface solution. The error in flow rate for the three-dimensional solution ($M = 60$) is less than 7% below the axisymmetric solution ($M = 64$).

## 7.4 Well problems

The three-dimensional BIEM formulation can be used, in principle, to find the shape of the water table in a groundwater basin where multiple wells exist. However, since the size of a groundwater basin is usually large and very fine elements are required to simulate accurately the water table near a well, it becomes economically not feasible to take this direct approach. An alternative is to simulate the flow field approximately by replacing wells with line sinks with appropriate intensities. By doing so accurate solutions can be obtained a few well diameters away from the wells (Dagan 1967b). When the solution is required in the immediate vicinity of a well, a local three-dimensional (or, perhaps, axisymmetric) BIEM can be used with the approximate solutions as far field boundary conditions.

In a groundwater basin with $N_w$ wells, which is replaced by $N_w$ vertical line sinks, the governing equation for the flow field becomes (see Ch. 1):

$$\nabla^2\Phi = -\sum_{j=1}^{N_W}\int_{z_j^B}^{z_j^T}\frac{Q_j^f}{z_j^T - z_j^B}\,\delta(x-x_j)\delta(y-y_j)\delta(z-z_j)\,dz_j \tag{7.24}$$

in which $Q_j^f$ is the constant withdrawal rate for the $j$th well. The vertical wells are assumed to be partially screened and are positioned at $(x_j, y_j, z_j;\ z_j^B < z_j < z_j^T\ j = 1, 2, 3, \ldots, N_w)$. The water levels in the wells never drop below the top of the screens, $z = z_j^T$ (see Fig. 7.14). Thus the lengths of the line sinks, $z_j^T - z_j^B$ remain the same throughout the analysis. If a well is entirely screened, a simple modification is required to define the length of the sink as a function of time, taking into account the seepage surface on the face of the well. The assumption that the flow into the well is uniform with respect to depth should not significantly affect the solution a short distance from the well. The integral on the right-hand side of (7.24) is further approximated as a summation of point sinks using Gaussian quadrature, i.e.

$$\begin{aligned}&\int_{z_j^B}^{z_j^T}\left(\frac{\cdot Q_j^f}{z_j^T - z_j^B}\right)\delta(x-x_j)\delta(y-y_j)\delta(z-z_j)\,dz_j\\ &= \sum_{k=1}^{n_p}\frac{Q_j^f\omega_k}{2}\,\delta(x-x_j)\delta(y-y_j)\delta(z-z_{jk}) + E\end{aligned} \tag{7.25}$$

where $z = z_{jk}$ is the vertical location of $k$th point sink along the $j$th well and is given by

$$z_{jk} = \frac{z_j^T - z_j^B}{2}\chi_k + \frac{z_j^T + z_j^B}{2} \tag{7.26}$$

The Gauss weights, $\omega_k$, are located at $\chi_k$, $k = 1, 2, 3, \ldots, n_p$.

The governing equation (7.24) can be rewritten as

$$\nabla^2\Phi = -\sum_{j=1}^{N_W}\sum_{k=1}^{n_p} Q_{jk}\,\delta(x-x_j)\delta(y-y_j)\delta(z-z_{jk}) \tag{7.27}$$

with

$$Q_{jk} = Q_j^f\omega_k/2 \tag{7.28}$$

the resulting integral equation is

$$-\alpha_i\Phi_i + \sum_{j=1}^{N_W}\sum_{k=1}^{n_p} Q_{jk}\,g(x_j, y_j, z_{jk}) = \iint_\Omega \left(\Phi\frac{\partial g}{\partial n} - g\frac{\partial \Phi}{\partial n}\right) dA \tag{7.29}$$

where

$$g(x_j, y_j, z_{jk}) = \{(x_i - x_j)^2 + (y_i - y_j)^2 + (z_i - z_{jk})^2\}^{-\frac{1}{2}} \tag{7.30}$$

The extra term on the left-hand side of (7.29) depends only on the strength of the source ($Q_{jk} > 0$) or sink ($Q_{jk} < 0$) and the distance between the $i$th base point and the location of the source (sink). The same numerical pro-

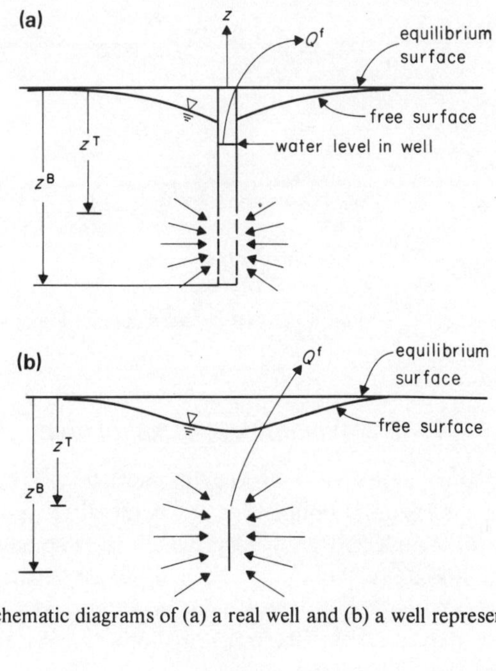


**Figure 7.14** Schematic diagrams of (a) a real well and (b) a well represented by a line sink.

cedure is used as that described in Section 7.3, except that the contribution due to sources and sinks is added to the known right-hand side of the discretized system of equations. Because the last known surface at the $k$th time level is used in the numerical calculations (including $g(x_j, y_j, z_{jk})$), both the governing equations and discretized boundary conditions are linear.

### 7.4.1 Single well solutions

Consider the case in which a line sink ($N_w = 1$) is placed in an aquifer of infinite horizontal extent and depth. Pumping begins at time zero with a flat water table at $z = 0$. Figure 7.15 shows the spacial axisymmetric BIEM solution of the free surface decline due to a vertical line sink of small strength ($Q_1^f = 0.02\pi$, $n_p = 8$, $x_1 = y_1 = 0$, $-2 < z < -1$) for selected time steps. Similar results are obtained for three-dimensional BIEM solutions, although the three-dimensional grid is much cruder than the axisymmetric grid. The same boundary conditions apply as in the recharge problem ((7.18)–(7.21)) with $W = 0$. Also included is the linearized theory for a line sink presented by Dagan (1967b). Because the line sink strength is so small, Dagan's (1967b) theory should be accurate. In the far field, Dagan's (1967b) theory is identical to the Theis solution (Bear 1972). The slight initial lag in the BIEM solutions is due to a rather coarse time step, and the discrepancy is reduced when a finer time step is chosen.

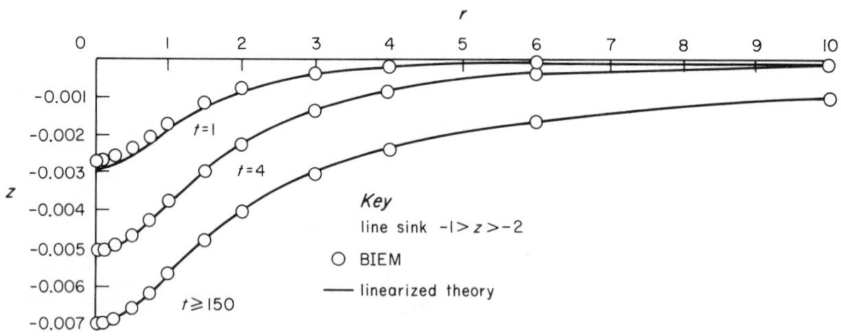


**Figure 7.15** The phreatic surface for a line sink representation of a well screened between $-1 > z > -2$.

### 7.4.2 A multiple well solution

Problems may also be analyzed for cases with multiple wells and aquifers of finite depth and/or horizontal boundaries at a finite distance from the well. In this section an example is presented in which the depth, $h$, is constant above an infinite horizontal boundary. Four wells are located at $(d, 0, z)$, $(0, d, z)$, $(-d, 0, z)$, and $(0, -d, z)$, where $d$ is a constant and $-2 < z < -1$ for all four wells (see Figure 7.16). Figure 7.17a shows the free surface profile

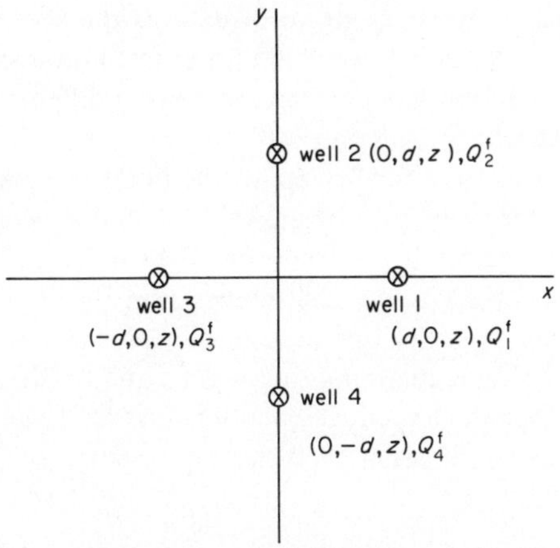


**Figure 7.16** Plan view of a well field consisting of four wells.

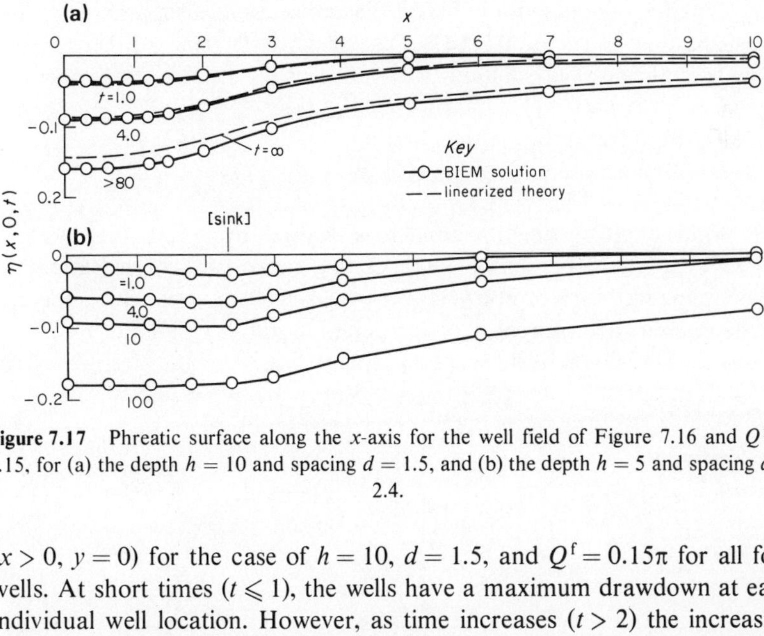


**Figure 7.17** Phreatic surface along the $x$-axis for the well field of Figure 7.16 and $Q^f =$ 0.15, for (a) the depth $h = 10$ and spacing $d = 1.5$, and (b) the depth $h = 5$ and spacing $d =$ 2.4.

($x > 0$, $y = 0$) for the case of $h = 10$, $d = 1.5$, and $Q^f = 0.15\pi$ for all four wells. At short times ($t \leqslant 1$), the wells have a maximum drawdown at each individual well location. However, as time increases ($t > 2$) the increasing interference of the wells causes a maximum drawdown at $(x, y) = (0, 0)$. (If the wells located at $(d, 0, z)$ and $(0, -d, z)$ were omitted, the maximum drawdown would be at the wells located at $(d, 0, z)$, $(-d, 0, z)$, and a 'bump' would exist between the two wells.) Also shown in Figure 7.17a is the linearized perturbation solution (Dagan 1967b) for the case of an infinitely deep aquifer.

For the relatively deep aquifer of Figure 7.17a ($h = 10$) the BIEM solution and Dagan's (1967b) solution agree well for times up to $t = 5$. For large times ($t > 10$), the solutions begin to diverge as the non-linear interference and effect of the finite depth become important. However, even for $t \to \infty$ Dagan's (1967b) solution has the same trend (but different magnitude) with the maximum drawdown at $(x, y) = (0, 0)$.

When the line sinks are further apart, the point of maximum drawdown should remain at the well for a longer period of time, if not always, as the magnitude of the interference is reduced. Figure 7.17b, in which the well distance from the origin is $d = 2.4$, shows that the maximum drawdown occurs at the wells for $t \leqslant 10$, a considerably longer time than in the previous example. The same pumping rate is used and the depth is decreased to $h = 5$. A major effect of the reduced depth is that for large times ($t > 2$), the free surface moves much faster than the case of Figure 7.17a with $h = 10$.

### 7.4.3 A non-symmetric solution

The two previous examples utilize a double symmetry which does not exploit the full potential of the BIEM. Therefore, we present another case in which the four values of $Q^{\mathrm{f}}$ are not the same. Referring to Figure 7.18, a fully three-dimensional solution is obtained using the following values:

$$\begin{array}{lll} d = 1.2 & Q_1^{\mathrm{f}} = +0.6\pi & Q_3^{\mathrm{f}} = -0.4\pi \\ h = \infty & Q_2^{\mathrm{f}} = -0.15\pi & Q_4^{\mathrm{f}} = -0.05\pi \end{array} \tag{7.31}$$

The primary effect on the aquifer is caused by a pumping well at $(x, y) = (1.2, 0)$ and a recharge well operating at two thirds of the strength of the pumping well located at $(0, -1.2)$. A weak recharge well which operates at one quarter of the strength of the pumping well is located at $(0, d)$. An even weaker recharge well operating at one twelfth of the strength of the

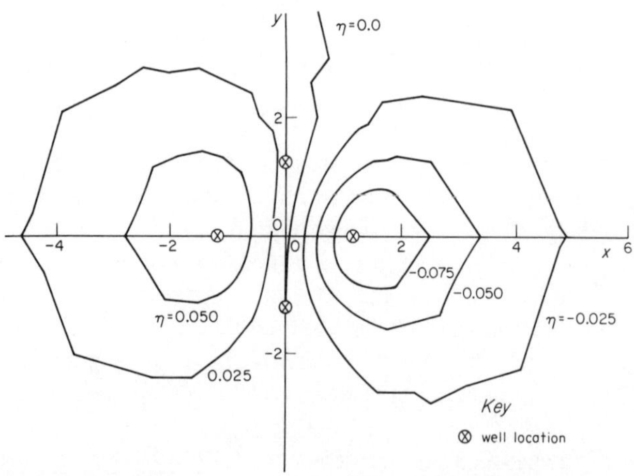


**Figure 7.18** Contour plot of the steady state solution for four wells with different outflow.

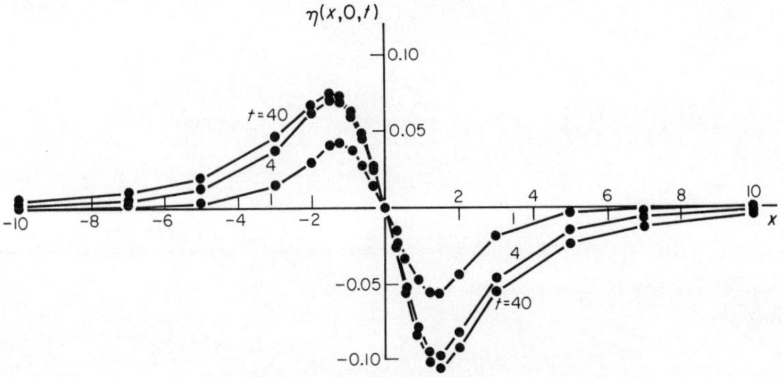


**Figure 7.19** Phreatic surface along the $x$-axis for various times in the case of four wells with different outflow.

pumping well is located at $(0, -d)$. Thus the net outflow is zero. The free surface contour lines in plan view in Figure 7.18 show that the steady state solution ($t > 80, h = \infty$) is basically a two well solution with a greater free surface deflection near the pumping well at (1.2, 0) than near the recharge well at (−1.2, 0). The line of $\eta = 0$ has swung clockwise from the $y$ axis due to the fact that the strength of the recharge well at (0, −1.2) is three times smaller than the strength of the well at (0, 1.2). The rather crude solution for $x > 2, y > 2$ is the result of the coarse grid used to represent the solution and can be made smoother by adding more elements. Previous results indicate that the solution for $\eta$ is much more accurate at the nodes than inside the elements. Figure 7.19 shows the distribution of the free surface elevation along the $x$ axis for selected time steps. Because the net outflow is zero, the volume of water above the $z = 0$ plane should be equal to the water withdrawn from below. Calculations show that these volumes agree within 0.5%.

# 8 Multiphase flows

In this chapter the BIEM is applied to the flow of two immiscible fluids in a porous medium. We assume that a sharp interface exists between the two fluids and, that when a movement of the interface occurs, one fluid completely displaces the other. The interfacial boundary becomes a free surface in the sense that the location of that boundary, and hence the position at which the non-linear boundary conditions must be applied, becomes a part of the solution. We consider only confined flow in this chapter. Volker (1980) has treated problems which not only have a moveable interface but also a free surface. Also we consider only two-dimensional problems, although three-dimensional calculation is simply a straightforward combination of the methods of this chapter and Chapter 7.

## 8.1 Theoretical formulation

Bear (1972) discusses the movement of immiscible fluids in porous media. The interfacial movement of a two-fluid system (which is easily extended to a multifluid system) is summarized below. As shown in Figure 8.1, the regions, designated $R_1$ and $R_2$, occupied by each fluid, are separated by the line $z = \xi(x, t)$.

Following the discussion presented in Chapter 1, we assume that Darcy's law is valid in a homogeneous porous medium. Thus the discharge velocity can be expressed as (cf. (1.7)):

$$q_{x_i} = -K_i \frac{\partial \Phi_i}{\partial x} \qquad q_{z_i} = -K_i \frac{\partial \Phi_i}{\partial z} \tag{8.1}$$

with

$$K_i = \frac{\rho_i g k}{\mu_i} \tag{8.2}$$

where the subscript $i$ denotes the flow region, $\rho$ is the density of fluid, $k$ is the intrinsic permeability, and $\mu$ is the dynamic viscosity of the fluid. As in

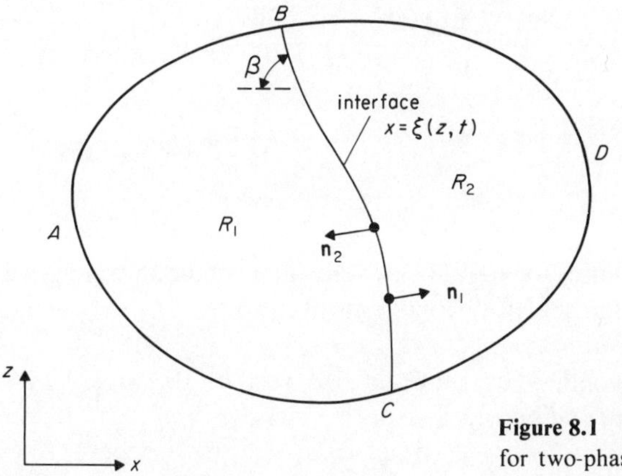


**Figure 8.1** The solution region for two-phase flow.

(1.8), $\Phi_i$ represents the piezometric head in the $i$th flow region, i.e. $\Phi = p_i/(\rho_i g) + z$. Using $L$ as the characteristic length scale, the following dimensionless variables are defined:

$$t^* = tK_1/nL, \; (\Phi^*, \eta^*, x^*, z^*) = (\Phi, \eta, x, z)/L \qquad (8.3)$$

where $n$ is the effective porosity. Equation (8.3) is equivalent to (1.38) for a single-fluid system. All quantities with one asterisk are non-dimensional; however, for brevity all asterisks will be dropped in the remainder of this chapter.

Use of the continuity equation, $\nabla \cdot \mathbf{q}_i = 0$, leads to

$$\nabla^2\Phi_i = 0 \qquad \text{in the flow regions } R_i \quad (i = 1, 2) \qquad (8.4)$$

The interfacial boundary conditions require the continuity of pressure across the interface, i.e.

$$s\Phi_2 - \Phi_1 = (s - 1)z \qquad \text{on } z = \eta \qquad (8.5)$$

with

$$s = \rho_2/\rho_1 \qquad (8.6)$$

Moreover, the interface is also a material surface. The total rate of change of the position of the interface following interfacial movement vanishes,

$$\frac{\mathrm{D}}{\mathrm{D}t}(z - \eta) = 0 \qquad \text{on } z = \eta \qquad (8.7)$$

Using the definition of the total derivative and Darcy's law, (8.1), we may rewrite (8.7) in the following form:

$$\frac{\partial \eta}{\partial t}=\frac{\partial \Phi_1}{\partial x}\frac{\partial \eta}{\partial x}-\frac{\partial \Phi_1}{\partial z}=\frac{s}{\alpha}\left(\frac{\partial \Phi_2}{\partial x}\frac{\partial \eta}{\partial x}-\frac{\partial \Phi_2}{\partial z}\right) \qquad \text{on } z=\eta \qquad (8.8)$$

where

$$\alpha = \mu_2/\mu_1 \qquad (8.9)$$

Equation (8.8) constitutes the non-linear, kinematic condition which must be satisfied along the interface. Both conditions (8.5) and (8.8) are to be applied on the unknown location $z=\eta$.

As in the corresponding case of the free surface flow (cf. (5.2)) the kinematic boundary condition (8.8) can be rewritten as

$$\frac{\partial \eta}{\partial t}=-\left[1+\left(\frac{\partial \eta}{\partial x}\right)^2\right]^{\frac{1}{2}}\frac{\partial \Phi_1}{\partial n_1}=\left[1+\left(\frac{\partial \eta}{\partial x}\right)^2\right]^{\frac{1}{2}}\left(\frac{s}{\alpha}\right)\frac{\partial \Phi_2}{\partial n_2} \qquad \text{on } z=\eta \qquad (8.10)$$

where $\mathbf{n}_1$ and $\mathbf{n}_2$ represent the unit outward normals on the interface from the regions $R_1$ and $R_2$ respectively (Fig. 8.1). The normal vectors are

$$\mathbf{n}_1=\frac{\nabla(z-\eta)}{|\nabla(z-\eta)|}=-\mathbf{n}_2 \qquad (8.11)$$

For some problems it is convenient to express the location of the interface in a slightly different way, e.g. $x=\xi(z,t)$, in which case the interfacial boundary conditions take the following form:

$$s\Phi_2-\Phi_1=(s-1)\xi \qquad \text{on } x=\xi(z,t) \qquad (8.12)$$

$$\frac{\partial \xi}{\partial t}=-\left[1+\left(\frac{\partial \xi}{\partial z}\right)^2\right]^{\frac{1}{2}}\frac{\partial \Phi_1}{\partial n_1}=\left[1+\left(\frac{\partial \xi}{\partial z}\right)^2\right]^{\frac{1}{2}}\left(\frac{s}{\alpha}\right)\frac{\partial \Phi_2}{\partial n_2} \qquad \text{on } x=\xi \qquad (8.13)$$

where

$$\mathbf{n}_1=-\mathbf{n}_2=\frac{\nabla(x-\xi)}{|\nabla(x-\xi)|} \qquad (8.14)$$

and

$$\sin\beta=\left[1+\left(\frac{\partial \xi}{\partial z}\right)^2\right]^{-\frac{1}{2}} \qquad (8.15)$$

with $\tan\beta = 1/(\partial\xi/\partial z)$ (Fig. 8.1). Other boundary conditions could be either the Dirichlet type, $\Phi$ given, or the Neumann type, $\partial\Phi/\partial n$ specified. In the transient problem initial conditions are also required. They consist of the initial location of the interface and the initial distribution of the potential $\Phi_i$ $(i = 1, 2)$.

For the steady state problems the time derivative term on the left-hand side of the kinematic boundary condition vanishes, which implies that the interface is a streamline. Although a steady state solution can always be obtained by extending the transient calculations for a long time (until there is no change in the solution), a more direct method is presented in the final section of this chapter.

## 8.2 The boundary integral equations

The integral representation of the boundary value problem can be written as follows (see (2.15)):

$$\alpha_1\Phi_1 = \int_{\Gamma_1}\left[\Phi_1\frac{\partial}{\partial n}(\ln r) - \ln r\frac{\partial\Phi_1}{\partial n}\right]\mathrm{d}s \tag{8.16a}$$

for region $R_1$ and

$$\alpha_2\Phi_2 = \int_{\Gamma_2}\left[\Phi_2\frac{\partial}{\partial n}(\ln r) - \ln r\frac{\partial\Phi_2}{\partial n}\right]\mathrm{d}s \tag{8.16b}$$

for region $R_2$. In the above equations, $\alpha_i$ measures the interior angle of the small sector excluding the base point $P_j$ from the region $R_i$, and $r$ is the distance from a base point on the boundary to a field point on the boundary.

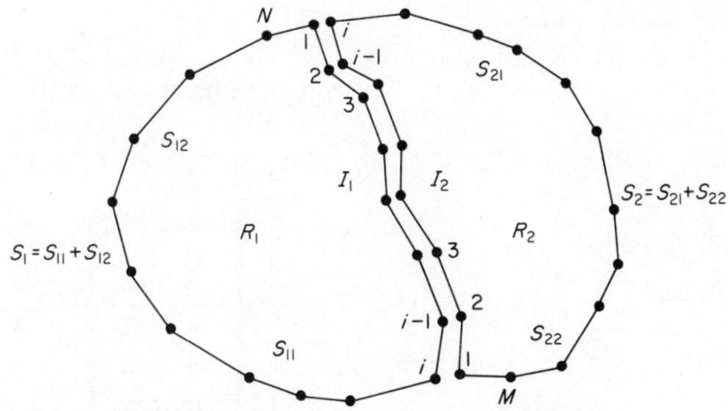


**Figure 8.2** Discretized region for the numerical solution.

A detailed discussion on the numerical discretization and integration of (8.16) is given in Section 2.2; only a brief summary of the procedure is given in this section. As shown in Figure 8.2, the boundaries $\Gamma_1$ and $\Gamma_2$ are discretized into a finite number of straight line segments. The interface is divided into $i-1$ segments and the interfacial nodal points are numbered twice, once in each of the two regions. In order not to lose generality, the boundary conditions along the external boundaries are described as follows (Fig. 8.2): along $S_{11}$ and $S_{21}$ the normal derivatives of the potential are given, and along $S_{12}$ and $S_{22}$ the potentials are specified. Applying the base point to each nodal point on the boundaries of $R_1$ and $R_2$, (8.16) can be integrated numerically. Assuming $N_1$ nodal points in $R_1$ and $N_2$ nodal points in $R_2$, a total of $N_1+N_2$ linear algebraic equations are obtained. Thus

$$\left[\begin{array}{ccc|ccc} H_{S_{11}} & H_{S_{12}} & H_{I_1} & & 0 & \\ \hline & 0 & & H_{S_{21}} & H_{S_{22}} & H_{I_2} \end{array}\right] \begin{Bmatrix} \Phi_{S_{11}} \\ \Phi_{S_{12}} \\ \Phi_{I_1} \\ \Phi_{S_{21}} \\ \Phi_{I_{22}} \\ \Phi_{I_2} \end{Bmatrix}$$

$$= \left[\begin{array}{ccc|ccc} G_{S_{11}} & G_{S_{12}} & G_{I_1} & & 0 & \\ \hline & 0 & & G_{S_{21}} & G_{S_{22}} & G_{I_2} \end{array}\right] \begin{Bmatrix} \partial\Phi_{S_{11}}/\partial n \\ \partial\Phi_{S_{12}}/\partial n \\ \partial\Phi_{I_1}/\partial n \\ \partial\Phi_{S_{21}}/\partial n \\ \partial\Phi_{S_{22}}/\partial n \\ \partial\Phi_{I_2}/\partial n \end{Bmatrix} \tag{8.17}$$

where $H$ and $G$ are coefficient matrices which depend only on the geometry of the flow domains; their detailed description can be found in Section 2.2. The subscripts appearing in (8.17) denote the part of the boundary as indicated in Figure 8.2.

Substituting the interfacial boundary conditions (8.12) and (8.13) into (8.17) and moving all the unknowns to the left-hand side of the resulting equations give

$$\left[\begin{array}{cc|cc|c|c} H_{S_{11}} & -G_{S_{12}} & & 0 & H_{I_1} & -G_{I_1} \\ \hline & 0 & H_{S_{21}} & -G_{S_{22}} & H_{I_2}/s & (\alpha/s)G_{I_2} \end{array}\right] \begin{Bmatrix} \Phi_{S_{11}} \\ \partial\Phi_{S_{12}}/\partial n \\ \partial\Phi_{S_{21}}/\partial n \\ \partial\Phi_{S_{22}}/\partial n \\ \partial\Phi_{I_1}/\partial n \\ \partial\Phi_{I_1}/\partial n \end{Bmatrix}$$

$$= \begin{bmatrix} G_{S_{11}} & -H_{S_{12}} & 0 & & \\ 0 & & G_{S_{21}} & -H_{S_{22}} & H_{I_2} \end{bmatrix} \begin{Bmatrix} \partial\Phi_{S_{11}}/\partial n \\ \Phi_{S_{12}} \\ \partial\Phi_{S_{21}}/\partial n \\ \Phi_{S_{22}} \\ (1/s - 1)z_{I_2} \end{Bmatrix} \tag{8.18}$$

On solving the system of linear equations, the potential and its normal derivative at the nodal points along the interface $I_1$ become known. From (8.12) and (8.13) $\Phi$ and its normal derivative along $I_2$ can be readily obtained. Finally, the new location of the interface can be found by rewriting (8.13) in the following finite difference form:

$$\xi^{k+1} = \xi^k + \frac{\Delta t}{\sin \beta^k}\left[\theta\left(\frac{\partial\Phi_1}{\partial n}\right)^k + (1-\theta)\left(\frac{\partial\Phi_1}{\partial n}\right)^{k+1}\right] \tag{8.19}$$

where the superscript $k$ denotes the $k$th time level (not to be confused with the intrinsic permeability) and $\theta$ is a weighting factor. The procedure can be repeated to find the time history of the movement of the interface.

## 8.3 Numerical examples for transient problems

Two transient interfacial flow problems are examined in this section. In both examples the numerical solutions are compared with experimental results.

### *8.3.1 Tilting of an interface in a Hele–Shaw model*

The first problem concerns the tilting of an interface between two fluids (distilled water and salt water) in a Hele–Shaw cell. The Hele–Shaw cell has a height of 4 cm (1.5 in) and length of 40 cm (16 in). The two fluids are placed in the apparatus and allowed to come to equilibrium with a barrier dividing them. At time $t = 0$ the barrier is removed so that two fluids come into contact. The heavier fluid tends to move underneath the lighter fluid. Thus, the location and the shape of the interface changes in time. The parameters for experimental runs are presented in Table 8.1. A more detailed discussion on the experimental set-up and procedure are found in Liu *et al.* (1981).

In Figure 8.3 the location of the toe of the interface, $l$, is plotted against the dimensionless time, $t$. A theoretical solution based on the Dupuit approximation (Keulegan 1954),

$$\frac{l}{H} = t^{\frac{1}{2}} \tag{8.20}$$

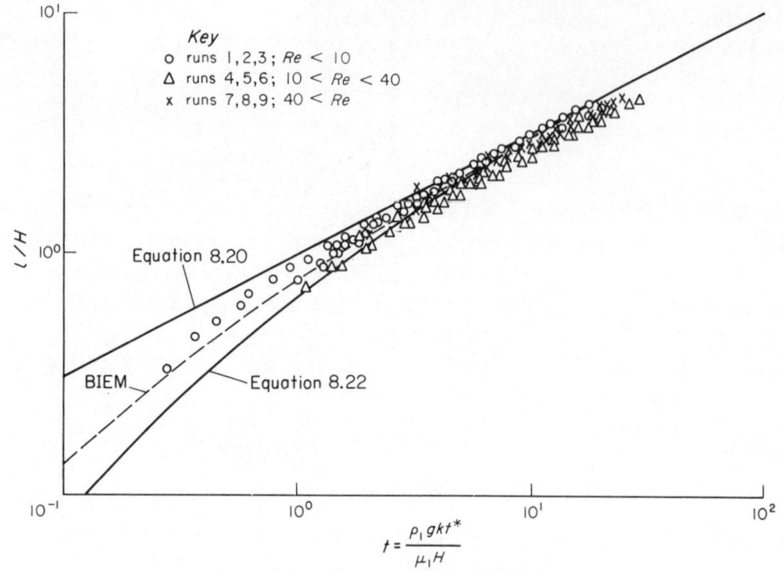


**Figure 8.3** Location of the moving toe.

is valid for large time and is compared to the experimental results in Figure 8.3. According to Gardner *et al.* (1962), the asymptotic solution for small time is

$$\frac{l}{H} = \frac{\sqrt{3}}{2} t \tag{8.21}$$

The interpolation formula encompassing (8.20) and (8.21) as given by Gardner *et al.* (1962),

$$\frac{l}{H} = \left( \frac{3t^2}{4 + 3t} \right)^{\frac{1}{2}} \tag{8.22}$$

is also plotted in Figure 8.3 for comparison. It is seen that $t^{\frac{1}{2}}$ behavior is well-exhibited by all the experimental data at large time. For small time all the experimental data fall consistently below the theoretical solution obtained by the Dupuit assumption, (8.20), yet above that of (8.22).

It is clear from Figure 8.3 that a more accurate description of the initial movement of the interface is needed. The BIEM is used to solve the same problem. In the numerical computations the interface is divided into 12 segments and each exterior boundary into 14–20 segments. The external boundaries require the no-flux condition. The matrix equation (8.18) for this particular problem is reduced to

$$\left[\begin{array}{c|c|c|c} H_{S_{11}} & 0 & H_{I_1} & -G_{I_1} \\ \hline 0 & H_{S_{21}} & H_{I_2}/s & (\alpha/s)G_{I_2} \end{array}\right] \begin{Bmatrix} \Phi_{S_{11}} \\ \Phi_{S_{21}} \\ \Phi_{I_1} \\ \partial\Phi_{I_1}/\partial n \end{Bmatrix} = [H_{I_2}] \left\{ \frac{1-s}{s} z_{I_2} \right\} \tag{8.23}$$

This problem is clearly a Neumann problem since nowhere on the boundary is a value of the potential specified. A constraint equation which specifies a reference potential at a nodal point is added to the above system by the Lagrangian multiplier method (e.g. Gallagher 1975). A reference node with zero potential is selected at the center of the solution region in order to maintain the symmetry of the problem.

An additional mathematical difficulty occurs when the interface intersects a horizontal impermeable boundary at a right angle (Fig. 8.4). It is obvious that the no-flux condition along the impermeable boundary

$$\frac{\partial\Phi_1}{\partial z} = \frac{\partial\Phi_2}{\partial z} = 0 \tag{8.24}$$

and the interfacial condition (8.12) cannot be satisfied simultaneously at the intersecting point. Linearized theoretical solutions by de Josselin de Jong (1960) and Verruijt (1980) for an initially vertical interface are singular at the toe (bottom) and the tip (top) of the interface. The actual initial interface used in the numerical computations is taken as an inclined straight line with a slope $l/H = 0.01$, so as to avoid the unwanted singularity.

Twenty-seven variable time steps are used in the computation for a time

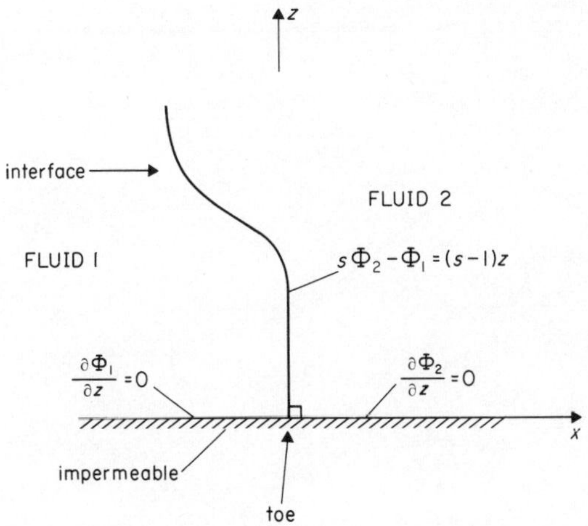


**Figure 8.4** Case where the interface intersects the boundary at a right angle.

**Table 8.1** Experimental parameters. Hele–Shaw model dimensions: spacing $b = 0.069$ in; height $H = 1.5$ in.

| *Run no.* | $\Delta\rho/\rho_1$ | *Temperature (°C)* | $\nu \times 10^5$ *($m^2/s$)* | $k = gb^2/12\nu$ *(m/s)* | $R = k\,\Delta\rho\, b/\rho\nu$ |
|---|---|---|---|---|---|
| 1 | 0.0008 | 25 | 0.090 | 2.783 | 4.3 |
| 2 | 0.0012 | 25 | 0.090 | 2.783 | 6.5 |
| 3 | 0.0015 | 24 | 0.092 | 2.728 | 7.8 |
| 4 | 0.0035 | 24.5 | 0.091 | 2.755 | 18.7 |
| 5 | 0.0055 | 22 | 0.097 | 2.597 | 25.9 |
| 6 | 0.007 | 24.5 | 0.091 | 2.755 | 37.1 |
| 7 | 0.009 | 24 | 0.092 | 2.728 | 46.8 |
| 8 | 0.010 | 24 | 0.092 | 2.728 | 52.0 |
| 9 | 0.017 | 24 | 0.092 | 2.728 | 88.4 |

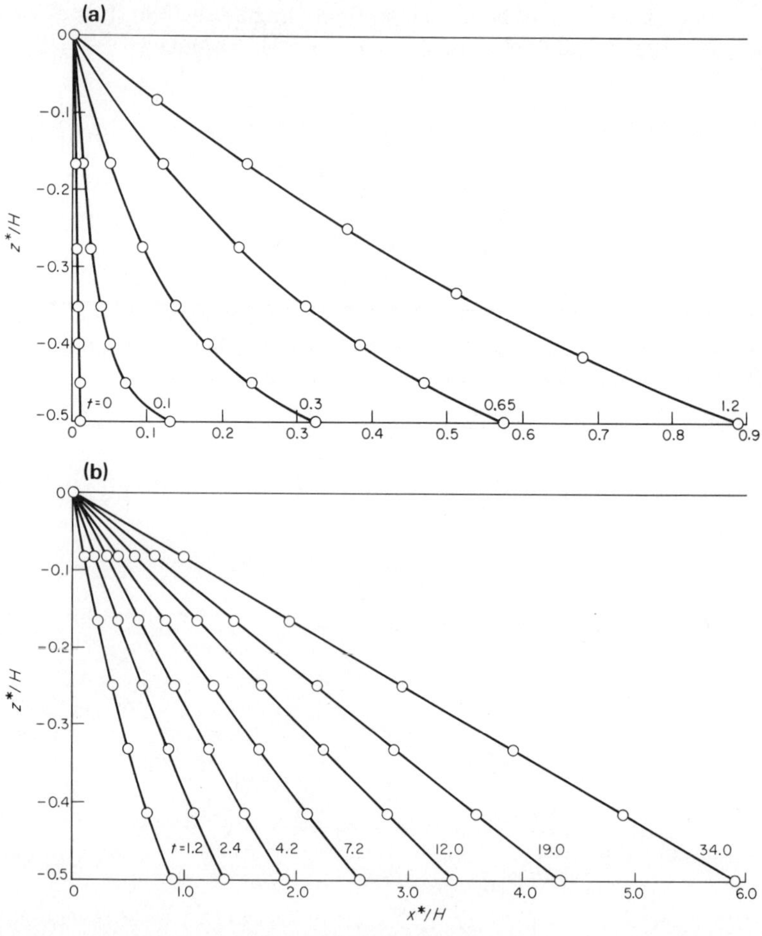


**Figure 8.5** The interface profile from the BIEM solution: (a) $t = 0$–$1.2$; (b) $t = 1.2$–$3.4$.

period of 34.0 (dimensionless). The numerical results of the history of the interfacial profile for run no. 1 (Table 8.1) are shown in Figure 8.5 for selected times. Only the lower half of the interface is plotted since the upper half is symmetrical. The interface becomes almost linear for $t > 1.0$. The same calculation was done for the parameters of run no. 9, producing a result almost the same as that shown in Figure 8.5.

The transient location of the toe (or the tip) of the interface is plotted in Figure 8.3. Good agreement with the experimental results is observed.

### 8.3.2 *A transient seawater intrusion problem*

The second problem studied herein is a typical seawater intrusion problem in a confined coastal aquifer. The aquifer is assumed to have a constant thickness, a horizontal bottom, and a known freshwater discharge into the sea. At time $t = 0$, the freshwater discharge is suddenly changed from that of the initial equilibrium value and the movement of the seawater wedge is observed. Bear and Dagan (1964) performed experimental studies of this problem in a Hele–Shaw cell. A definition sketch of their model is produced in Figure 8.6.

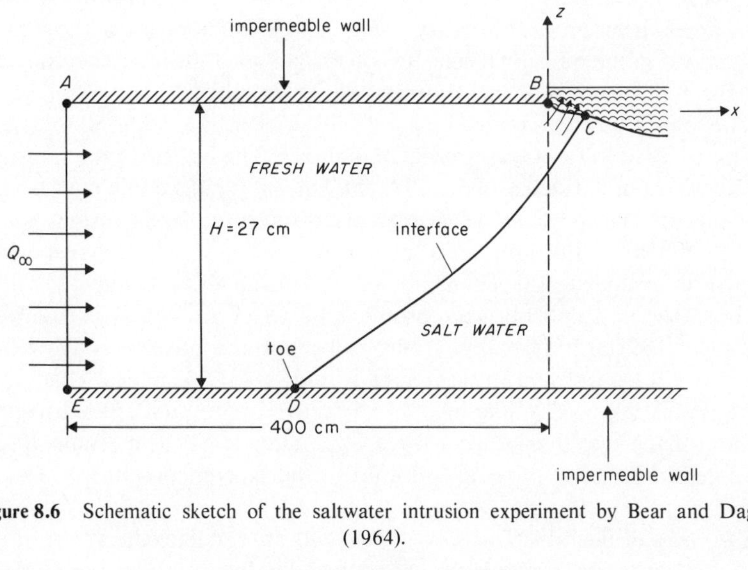


**Figure 8.6** Schematic sketch of the saltwater intrusion experiment by Bear and Dagan (1964).

The numerical procedure for this problem follows that described in the previous sections. A further complication involved here is the treatment of two boundary points with singular behavior. As observed in Figure 8.7, point $B$ may be identified as a cut-off wall type, i.e., an $r^{-\frac{1}{2}}$ singularity. Singular elements, discussed in Section 3.3, can be used to improve the accuracy of computation.

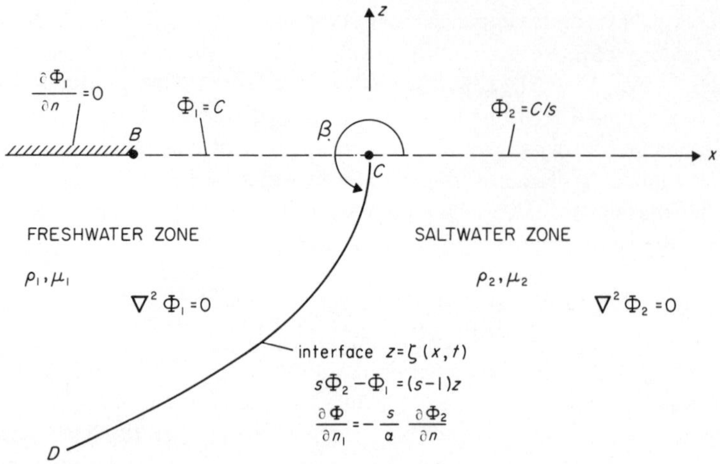


**Figure 8.7** The outflow tip for the saltwater intrusion problem.

The situation at point $C$ is different. Since $\partial\Phi_1/\partial x$ and $\partial\Phi_2/\partial x$ on each side of point $C$ vanish for all time, the $x$-component velocity should always disappear at that point. The interface should intersect the outflow face at a right angle at all times (Liu *et al.* 1981). The location of the outflow tip is determined in the numerical scheme by extrapolating the interface using the nearby nodes such that the above condition is fulfilled.

The interface is represented by 11–12 linear elements. Some 30 to 35 elements are used to form the external boundaries. The far end of the saltwater aquifer is cut at a distance of $x = 200$ cm. Since it is judged that the seaward flow has only negligible influence at such a distance, either a constant potential condition or the no-flux condition is applied there. The freshwater aquifer is extended 500 cm inland and a constant flux condition is prescribed. The initial steady state profile is obtained through a full transient treatment, i.e. starting from an arbitrary location the interface is allowed to move until the movement ceases. At time $t = 0$ a sudden change of discharge is imposed at the inland end of the aquifer and the subsequent motion of the interface is recorded. In the numerical scheme some 40–50 time steps are used to cover the duration of the experiments (Bear & Dagan 1964).

Two runs of Bear and Dagan's experiments are simulated here, one for an advancing saltwater wedge (exp. 1) and the other for a receding one (exp. 3). The relevant parameters for the two runs are given in Table 8.2. The time histories of the interfaces are shown in Figures 8.8 and 8.9 in comparison with the experimental results by Bear and Dagan (1964). The numerical results are not in good agreement with the experimental ones. However, comparisons between other theoretical and numerical results (Shamir & Dagan 1971, Sá da Costa & Wilson 1979) and the experimental ones (Bear

**Table 8.2** Parameters of Bear and Dagan's (1964) experiments. $H$ (Hele–Shaw model depth) = 27 cm; $K$ (permeability) = 69 cm/s; $\rho_1$ (freshwater density) = 1.001 g/cm$^3$; $\rho_2$ (saltwater density) = 1.030 g/cm$^3$.

| *Exp. no.* | $Q(t=0^-)$ $(cm^2/s)$ | $Q(t=0^+)$ $(cm^2/s)$ |
|---|---|---|
| 1 | 19.1 | 0 |
| 3 | 3.9 | 18.8 |

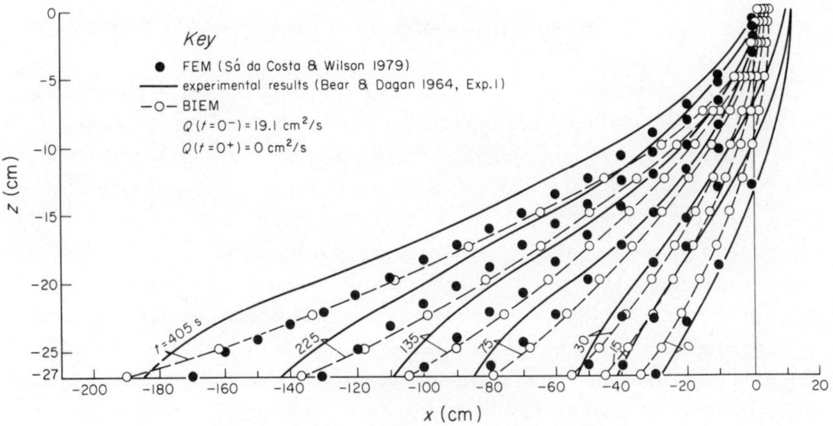


**Figure 8.8** The advancing saltwater wedge.

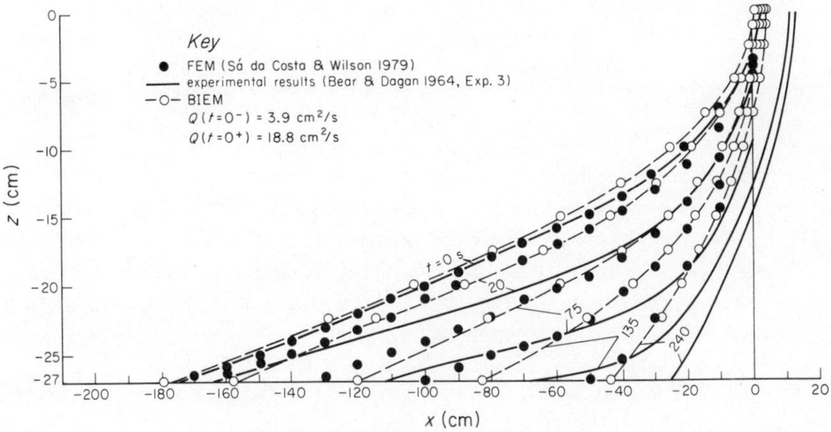


**Figure 8.9** The retreating saltwater wedge.

& Dagan 1964) were not close either. There is reason to believe that the experimental error may be significant.

A further check of credibility of the numerical solution may be done by comparing the outflow surface length, $\overline{BC}$ in Figure 8.7, with that obtained from an analytical solution for semi-finite aquifer by Glover (1959). It is found that under the steady state condition the length based on the numerical computation is always within 1% of that of the analytical solution.

## 8.4 Approximate solutions for steady-state problems

The steady state solution can be obtained by solving the problem as if it were transient. On the other hand, an iteration scheme can be developed to solve the steady state problem directly. Equation (8.18) can be used to solve for $\Phi$ and $\partial\Phi/\partial n$ on the interface if its estimated position is given. An adjusted location of the interface is found by satisfying the boundary condition $\partial\Phi/\partial n = 0$ along the interface; i.e. the interface must be a streamline for steady state flows. Equation (8.18) is solved again and the procedure is repeated until all boundary conditions are satisfied.

The methods mentioned above involve either time stepping or iterations, either of which could be time consuming for some problems. An alternative method is now discussed.

For steady state problems, the interfacial boundary conditions, (8.12) and (8.13) can be reduced to

$$s\Phi_2 - \Phi_1 = (s - 1)z \qquad \text{on } x = \xi(z) \tag{8.25a}$$

$$\frac{\partial\Phi_1}{\partial n_1} = \frac{\partial\Phi_2}{\partial n_2} = 0 \qquad \text{on } x = \xi(z) \tag{8.25b}$$

For saltwater intrusion problems such as the one shown in Figure 8.10 we can assume that in the steady state the velocity in the ocean (saltwater region) is negligible. Therefore, the pressure is hydrostatic in the saltwater region, which implies that $\Phi_2 = 0$. This is the well-known Ghyben–Herzberg approximation (Bear 1972). The dynamic interfacial boundary condition can be approximated as

$$\Phi_1 = (1 - s)z \qquad \text{on } x = \xi(z) \tag{8.26}$$

where

$$s = \gamma_s/\gamma_f \tag{8.27}$$

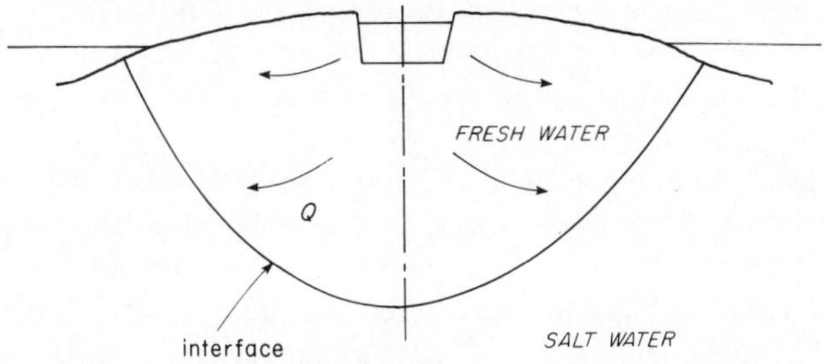


**Figure 8.10** Definition sketch of a problem with an interface separating fresh water from salt water.

and $\gamma_s$ is the specific weight of salt water and $\gamma_f$ the specific weight of fresh water. Hence, the two-fluid problem has been transformed into a single-fluid system for the freshwater region only.

For the convenience of discussion, we redefine the variables in the following forms:

$$(\bar{x}, \bar{z}) = (2N/Q)\,(x, z) \tag{8.28}$$

$$(\bar{\Phi}, \bar{\psi}) = (2/Q)\,(-\Phi, \psi) \tag{8.29}$$

with

$$N = K(s - 1)$$

in which $Q$ represents the discharge from the canal per unit width of the aquifer and $\bar{\psi}$ is the stream function which satisfies

$$u = \frac{\partial \bar{\Phi}}{\partial \bar{x}} = -\frac{\partial \bar{\psi}}{\partial \bar{z}} \qquad w = \frac{\partial \bar{\Phi}}{\partial \bar{z}} = \frac{\partial \bar{\psi}}{\partial \bar{x}} \tag{8.30}$$

The boundary conditions for the problem are shown in Figure 8.11. The BIEM can be applied directly to the physical problem in terms of either $\bar{\Phi}$ or $\bar{\psi}$. Of course an iteration scheme is still needed. To avoid the iterations, we can map the problem onto the complex potential plane $(\bar{\Phi}, \bar{\psi})$, since

$$\frac{\partial^2 \bar{x}}{\partial \bar{\Phi}^2} + \frac{\partial^2 \bar{x}}{\partial \bar{\psi}^2} = 0 \qquad \frac{\partial^2 \bar{z}}{\partial \bar{\Phi}^2} + \frac{\partial^2 \bar{z}}{\partial \bar{\psi}^2} = 0 \tag{8.31}$$

in which $\bar{x}\,(\bar{\Phi}, \bar{\psi})$ and $\bar{z}\,(\bar{\Phi}, \bar{\psi})$ satisfy the Cauchy–Riemann conditions; thus

$$\frac{\partial \bar{x}}{\partial \bar{\Phi}} = \frac{\partial \bar{z}}{\partial \bar{\psi}} \qquad \frac{\partial \bar{x}}{\partial \bar{\psi}} = -\frac{\partial \bar{z}}{\partial \bar{\Phi}} \tag{8.32}$$

The physical problem described in Figure 8.11 is transformed into a problem for the elevation $\bar{z}$ as shown in Figure 8.12. The boundaries are all straight lines. The BIEM can be readily applied to solve for $\bar{z}$, which gives the location of interface directly.

Numerical computation is carried out for the case $\Phi_B = -2.6$ and $\Phi_C = -4.0$. Only 32 points are used in the calculation. The results of the interface location are shown in Figure 8.13. The computed width of the freshwater canal and that of the face *OA* are compared in Table 8.3 with numerical results obtained by Charmonman (1966) and the analytical results by Hewson-Browne (1974). It is seen that the agreement is quite good. The efficiency of the solution in the transformed phase is obvious.

The same approach can be used to study other two-dimensional, free surface flow problems, e.g. seepage flow through an earth dam. More detailed discussion can be found in Liu and Liggett (1978).

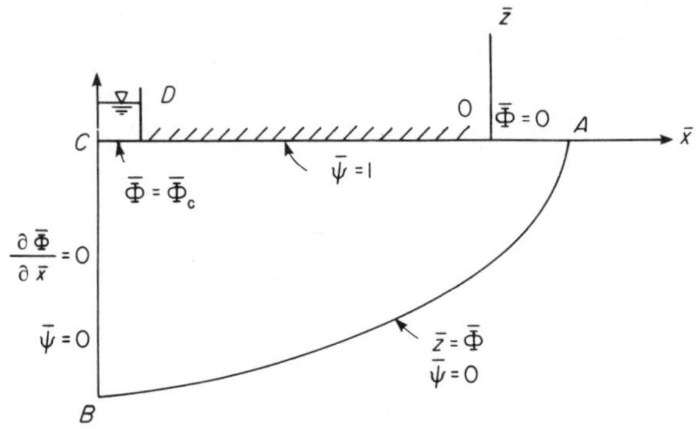


**Figure 8.11** Simplified model of the problem shown in Figure 8.10.

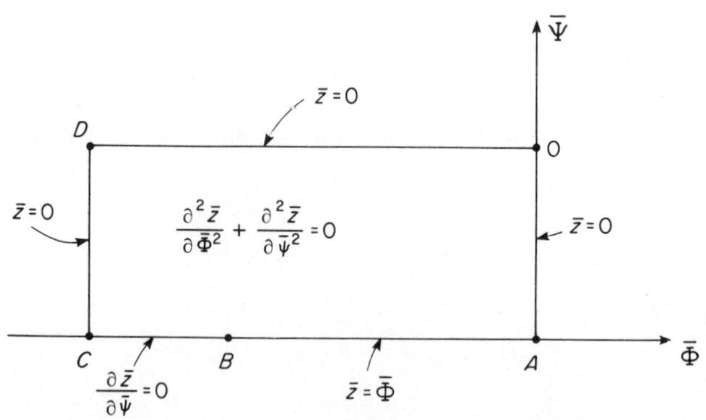


**Figure 8.12** Formulation of the problem of Figure 8.11 into the complex potential plane.

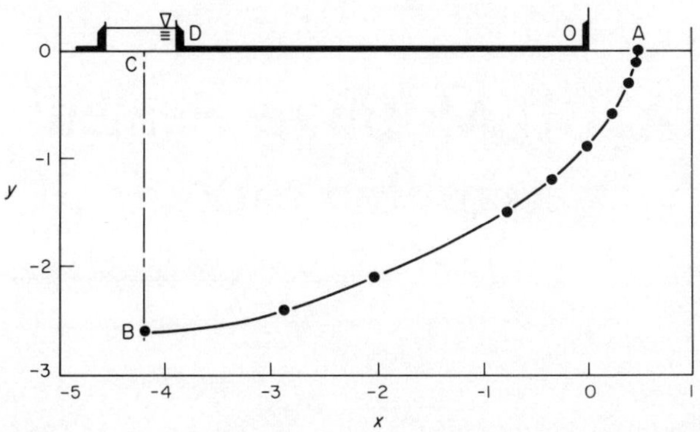


**Figure 8.13** Location of the interface between fresh water and salt water.

**Table 8.3** Results of the numerical computations for the case $\Phi_B = -2.6$ and $\Phi_C = -4.0$.

| *Method* | $x_A$ | $x_D$ | $x_C$ | *Total grid points* | *Iteration* | *Initial guess* |
|---|---|---|---|---|---|---|
| analytical solution | 0.499 | –3.970 | –4.311 | | | |
| successive over-relaxation | 0.499 | –3.893 | –4.214 | 1600 | yes | yes |
| BIEM | 0.433 | –3.855 | –4.192 | 32 | no | no |

# 9 A BIEM–finite element combination

The finite element method (FEM) has a history of successful use in groundwater flow problems (Pinder & Gray 1977). It is an extremely powerful technique which can solve problems with non-constant coefficients and a non-linear governing equation. Although the BIEM can also solve non-linear problems (Ch. 10) and problems with non-constant coefficients (Ch. 4), the FEM may be a better choice in some cases. In unsaturated flow, for example, the effective hydraulic conductivity can change several orders of magnitude over a short distance, thus severely testing any numerical method.

In many groundwater situations there is a wide area of Darcy flow combined with regions of non-linear flow (say unsaturated or non-Darcy, see Ch. 1). It would seem most appropriate to take advantage of the BIEM with its efficiency, ease of data input, and ability to handle infinite regions and singularities for the Darcy part of the region and to use the FEM for the difficult area. In this chapter we describe a BIEM–FEM combination for use in such circumstances.

## 9.1 The finite element method

It is not our intention to describe the FEM, as that is done in several books (Zienkiewicz 1971, Gallagher 1975, Segerlind 1976). Instead, this section serves as a reminder to those who are familiar with the FEM and as an outline to those who are not.

### *9.1.1 The approximation procedure*

Although the FEM is often considered in terms of a variational formulation of the problem at hand, only the method of weighted residuals will be considered here. The method of weighted residuals, although not as mathematically rigorous, can be applied more generally and for many problems leads to precisely the same results. First, we assume that a governing equation can be written in the form

$$L(\phi) = f \tag{9.1}$$

with boundary conditions

$$\phi = \phi_A \quad \text{on } \Gamma_1 \qquad \text{and} \qquad \frac{\partial \phi}{\partial n} = q \quad \text{on } \Gamma_2 \tag{9.2}$$

where $L$ is a differential operator and the boundary consists of two parts, $\Gamma = \Gamma_1 + \Gamma_2$, where each of the conditions (9.2) is applied (Fig. 9.1). Although $\phi$ is an unknown, it is approximated by interpolating functions (also called basis function, trial functions, or shape functions) which have unknown coefficients. As an example consider the one-dimensional case shown in Figure 9.2. The linear interpolation functions approximate $\phi$ as

$$\hat{\phi} = \frac{x_2 - x}{x_2 - x_1}\phi_1 + \frac{x - x_1}{x_2 - x_1}\phi_2 \qquad x_1 \leqslant x \leqslant x_2 \tag{9.3a}$$

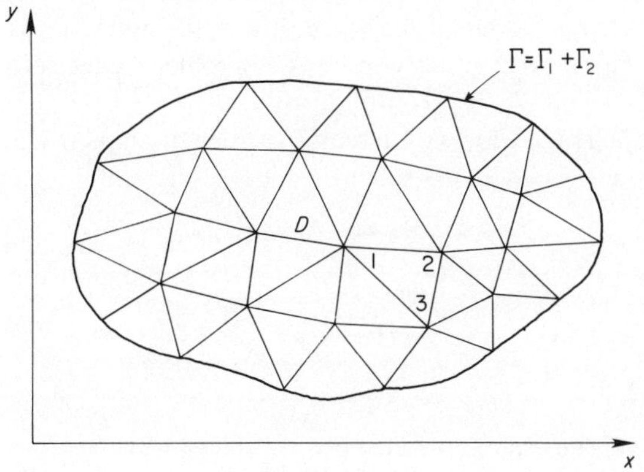


**Figure 9.1** The division of a region into triangular elements.

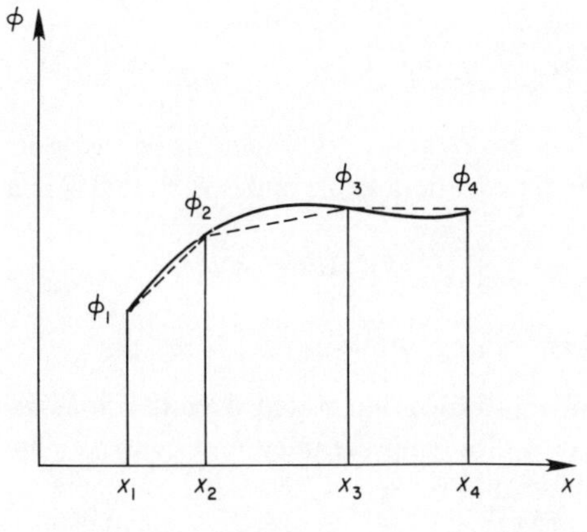


**Figure 9.2** The approximation of $\Phi$ by linear interpolation.

$$\hat{\phi} = \frac{x_3 - x}{x_3 - x_2}\phi_2 + \frac{x - x_2}{x_3 - x_2}\phi_3 \qquad x_2 \leqslant x \leqslant x_3 \tag{9.3b}$$

$$\hat{\phi} = \frac{x_4 - x}{x_4 - x_3}\phi_3 + \frac{x - x_3}{x_4 - x_3}\phi_4 \qquad x_3 \leqslant x \leqslant x_4 \tag{9.3c}$$

Equations (9.3) can be written as

$$\hat{\phi} = \sum_{i=1}^{n} N_i^e \phi_i \tag{9.4}$$

in which $n = 2$ and

$$N_1^e = \frac{x_{(2)} - x}{x_{(2)} - x_{(1)}} \quad N_2^e = \frac{x - x_{(1)}}{x_{(2)} - x_{(1)}} \qquad x_{(1)} \leqslant x \leqslant x_{(2)} \tag{9.5}$$

where $x_{(1)}$ and $x_{(2)}$ are the first and second nodes of an element and where $N_i^e = 0$ outside of the $e$th element. Thus if the value of $\phi$ at the nodes, $\phi_i$, is known, then the approximation gives $\phi$ everywhere. The problem reduces to finding $\phi$ at the nodes.

The same idea can be extended to two dimensions using triangular elements. Equation (9.4) applies with $n = 3$ and (Fig. 9.1)

$$N_1^e = [x_2y_3 - x_3y_2 + (y_2 - y_3)x + (x_3 - x_2)y]/2A^e$$

$$N_2^e = [x_3y_1 - x_1y_3 + (y_3 - y_1)x + (x_1 - x_3)y]/2A^e$$

$$N_3^e = [x_1y_2 - x_2y_1 + (y_1 - y_2)x + (x_2 - x_1)y]/2A^e$$

in which $A^e$ is the area of the triangular element.

Equation (9.1) is then written

$$L(\hat{\phi}) - f = R \tag{9.6}$$

in which $R$ is the error that has been made by replacing $\phi$ with its approximation, $\hat{\phi}$. We now seek to choose the $\hat{\phi}$ at the nodes such that $R$ is a minimum in some sense.

### 9.1.2 *The integration*

In order to minimize the residual $R$, (9.6) is integrated over the area. A method of minimization consists of multiplying the integrand by a 'weighting function' and setting the result equal to zero,

$$\int_D |L(\hat{\phi}) - f| w \, dA = 0 \tag{9.7}$$

The weighting function, $w$, can be any one of a number of functions. For example, we could choose a series of weighting functions

$$w_i = 1 \quad \text{at } (x_i, y_i) \qquad i = 1, 2, \ldots, n \tag{9.8}$$

in which the residual would be forced to zero at each of the points $(x_i, y_i)$ (point collocation). Of course there must be at least as many parameters (the $\phi_i$) as there are points $(x_i, y_i)$ in which case this formulation leads to $n$ simultaneous equations in the $\phi_i$.

It is common, however, to choose the interpolating functions as weighting functions so that

$$\int_D \left[ L\left( \sum_{i=1}^{n} N_i^e \phi_i \right) - f \right] N_j^e \, dA = 0 \qquad j = 1, 2, \ldots, n \tag{9.9}$$

Thus there are $n$ such equations, each with a different weighting function. This process is called the 'Galerkin's method'. For linear, self-adjoint differential operators (Greenberg 1971) it gives exactly the same result as a variational formulation.

Consider the two-dimensional version of (1.64): the differential operator is

$$L = \frac{\partial}{\partial x}\left( K \frac{\partial}{\partial x} \right) + \frac{\partial}{\partial y}\left( K \frac{\partial}{\partial y} \right) - \bar{C} \frac{\partial}{\partial t}; \; \bar{C} = \frac{\partial C}{\partial \Phi} \tag{9.10}$$

and the integral is

$$\sum_{i=1}^{n} \left\{ \Phi_i \int_A |\nabla \cdot (K \nabla N_i^e)| N_j^e \, dA - \int_A \bar{C} N_i^e N_j^e \frac{\partial \Phi_i}{\partial t} \, dA \right\} = 0$$

$$j = 1, 2, \ldots, n \tag{9.11}$$

The divergence theorem, (2.1), gives

$$\int_A \nabla \cdot \mathbf{q} \, dA = \int_\Gamma \mathbf{q} \cdot \mathbf{n} \, ds \tag{9.12}$$

Let $\mathbf{q} = U \, \nabla V$; then

$$\nabla \cdot \mathbf{q} = U \, \nabla^2 V + \nabla U \cdot \nabla V \tag{9.13}$$

and

$$\int_A U\,\nabla^2 V\,\mathrm{d}A = \int_\Gamma U\frac{\partial V}{\partial n}\,\mathrm{d}s - \int_A \nabla U\cdot\nabla V\,\mathrm{d}A \qquad (9.14)$$

which is Green's first identity (Green's theorem). Applying Green's theorem to the first integral of (9.11) yields

$$\sum_{i=1}^{n}\Phi_i\left\{\int_\Gamma KN_j^e\frac{\partial N_i^e}{\partial n}\,\mathrm{d}s - \int_\Gamma (K\,\nabla N_j^e\cdot\nabla N_i^e)\,\mathrm{d}s\right\}$$

$$-\sum_{i=1}^{n}\frac{\partial\Phi_i}{\partial t}\int_A \bar{C}N_i^eN_j^e\,\mathrm{d}A = 0 \qquad j = 1, 2, \ldots, n \qquad (9.15)$$

In applying Green's theorem the derivatives of $K$ have disappeared and only first order derivatives of the interpolation functions appear in the equation, thus making the problem more tractable. The first boundary integral in (9.15) can be rewritten as

$$\sum_{i=1}^{n}\Phi_i\int_\Gamma KN_j^e\frac{\partial N_i^e}{\partial n}\,\mathrm{d}s = \int_\Gamma N_j^e\left[K\sum_{i=1}^{n}\frac{\partial}{\partial n}(\Phi_iN_i^e)\right]\mathrm{d}s = -\int_\Gamma N_j^eq\,\mathrm{d}s$$

$$j = 1, 2, \ldots, n \qquad (9.16)$$

in which $q$ is the outflow velocity across the element boundary. Finally, (9.15) can be written in matrix form as

$$|F|\left\{\frac{\partial\Phi}{\partial t}\right\} + |D|\{\Phi\} = \{Q\} \qquad (9.17)$$

in which the elements of the matrices are

$$D_{ij} = \int_A K\left[\frac{\partial N_i^e}{\partial x}\frac{\partial N_j^e}{\partial x} + \frac{\partial N_i^e}{\partial y}\frac{\partial N_j^e}{\partial y}\right]\mathrm{d}A \qquad (9.18a)$$

$$F_{ij} = \int_A \bar{C}N_i^eN_j^e\,\mathrm{d}A \qquad (9.18b)$$

$$Q_i = -\int_\Gamma N_i^eq\,\mathrm{d}s \qquad (9.18c)$$

Notice that $F_{ij} = 0$ for saturated flow because the differential operator (9.10) does not contain a time derivative in that case.

### 9.1.3 *Time stepping*

Although finite elements can be created in time, it is much more common to use finite differences. There are a number of ways in which (9.17) can be discretized; we chose the following:

$$|F|^{k+\frac{1}{2}} \frac{\{\Phi\}^{k+1} - \{\Phi\}^{k}}{\Delta t} = \{Q\}^{k+\frac{1}{2}} - \theta |D|^{k+\frac{1}{2}} \{\Phi\}^{k+1} - (1-\theta)[D]^{k+\frac{1}{2}}\{\Phi\}^{k} \tag{9.19}$$

where $\theta$ is a weighting factor which may take on any value between zero and unity and $k$ is the time level ($t = \Sigma_{i=1}^{k} \Delta t_i$). The unknowns are the $\Phi$ at the advanced time level, $\{\Phi\}^{k+1}$. Thus (9.19) is rewritten

$$|M|\{\Phi\}^{k+1} = |P|\{Q\} + \{S\} \tag{9.20}$$

in which

$$|M| = |F|^{k+\frac{1}{2}} + \theta\ \Delta t\ |D|^{k+\frac{1}{2}} \tag{9.21a}$$

$$|P| = \Delta t\ [1] \tag{9.21b}$$

$$\{S\} = (|F|^{k+\frac{1}{2}} - (1 - \theta)\ \Delta t\ |D|^{k+\frac{1}{2}})\{\Phi\}^{k} \tag{9.21c}$$

Matrices evaluated at time level $k + \frac{1}{2}$ should be

$$|\quad|^{k+\frac{1}{2}} = \tfrac{1}{2}(|\quad|^{k} + |\quad|^{k+1}) \tag{9.22}$$

If the problem is non-linear (i.e. if $C$ and $K$ of (1.64) are really functions of $\Phi$), then $|F|^{k+1}$ and $|D|^{k+1}$ are unknown. Thus the solution is iterative with $|F|^{k+1}$ and $|D|^{k+1}$ taken first equal to their respective values at time level $k$ and, after the solution is found for $\Phi^{k+1}$, these values are revised and the problem is solved again for a better value of $\Phi^{k+1}$. When this value no longer changes (i.e. has converged), the solution passes on to the next time step.

Normally the weighting factor, $\theta$, would be taken to be 0.5 in order to have a fully centered difference scheme of second order accuracy in $\Delta t$. A difficulty occurs, however, if any part of the flow region is saturated. In saturated flow the time derivative of (9.17) disappears. The governing equation is a parabolic type for unsaturated flow and an elliptic type for saturated flow. A change in the boundary conditions in the saturated flow has an instantaneous effect on the entire saturated region; the solution is no longer a continuous function of time and $\Phi^{k+1}$ is independent of $\Phi^k$. At the sacrifice of some accuracy this problem can be solved by taking $\theta = 1$ (Neuman 1973) which works for both saturated and unsaturated flow.

### 9.1.4 Boundary conditions

For Neumann type boundaries the evaluation of the integrals is straightforward. All of the nodes, both interior and boundary nodes, have a single unknown, $\Phi$. The boundary integral of (9.18c) cancels along adjacent interior elements since the outflow from one interior element is the inflow to its neighbor. Along the exterior boundaries this integral is not necessarily zero but can be evaluated from the specified flow rate out of the domain. Thus the integral can be calculated explicitly, leading to the vector $\{Q\}$ which is the forcing function of the system.

For Dirichlet boundary conditions all of the $\Phi$ at the boundary nodes are known. A common way to enforce this condition is to first set up the matrices as previously described, then in (9.20) the rows of the $[M]$ matrix which correspond to a boundary node are made zero except for a value of unity on the main diagonal. The right-hand side of (9.20) is changed so that the row corresponding to a boundary node row contains the given boundary value. Thus the solution of the set of equations will always return the proper (given) value of $\Phi$ on the boundary. Since the outflow is unknown, the boundary integral of (9.18c) cannot be evaluated; however, it is unnecessary to do so since those terms in (9.20) have been replaced. After all of the $\Phi$ are found, the outflow velocity at any node, $i$, is obtained from

$$q = -K \sum_{i=1}^{n} \Phi_i \frac{\partial N_i^e}{\partial n} \tag{9.23}$$

The quantity of outflow across the exterior boundary of an exterior element can be found from integrating (9.23):

$$Q^e = -K \sum_{i=1}^{n} \Phi_i \int_{\Gamma^e} \frac{\partial N_i^e}{\partial n}\, ds \tag{9.24}$$

in which $\Gamma^e$ is the element exterior boundary.

Mixed Dirichlet and Neumann conditions form a problem which is a straightforward combination of those described above. The unknowns are the $\Phi$ at interior nodes and those on Neumann boundaries. The boundary integral of (9.18c) is evaluated at Neumann boundaries and the proper value of $\Phi$ is enforced at Dirichlet boundaries.

## 9.2 Combining the FEM with the BIEM

Consider the domain $D = B + F$ (Fig. 9.3) in which the region is divided into a portion in which the solution is to be accomplished by the BIEM ($B$) and a part where the FEM is to be applied ($F$). On the exterior boundary the potential, $\Phi$, or its normal derivative, $\partial\Phi/\partial n$, is given. Thus the boundary is divided as follows:

$b_h$ = the boundary of $B$ where $\Phi$ is prescribed (Dirichlet)
$b_q$ = the boundary of $B$ where $\partial\Phi/\partial n$ is prescribed (Neumann)
$f_h$ = the boundary of $F$ where $\Phi$ is prescribed (Dirichlet)
$f_q$ = the boundary of $F$ where $\partial\Phi/\partial n$ is prescribed (Neumann)
$c$ = the common boundary between $B$ and $F$

The BIEM is written in the form of (2.34)

$$|R|\{\Phi\} = |L|\left\{\frac{\partial\Phi}{\partial n}\right\} \tag{9.25}$$

which can be partitioned to form

$$|R_q \quad -L_h \quad -L_c \quad R_c| \begin{Bmatrix} \Phi_{bq} \\ \partial\Phi_{bh}/\partial n \\ \partial\Phi_{bc}/\partial n \\ \Phi_{bc} \end{Bmatrix} = |L_q \quad -R_h| \begin{Bmatrix} \partial\Phi_{bq}/\partial n \\ \Phi_{bh} \end{Bmatrix} \tag{9.26}$$

in which the subscript h designates the $b_h$ part, the subscript q designates the $b_q$ part and the subscript c designates the common boundary part. On all

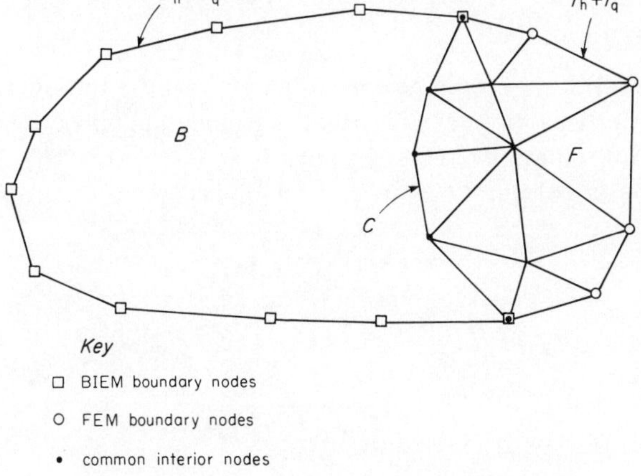


**Figure 9.3** The combination BIEM–FEM region.

the $\Phi$ the subscript b is added to emphasize that this quantity applies to the $b = b_h + b_q$ portion of the boundary and later to differentiate it from a similar quantity on the FEM boundary. Equation (9.26) has been arranged with the unknowns on the left-hand side. By definition $\partial\Phi_{bq}/\partial n$ and $\Phi_{bh}$ are prescribed boundary conditions. The potential and its normal derivative are, of course, unknown on the common boundary; these values have intentionally been placed last on the left-hand side of (9.26).

In the FEM region (9.20) is rewritten as

$$\lfloor -P_c \quad M_c \quad M_q \quad M_i \rfloor \begin{Bmatrix} Q_{fc} \\ \Phi_{fc} \\ \Phi_{fq} \\ \Phi_i \end{Bmatrix} = \lfloor P_f \quad -M_h \rfloor \begin{Bmatrix} Q_{fq} \\ \Phi_{fh} \end{Bmatrix} + \{S\} \quad (9.27)$$

in which the meanings of the subscripts are the same as for (9.26) except that f is added to the $\Phi$ in order to designate that it now applies to the $f = f_h + f_q$ part of the boundary. The interior nodes in the FEM region are designated by the subscript $i$. The unknowns at the common boundary nodes have been placed first on the left-hand side.

The unknown potential and flow rate along the common boundary must be the same whether viewed from the BIEM region or the FEM region. Thus

$$\{\Phi_{bc}\} = \{\Phi_{fc}\} = \{\Phi_c\} \quad (9.28)$$

$$\{Q_{fc}\} = -K\lceil l \rfloor\{\partial\Phi_{bc}/\partial n\} \quad (9.29)$$

in which $\lceil l \rfloor$ is a diagonal matrix of the average length of the two line segments joining a common boundary node (Fig. 9.4), i.e.

$$l_i = \tfrac{1}{2}(l_{i,i+1} + l_{i-1,i}) \quad (9.30)$$

Equation (9.29) comes from the approximation that the outflow corresponding to a node is the normal velocity at that node multiplied by $l_i$ and it is consistent with linear interpolation functions.

Equations (9.26) and (9.27) can be combined to form

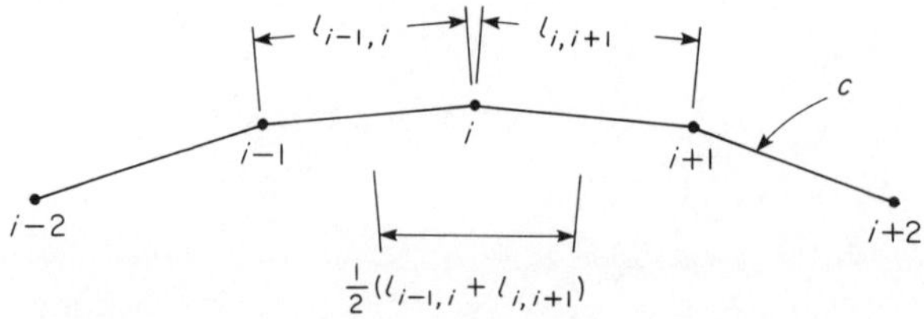


**Figure 9.4** The elements of $\lceil l \rfloor$ along the common boundary.

$$\begin{bmatrix} R_q & -L_h & L_c & -R_c & 0 & 0 \\ 0 & 0 & -\hat{P}_c & M_c & M_q & M_i \end{bmatrix} \begin{Bmatrix} \Phi_{bq} \\ \partial\Phi_{bh}/\partial n \\ \partial\Phi_c/\partial n \\ \Phi_c \\ \Phi_{fq} \\ \Phi_i \end{Bmatrix} = \begin{Bmatrix} S_1 \\ S_2 \end{Bmatrix} \tag{9.31}$$

in which

$$|\hat{P}_c| = -K\lceil l \rfloor |P_c| \tag{9.32a}$$

$$\{S_1\} = |L_q \quad -R_h| \begin{Bmatrix} \partial\Phi_{bq}/\partial n \\ \Phi_{bq} \end{Bmatrix} \tag{9.32b}$$

$$\{S_2\} = \{S\} \tag{9.32c}$$

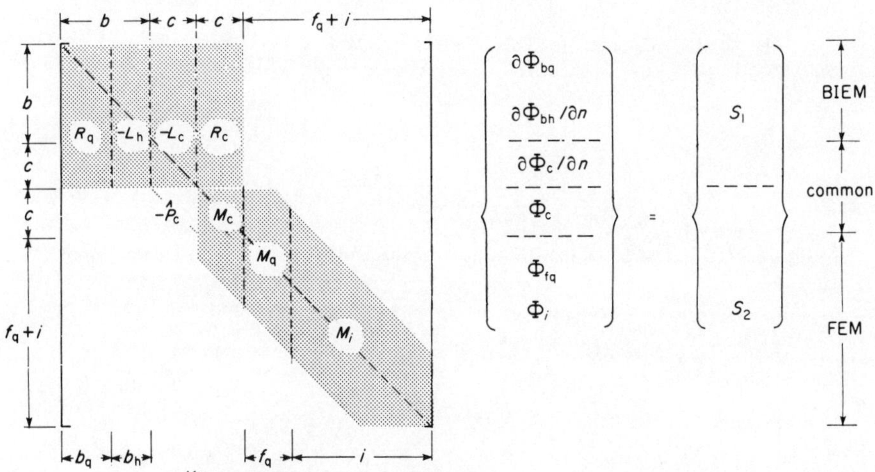


*Key*
*b* number of boundary nodes in the BIEM region
*c* number of nodes on the BIEM–FEM division line
*f* number of boundary nodes in the FEM region
*i* number of interior nodes in the FEM region

**Figure 9.5** The combined BIEM–FEM coefficient matrix.

Figure 9.5 shows a schematic diagram of the assembly of (9.31). The blank portions in the figure represent zeros. In practice the purely FEM part ($M_q$, $M_i$) need not be assembled in the order shown but can be arranged so as to minimize the band width of the FEM part of the matrix.

## 9.3 Reduction of the equations

Equations (9.31) can now be solved for the unknowns at all nodes, which consist of the exterior boundary, the common boundary, and the interior of the FEM region. As assembled, however, these equations are relatively inefficient to solve. Normally the FEM unknowns are solved by special routines that take advantage of the banded nature of the coefficient matrix. The well-populated BIEM prevents the use of such a solver, or at least the solver may have to operate on a wider band width if the situation is as appears in Figure 9.5. If the solution is to be done only once (i.e. a single-step problem) the computing time for the BIEM–FEM may be more than if the FEM is used alone. Of course there are other advantages in the combined method such as data handling, the solution in infinite regions, and the solution in the presence of singularities, all of which are easier in the BIEM than in the FEM.

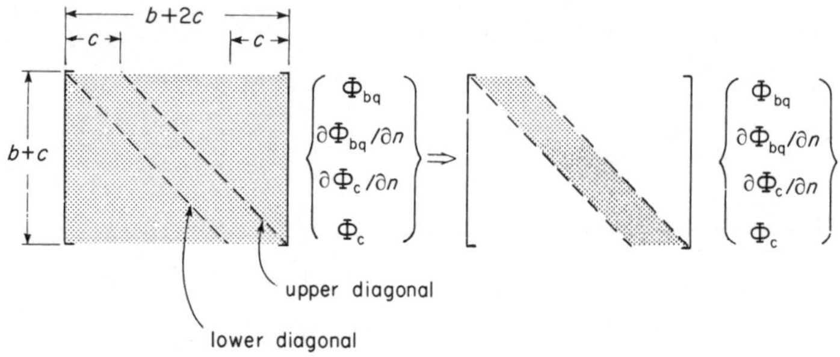


**Figure 9.6** Reduction of the BIEM part of the coefficient matrix to banded form.

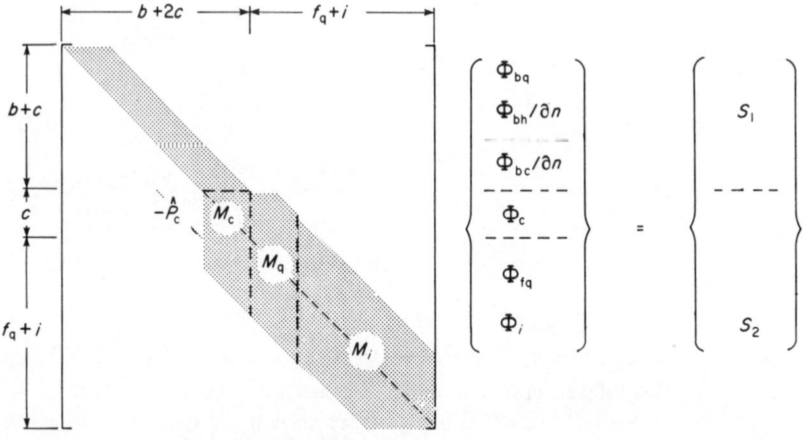


**Figure 9.7** The reduced BIEM–FEM equations.

For time dependent and/or iterative problems (i.e. multistep problems) the solution of (9.31) can be made very efficient if the BIEM portion is confined to the linear (Darcy flow) part of the domain. In that case the BIEM portion of the matrix does not change from one step to the next. The BIEM part can be reduced to banded form (Fig. 9.6) by zeroing (by Gaussian elimination, for example) the elements below the lower diagonal and above the upper diagonal. The result is a diagonalized set of equations such as is shown in Figure 9.7. Not only can a banded equation solver be used but the non-zero BIEM part is all above the main diagonal of the combined matrix (Fig. 9.7); thus, the solver begins with the common boundary part and the solution to the BIEM part is obtained by back substitution only.

In a multistep problem the coefficient matrix is first reduced to the form shown in Figure 9.7. Although this first step may require slightly more time than if the matrix were solved directly, subsequent steps become very efficient.

## 9.4 Numerical examples

Problems which can use the BIEM–FEM to advantage are those that require iteration or time stepping or both and that have both a linear (Darcy) region and a non-linear region. In this section we present three such cases. Two involve unsaturated flow where the effective permeability is dependent on the moisture content; the third involves non-Darcy flow in which the head gradient is a non-linear function of the velocity.

### 9.4.1 A drawdown problem

A problem similar to those treated in Section 5.3 is solved here with the difference that unsaturated flow is considered above the line of the phreatic surface (line of atmospheric pressure). The block shown in Figure 9.8 is initially saturated with the tailwater level at height $T_h$. At time $t = 0$ the tail water is lowered suddenly to height $T_w$ and the phreatic surface begins to fall. The zone between the phreatic surface and the initial water height becomes unsaturated although water continues to drain from this zone. The boundary and initial conditions are

$$\Phi = T_h \qquad \text{in } B + F \quad t \leqslant 0 \tag{9.33a}$$

$$\frac{\partial \Phi}{\partial z} = 0 \qquad \text{along } AB \text{ and } DE \tag{9.33b}$$

$$\frac{\partial \Phi}{\partial x} = 0 \qquad \text{along } BCD \tag{9.33c}$$

$$\Phi = T_w \quad \text{along } EF \quad t > 0 \tag{9.33d}$$

$$\Phi = z \quad \text{along } FE \quad t > 0 \tag{9.33e}$$

$$\frac{\partial \Phi}{\partial x} = 0 \quad \text{along } GA \quad t > 0 \tag{9.33f}$$

Discretization of the region is shown in Figure 9.9. The common boundary between the BIEM and FEM regions is chosen so that the flow is saturated along that boundary for the duration of the run. For maximum efficiency the FEM region should be as small as possible, but there has been no effort here to crowd the line of the lowest phreatic surface.

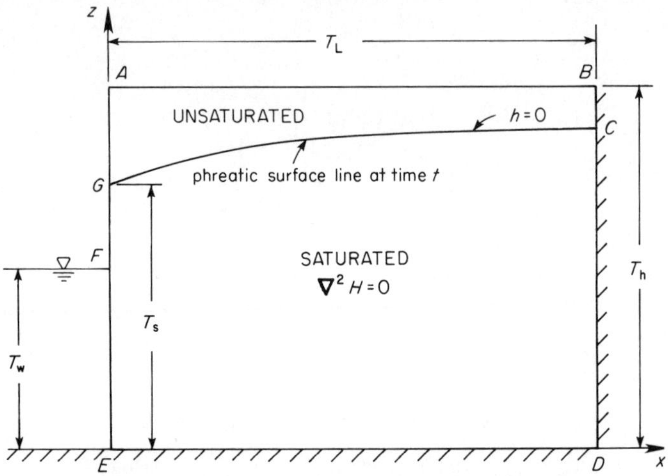


**Figure 9.8** The configuration of the drawdown problem.

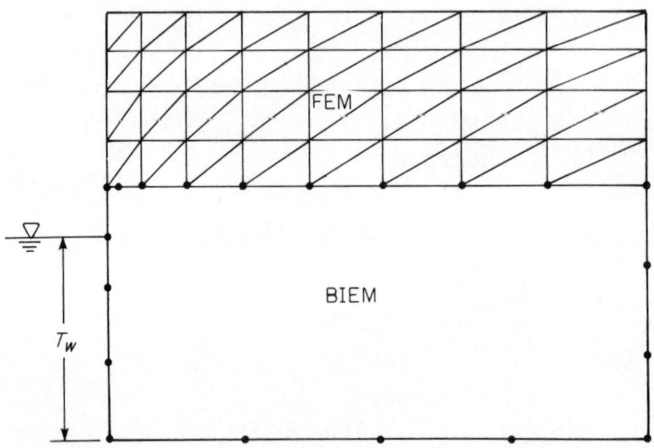


**Figure 9.9** Discretization of the region for the drawdown problem.

Richards' (1931) equation is to be used for the solution in the unsaturated zone (see Sec. 1.8). The capillary pressure and hydraulic conductivity are given by (Verma & Brutsaert 1971)

$$C(h_c) = \frac{\alpha\beta n |h_c|^{\beta-1}}{(\alpha + |h_c|^{\beta})^2} \tag{9.34}$$

$$K(h_c) = K_0 \left(\frac{\alpha}{\alpha + |h_c|^{\beta}}\right)^{\eta} \tag{9.35}$$

in which $\alpha$, $\beta$, and $\eta$ are constants for any particular medium, $n$ is the saturated porosity and $K_0$ is the saturated hydraulic conductivity. The problem was solved dimensionlessly with

$$t_* = \frac{tK_0}{nT_h} \quad x_* = \frac{x}{T_h} \quad z_* = \frac{z}{T_h} \quad q_* = \frac{q}{K_0} \quad \alpha_* = \frac{\alpha}{T_h^{\beta}} \tag{9.36}$$

A solution was obtained for the following values of the dimensionless variables:

$$T_L^* = 1 \quad T_w^* = 0.5 \quad \alpha_* = 0.211 \quad \beta = 1.5 \quad \eta = 3.0 \quad K_0 = 1 \quad n = 0.2 \tag{9.37}$$

A summary of the results is shown in Table 9.1 and Figure 9.10. Three different BIEM–FEM runs are compared with the finite difference solution of Verma and Brutsaert (1971) and a BIEM solution which uses the methods of Chapter 5 and does not consider the unsaturated region. The FEM and BIEM–FEM solutions with the same numbered subscript are roughly comparable discretizations, although it appears that the number of boundary nodes for a BIEM calculation can be fewer than for an equally accurate FEM calculation.

As indicated in Table 9.1 and Figure 9.10 all calculations, except the finite difference, had good accuracy. The finite difference calculation was done by Verma (1969) years earlier and thus it may be unfair to compare it to present day methods. The difference between the BIEM calculation and all of the others indicates the relative importance of the unsaturated zone for this problem.

Computing times and total number of nodes are shown for the FEM and BIEM–FEM calculations in Table 9.1. All computing (except finite differences) was done on Cornell's IBM 370/168 using a FORTRAN H compiler. The total nodes shown in the BIEM–FEM case are FEM nodes plus BIEM nodes (where the common boundary nodes are included in both,

**Table 9.1** Comparison of solutions for the drawdown problem.

| | | $FEM_1$ | $BIEM–FEM_1$ | $FEM_2$ | $BIEM–FEM_2$ | $FEM_3$ | $BIEM–FEM_3$ | $FD$ | $BIEM$ |
|---|---|---|---|---|---|---|---|---|---|
| | *no. nodes* | 117 | 50 + 20 = 70 | 81 | 40 + 16 = 56 | 40 | 30 + 12 =42 | 81 | |
| | *no. iterations* | 43 | 42 | 43 | 43 | 42 | 42 | | |
| *computing* | *init. time* | 0.20 | 0.25 | 0.14 | 0.20 | 0.10 | 0.15 | | |
| *time* | *total time* | 8.28 | 4.89 | 4.97 | 3.84 | 2.40 | 2.78 | | |
| *(CPU seconds)* | *time/iteration* | 0.1879 | 0.1105 | 0.1123 | 0.0847 | 0.0548 | 0.0612 | | |
| phreatic | $x_* = 0$ | 0.750 | 0.750 | 0.750 | 0.750 | 0.750 | 0.750 | 0.625 | 0.747 |
| surface | 0.125 | 0.808 | 0.808 | 0.810 | 0.807 | 0.804 | 0.803 | | 0.791 |
| elevation at | 0.250 | 0.847 | 0.847 | 0.848 | 0.844 | 0.850 | 0.843 | 0.817 | 0.821 |
| time $t_* = 0.013$ | 0.375 | 0.869 | 0.869 | 0.869 | 0.868 | 0.865 | 0.866 | | 0.847 |
| for various | 0.500 | 0.882 | 0.882 | 0.882 | 0.883 | 0.882 | 0.880 | 0.870 | 0.863 |
| distances | 0.750 | 0.897 | 0.897 | 0.897 | 0.897 | 0.901 | 0.896 | 0.895 | 0.887 |
| | 1.000 | 0.901 | 0.901 | 0.901 | 0.901 | 0.908 | 0.900 | 0.902 | 0.893 |
| phreatic | $x_* = 0$ | 0.500 | 0.500 | 0.500 | 0.500 | 0.500 | 0.500 | 0.500 | |
| surface | 0.125 | 0.539 | 0.538 | 0.538 | 0.535 | 0.533 | 0.531 | | |
| elevation at | 0.250 | 0.559 | 0.559 | 0.559 | 0.556 | 0.554 | 0.566 | 0.551 | |
| time $t_* = 0.518$ | 0.375 | 0.575 | 0.574 | 0.575 | 0.573 | 0.577 | 0.572 | | |
| for various | 0.500 | 0.587 | 0.586 | 0.586 | 0.585 | 0.586 | 0.584 | 0.577 | |
| distances | 0.750 | 0.603 | 0.602 | 0.601 | 0.601 | 0.602 | 0.601 | 0.592 | |
| | 1.000 | 0.607 | 0.607 | 0.607 | 0.606 | 0.608 | 0.606 | 0.597 | |
| outflow rate | $t_* = 0.013$ | 0.451 | 0.451 | 0.450 | 0.452 | 0.448 | 0.455 | 0.373 | 0.409 |
| | 0.518 | 0.117 | 0.117 | 0.117 | 0.116 | 0.116 | 0.114 | 0.114 | |

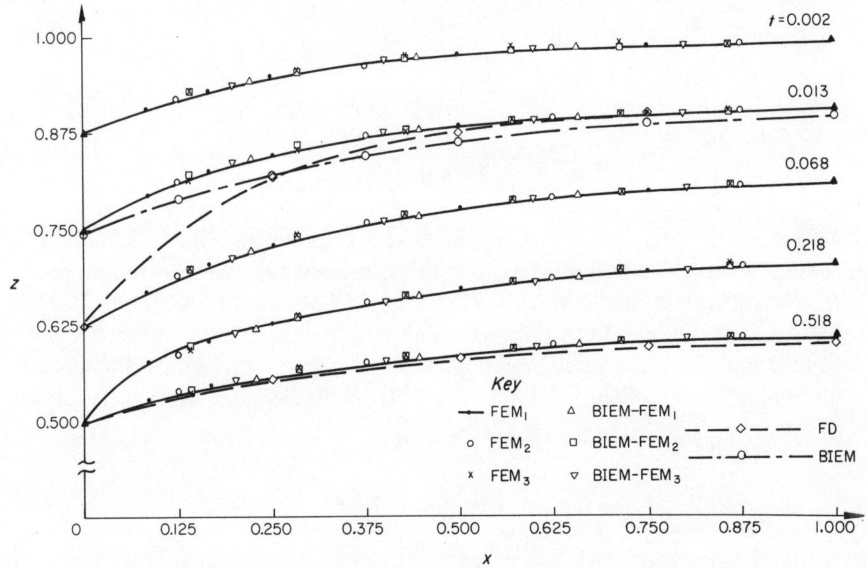


**Figure 9.10** The phreatic surface at various times in the drawdown problem.

i.e. counted twice). The number of iterations (time steps plus intermediate iterations) is indicated in the second row of Table 9.1. The third row shows the computing time required for the initial solution which includes the reduction of the BIEM portion of the matrix. The fourth row is the total time required for each iteration excluding the initial solution and other calculations performed only once during operation of the program. (Thus the time per iteration is not the total time divided by the number of iterations.) The solution is shown in subsequent rows.

The time saving of the BIEM–FEM over the FEM is apparent from Table 9.1. The greatest economy is realized in the cases with the most nodes or the finest discretization, a fact which has obvious implications for large field problems. In this example the BIEM and FEM regions are of comparable size, but the BIEM–FEM should gain in computational efficiency over the FEM if the FEM part can be made a smaller part of the total.

### 9.4.2 *A recharge problem*

In this example we have chosen a problem which has been solved both experimentally and numerically by Vauclin *et al.* (1979). The configuration is shown in Figure 9.11. Initially the free water surface is horizontal at the level of the tail water. At time $t = 0$ a recharge rate of $W = 14.8$ cm/h is applied over an area of 50 cm. The half length of the region is 300 cm, the height is 200 cm, and the tailwater elevation is 65 cm. The moisture capacity and hydraulic conductivity are

$$C(h_c) = \frac{3400|h_c|^{1.9}}{(40\,000 + |h|^{2.9})^2} \tag{9.38}$$

$$K(h_c) = \frac{104.65 \times 10^6}{2.99 \times 10^6 + |h_c|^5} \tag{9.39}$$

One of the BIEM–FEM discretizations is shown in Figure 9.12. The results of the calculation are as shown in Figure 9.13. The finite difference results are from Vauclin *et al.* (1979) who used 396 nodes. Although all of the results are good, both the FEM and the BIEM–FEM are closer to the experiment than the finite differences even though the finite difference discretization was finer. The FEM and BIEM–FEM$_1$ (Table 9.2) showed

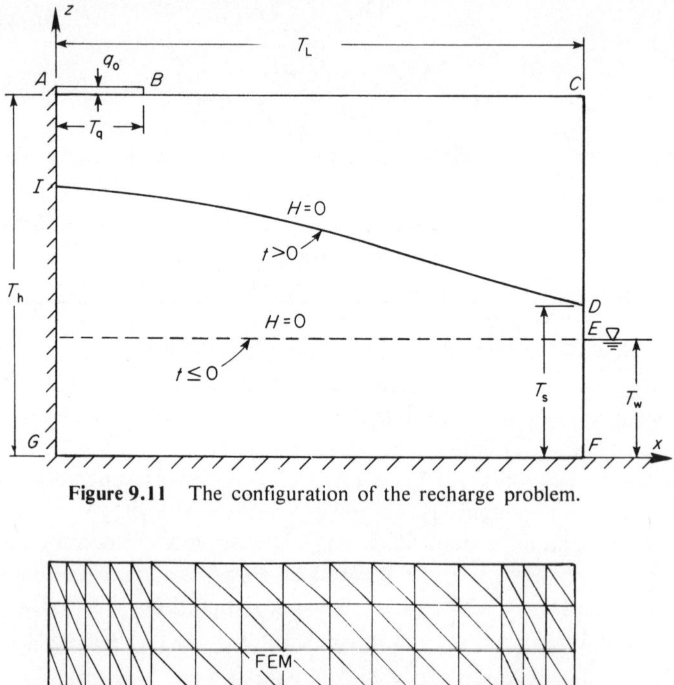


**Figure 9.11** The configuration of the recharge problem.

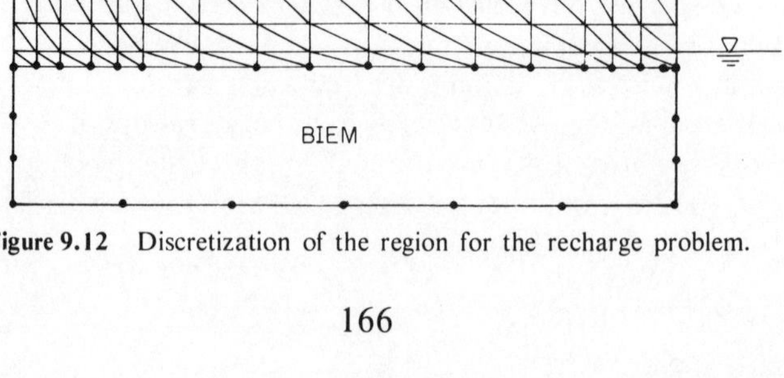


**Figure 9.12** Discretization of the region for the recharge problem.

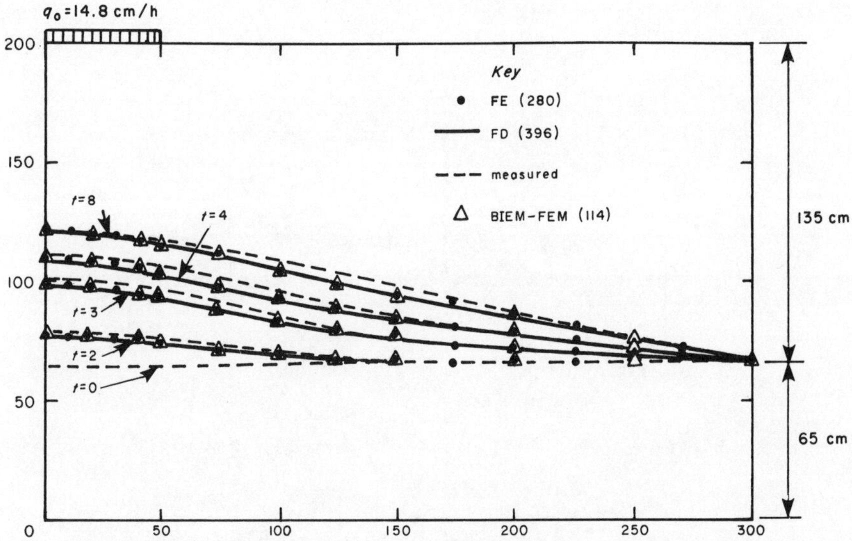


**Figure 9.13** The phreatic surface for various times in the recharge problem.

**Table 9.2** Computing times (CPU) for the recharge problem.

| | *FEM* | *BIEM–FEM$_1$* | *BIEM–FEM$_2$* | *FD* |
|---|---|---|---|---|
| no. of nodes | 280 | 162 + 38 = 200 | 88 + 26 = 114 | 396 |
| initial time (s) | 0.13 | 0.47 | 0.24 | |
| total time (s) | 146.73 | 121.27 | 51.78 | |
| no. iterations | 200 | 205 | 195 | |
| time/iteration (s) | 0.7319 | 0.5893 | 0.2643 | |

almost identical results to three or four significant figures. The BIEM–FEM$_2$ showed minor differences of the order of 0.1%.

The relative computing times for the FEM and BIEM–FEM are shown in Table 9.2. Although the initial step for the BIEM–FEM is considerably longer than for the FEM, the overall computational time is much less.

### 9.4.3 *Non-Darcy flow at the tip of a cut-off wall*

The problem to be used as an example of non-Darcy flow is essentially the same as that shown in Figure 3.4. Due to symmetry only the right half of the box is used as indicated in Figure 9.14. The discretization for the FEM and BIEM–FEM solutions are indicated in Figure 9.15.

The relationship between the velocity and the gradient of the potential (see (1.63)) is taken to be

$$\nabla\Phi = -(0.938 + 0.179|v|)\mathbf{v} \tag{9.40}$$

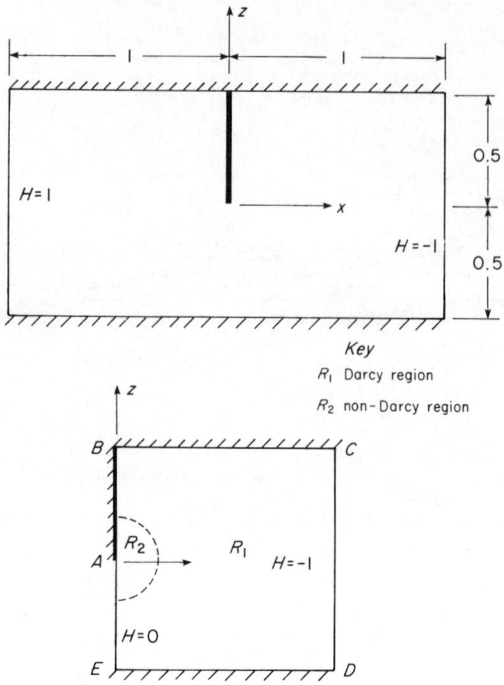


**Figure 9.14** Flow around the tip of a cut-off wall. The full region (top) and the problem as solved due to symmetry (bottom).

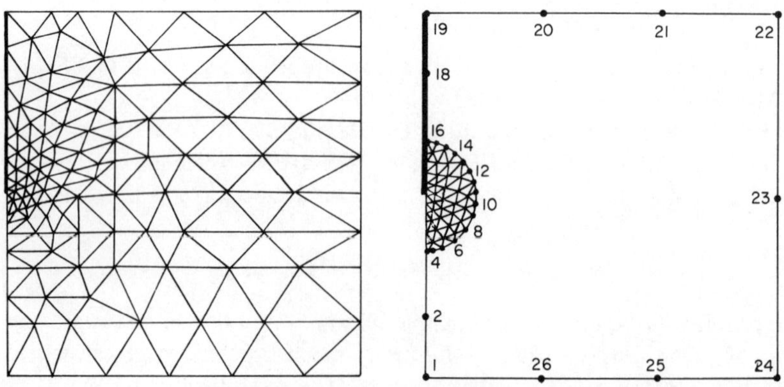


**Figure 9.15** Discretization for the flow around a cut-off wall – the FEM discretization (left) and the BIEM–FEM discretization (right).

or, expressed in terms of the hydraulic conductivity,

$$K(v) = (0.938 + 0.179|v|)^{-1} \tag{9.41}$$

Because the effective permeability is a function of the velocity, the solution must be iterative. In order to begin we assume Darcy flow throughout and obtain an initial solution. Subsequent iterations use (9.41) for the hydraulic conductivity. Iteration is continued until there is no change in the results.

The non-Darcy and Darcy regions are arbitrarily divided as shown in Figure 9.14. Using (9.40) the gradient of the potential is not exactly matched on the boundary of the two regions. Either the equation for hydraulic conductivity or the region could be changed so that the potential gradient does match, but since it is close in any case such a refinement seems unnecessary.

In the initial solution, where Darcy flow was assumed throughout, there was a considerable time disadvantage in using the BIEM–FEM over the FEM. When these solutions were compared to the analytic solution (Table 3.2), the BIEM–FEM was more accurate than the FEM; thus the question of time *vs.* accuracy is still open. After several iterations the BIEM–FEM regains its time advantage. Since the number of iterations is small, the time advantage of the BIEM–FEM is also small. Thus we conclude that its primary advantage lies elsewhere (i.e. data handling, grid specification, singularities, infinite regions) unless there are many iterations. The comparison of the BIEM–FEM with the FEM is shown in Table 9.3.

**Table 9.3** The FEM/BIEM–FEM comparison to the cut-off wall solution.

| | | *FEM* | *BIEM–FEM* |
|---|---|---|---|
| no. of nodes | | 125 | 51 + 26 = 77 |
| total time (s) | | 1.56 | 1.36 |
| no. of iterations | | 7 | 6 |
| time/iteration (s) | | 0.2014 | 0.1600 |
| solution along | $z = 0$ | $-H = 0$ | $-H = 0$ |
| the cut-off wall | 0.01414 | 0.0858 | 0.1117 |
| | 0.04243 | 0.1995 | 0.2047 |
| | 0.14142 | 0.3441 | 0.3442 |
| | 0.28284 | 0.4310 | 0.4250 |
| | 0.50000 | 0.4741 | 0.4734 |
| solution along | $x = 0.2$ | $-H = 0.5078$ | $-H = 0.5247$ |
| the boundary | 0.4 | 0.6010 | 0.5985 |
| *BC* | 0.6 | 0.7285 | 0.7171 |
| | 0.8 | 0.8681 | 0.8450 |
| flow through boundary *CD* | | 0.9555 | 0.8040 |

# 10 General groundwater problems

Up to this point we have used the BIEM to solve a number of specific problems in flow through porous media. These applications have involved, primarily, Laplace's equation. There are, however, other groundwater problems which do not fit the mode of the solutions previously presented. Fortunately, the BIEM can be made considerably more flexible. Our goal is to solve a general case such as that illustrated in Figure 10.1. The total groundwater basin may consist of several aquifers separated by aquitards. The aquifers may be confined or unconfined, isotropic or anisotropic. There may be steady or unsteady flow. In this chapter we consider several of the elements separately and then put them together to form a calculation method which can be applied to complex cases.

Unsteady flow in confined aquifers is an important groundwater case. The compressibility of the aquifer cannot be ignored, leading to a storativity term in the governing equation. Instead of Laplace's equation we are required to solve the diffusion equation. The solution to leaky aquifers involves many of the same considerations; these problems are therefore considered next.

In general any steady state groundwater problem of saturated, Darcy flow is a solution of the three-dimensional Laplace equation. Historically it has been impossible, at least from a practical point of view, to obtain such solutions. Even using the BIEM, as in Chapter 7, three-dimensional solutions become large and require a considerable amount of computer time and human data preparation. We often use a two-dimensional approximation which serves well for many flows. Indeed, it is seldom that we know the constitutive properties of an aquifer with sufficient precision that the accuracy of a three-dimensional solution is justified except in a local situation. Thus there are approximations which are commonly used to decrease the dimensionality of groundwater problems. In this chapter the development of a general method of computation is presented. This method uses common approximations to reduce a basically three-dimensional situation to a two-dimensional calculation, then the BIEM is used to further reduce the computational dimensionality to one dimension. In doing so many of the three-dimensional features of the aquifer are preserved.

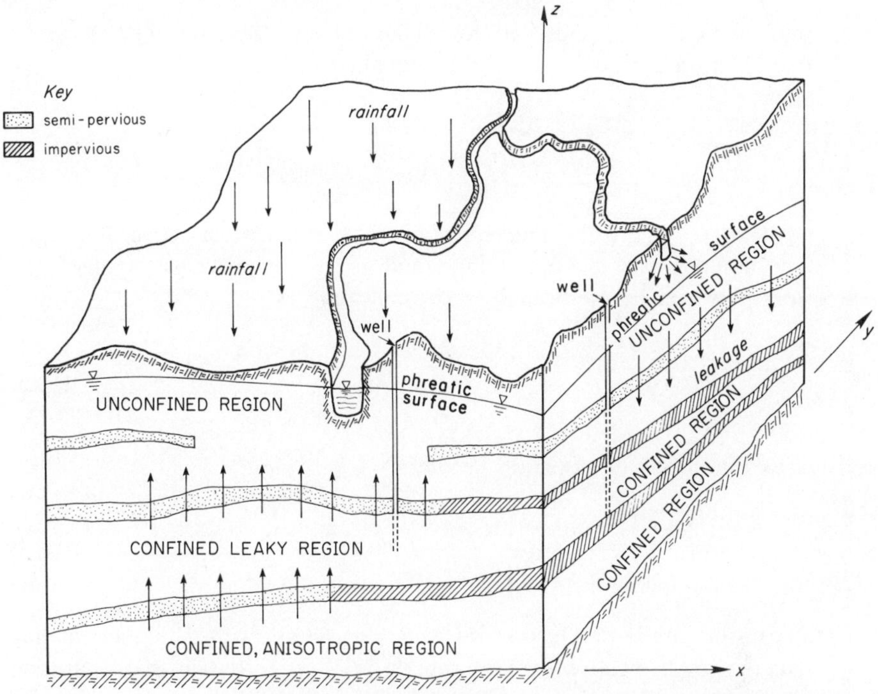


**Figure 10.1** A groundwater basin composed of confined, unconfined leaky, layered and anisotropic aquifers.

## 10.1 Unsteady flow in a confined aquifer

Equation (1.13) for unsteady flow in a confined aquifer is equivalent to the heat equation which governs transient heat flow in a two-dimensional solid (Rizzo & Shippy 1970). In contrast to the equations which have been solved in the previous chapters, that equation contains a time derivative. If the time derivative is removed by means of a Laplace transform, it takes on the same form as (1.61) which governs flow in leaky aquifers. In this section, therefore, we consider the solution of problems in a compressible medium. Two methods are given: (1) the use of the Laplace transform to remove the time derivative and (2) direct integration in time and space using the BIEM.

We repeat for convenience (1.13) which governs unsteady flow in a confined aquifer:

$$S_e \frac{\partial \Phi}{\partial t} = \nabla \cdot (\boldsymbol{K} \cdot \nabla \Phi) \tag{10.1}$$

in which $S_e$ is the specific storage and $\boldsymbol{K}$ is the coefficient of permeability. If we assume that the Dupuit approximation is valid (see Sec. 1.7) and write the equation in the stretched co-ordinate system (Sec. 4.4), (10.1) can be reduced to a two-dimensional flow equation

$$S\frac{\partial\Phi}{\partial t}=T_{**}\left(\frac{\partial^2\Phi}{\partial x_{**}^2}+\frac{\partial^2\Phi}{\partial y_{**}^2}\right) \tag{10.2}$$

in which $S = S_e b$ is the aquifer's storativity. Of course, the preceding equation assumes an incompressible fluid in a compressible aquifer. The storativity can be removed by redefining time as

$$t_*=t\frac{T}{SL^2} \tag{10.3}$$

Thus the governing equation becomes

$$\frac{\partial\Phi}{\partial t_*}=\nabla_{**}^2\Phi \tag{10.4}$$

For further use in this section we drop the asterisks with the understanding that the result must be scaled properly in order to obtain the solution in physical co-ordinates.

The same boundary conditions apply to flow in confined aquifers as to other problems in porous media, namely

$$\Phi=\Phi_b \qquad \text{on } \Gamma_1 \tag{10.5}$$

or

$$\frac{\partial\Phi}{\partial n}=-q \qquad \text{on } \Gamma_2 \tag{10.6}$$

where $q$ is the outflow through the boundary.

### *10.1.1 The Laplace transform solution*

Equation (10.4) can be reduced to an equation which contains only space variables through use of the Laplace transform, which is defined as

$$\tilde{\tilde{\Phi}}(x,y;s)=\int_0^\infty \Phi e^{-st}\,dt \tag{10.7}$$

in which $\tilde{\tilde{\Phi}}$ denotes the transformed variable and $s$ is the transformation parameter. The transformation of the right-hand side of (10.4) is

$$\int_0^\infty \nabla^2\Phi\, e^{-st}\, dt = \nabla^2\tilde{\tilde{\Phi}} \tag{10.8}$$

For the left-hand side we integrate by parts:

$$\int_0^\infty \frac{\partial\Phi}{\partial t} e^{-st}\, dt = \Phi e^{-st}\Big|_0^\infty + s\int_0^\infty \Phi e^{-st}\, dt = -\Phi(x, y, 0) + s\tilde{\tilde{\Phi}}(x, y; s) \tag{10.9}$$

in which $\Phi_0 = \Phi(x, y, 0)$ is the initial value of $\Phi$. Thus the differential equation becomes

$$\left(\frac{\partial^2}{\partial x^2} + \frac{\partial^2}{\partial y^2}\right)\tilde{\tilde{\Phi}} - s\tilde{\tilde{\Phi}} + \Phi_0 = 0 \tag{10.10}$$

We now consider the homogeneous part of (10.10)

$$\nabla^2\tilde{\tilde{\Phi}} - s\tilde{\tilde{\Phi}} = 0 \tag{10.11}$$

and seek a free space Green function which satisfies this equation in the same way that $\ln r$ satisfies Laplace's equation. Greenberg (1971) gives the free space Green function as

$$G = -K_0(s^{\frac{1}{2}}r) \tag{10.12}$$

in which $r$ has the same meaning as before and $K_0$ is a modified Bessel function of the second kind and of order zero. Equation (10.12) represents a singular solution to (10.11) as $K_0$ is singular for the parameter $s^{\frac{1}{2}}r \to 0$,

$$G \to \ln r + \tfrac{1}{2}\ln s \qquad \text{for } s^{\frac{1}{2}}r \to 0 \tag{10.13}$$

In (10.13) we see that for small $s$, $G$ behaves as $\ln r$, the two-dimensional Green function for Laplace's equation.

The boundary conditions (10.5) and (10.6) must be transformed. If $\Phi_b$ and $q$ are constants in time

$$\tilde{\tilde{\Phi}} = \Phi_b/s \qquad \text{on } \Gamma_1 \tag{10.14}$$

and

$$\partial\tilde{\tilde{\Phi}}/\partial n = -q/s \qquad \text{on } \Gamma_2 \tag{10.15}$$

The Green's formula for (10.11) is (Courant & Hilbert 1953, p. 280)

$$\int_D [VL(U) - UL(V)]\, \mathrm{d}A = \int_\Gamma \left( V\frac{\partial U}{\partial n} - U\frac{\partial V}{\partial n} \right) \mathrm{d}\lambda \tag{10.16}$$

where

$$L = \frac{\partial^2}{\partial x^2} + \frac{\partial^2}{\partial y^2} - s \tag{10.17}$$

is the differential operator. With the identity $U = G$ and $V = \tilde{\tilde{\Phi}}$, and since $L(G) = 0$, equation (10.10) gives

$$\int_\Gamma \left( \tilde{\tilde{\Phi}} \frac{\partial G}{\partial n} - G\frac{\partial \tilde{\tilde{\Phi}}}{\partial n} \right) \mathrm{d}\lambda = \int_D G\Phi_0 \, \mathrm{d}A \tag{10.18}$$

Since $G$ is a modified Bessel function as in (10.12), the derivative is (Janke & Emde 1945)

$$\frac{\partial}{\partial n} K_0(s^{\frac{1}{2}}r) = -s^{\frac{1}{2}} K_1(s^{\frac{1}{2}}r) \frac{\partial r}{\partial n} \tag{10.19}$$

in which $K_1$ is a modified Bessel function of the second kind and of order one. Thus (10.18) becomes

$$\int_\Gamma \left[ \tilde{\tilde{\Phi}} s^{\frac{1}{2}} K_1(s^{\frac{1}{2}}r) \frac{\partial r}{\partial n} + K_0(s^{\frac{1}{2}}r) \frac{\partial \tilde{\tilde{\Phi}}}{\partial n} \right] \mathrm{d}\lambda = -\int_D K_0(s^{\frac{1}{2}}r)\Phi_0 \, \mathrm{d}A \tag{10.20}$$

The area integral depends on the initial conditions and needs to be evaluated only once. The solution for any point in the solution domain can be found in the same way as described in Chapter 3 and is

$$\tilde{\tilde{\Phi}} = \frac{1}{2\pi} \int_\Gamma \left[ \tilde{\tilde{\Phi}} s^{\frac{1}{2}} K_1(s^{\frac{1}{2}}r) \frac{\partial r}{\partial n} + K_0(s^{\frac{1}{2}}r) \frac{\partial \tilde{\tilde{\Phi}}}{\partial n} \right] \mathrm{d}\lambda + \frac{1}{2\pi} \int_D K_0(s^{\frac{1}{2}}r)\Phi_0 \, \mathrm{d}A \tag{10.21}$$

Therefore, after the initial evaluation of the area integral, an internal evaluation of $\tilde{\tilde{\Phi}}$ consist of simply a contour integration around the solution domain. As in the potential flow case only half the boundary data are known in a well-posed problem and thus (10.20) needs to be used initially to find the remaining half. The boundary data come from (10.18) and/or

(10.21). The boundary discretization and numerical methods are exactly those used for Laplace's equation, the difference being that the Green function of (10.12) must be used in place of $\ln r$ and, if $\Phi_0$ is not equal to zero, the solution domain must be discretized for the once only evaluation of the area integral.

The numerical method requires frequent evaluation of $K_0$ and $K_1$. This can be done to a relatively high degree of accuracy by using a polynomial approximation (see, e.g. Abramowitz & Stegun 1974, p. 379). Most computer libraries have standard subroutines for the evaluation of Bessel functions; for example, the International Mathematical and Statistical Library (IMSL) contains FORTRAN routines that can be used directly.

The integration must be done numerically and the best choice is Gaussian quadrature (App. A.5). However, care must be taken with the logarithmic singularity in the modified Bessel function when the base point is at one end of the element. In that case $\ln r$ can be added to the modified Bessel function to produce a non-singular integrand, then the $\ln r$ parts integrated separately as follows:

$$\int K_0(s^{\frac{1}{2}}r)\frac{\partial\tilde{\tilde{\Phi}}}{\partial n}\,\mathrm{d}\lambda = \int [K_0(s^{\frac{1}{2}}r) + \ln r]\frac{\partial\tilde{\tilde{\Phi}}}{\partial n}\,\mathrm{d}\lambda - \int \ln r\frac{\partial\tilde{\tilde{\Phi}}}{\partial n}\,\mathrm{d}\lambda \tag{10.22}$$

The last integral can be done either analytically, as in Chapter 2, or by special Gaussian quadrature (see App. A.5.3).

### 10.1.2 Inversion of the transform

The last section presents the solution to the transformed equation but we need to invert the transform in order to return to physical variables. There are a number of techniques to invert the Laplace transform. We will use the method of Schapery (1962). It assumes that the solution can be expanded in a series

$$\Phi(x, y, t) = \Phi_s(x, y) + \sum_{i=1}^{M} a_i \mathrm{e}^{-b_j t} \tag{10.23}$$

in which $\Phi_s$ is the steady state solution which is obtained at $t = \infty$ and $a_i$ and $b_i$ are constants which must be determined. Equation (10.23) is transformed to produce

$$\tilde{\tilde{\Phi}}(x, y; s) = \frac{\Phi_s}{s} + \sum_{i=1}^{M} \frac{a_i}{b_i + s} \tag{10.24}$$

We now choose, somewhat arbitrarily, $M$ values of $b_i$ and $M$ values of $s$, say $s_i$. Although it is not necessary to do so, the $b_i$ are usually chosen equal to the $s_i$ for lack of a better choice:

$$b_i = s_i \qquad i = 1, 2, \ldots, M \tag{10.25}$$

Then (10.24) becomes $M$ different equations, one for each of the $s$ values

$$\tilde{\tilde{\Phi}}(x, y; s_j) = \frac{\Phi_s}{s_j} + \sum_{i=1}^{M} \frac{a_i}{s_i + s_j} \qquad j = 1, 2, \ldots, M \tag{10.26}$$

Equation (10.26) represents $M$ simultaneous equations for $M$ values of $a_i$. At any point $x$, $y$ and parameter $s$ the BIEM solution gives us a number for $\tilde{\tilde{\Phi}}$. We can also find the steady state solution, $\Phi_s$, which is simply the transformed solution for $s = 0$. Since we have chosen the $s$ values, the only remaining unknowns in (10.26) are the $a_i$. It is necessary to solve the BIEM problem $M$ times in order to find $M$ values of $\Phi$, one for each $s$.

Although the BIEM needs to be used only $M$ times, we need the solution of (10.26) for each point $(x, y)$. If there are $N$ such points, the complete solution requires that we solve the set of simultaneous equations $M$ times, thus producing a total of $M \times N$ of the $a_i$. However, the coefficients of the simultaneous equations are the same for each point. That is, (10.26) can be written

$$|P|\{a\} = \left\{\tilde{\tilde{\Phi}} - \frac{\tilde{\tilde{\Phi}}_s}{s}\right\} \tag{10.27}$$

where

$$|P| = \begin{bmatrix} (s_1 + s_1)^{-1} & (s_1 + s_2)^{-1} & \ldots & (s_1 + s_M)^{-1} \\ (s_2 + s_1)^{-1} & (s_2 + s_2)^{-1} & \ldots & (s_2 + s_M)^{-1} \\ \cdot & \cdot & & \cdot \\ \cdot & \cdot & & \cdot \\ \cdot & \cdot & & \cdot \\ (s_M + s_1)^{-1} & (s_M + s_2)^{-1} & \ldots & (s_M + s_M)^{-1} \end{bmatrix} \tag{10.28}$$

Since $|P|$ does not vary from point to point, this matrix needs to be inverted only once. A usual choice of $M$ is from five to twelve, so the inversion of the matrix is not a great burden. Once the matrix is inverted the $a_i$ are found from a simple matrix multiplication.

The above analysis is written for $\Phi$ as the variable of interest. It could as well have been written for $\partial\Phi/\partial n$. Equation (10.23) would become

$$\frac{\partial}{\partial n}\Phi(x, y, t) = \frac{\partial \Phi_s}{\partial n} + \sum_{i=1}^{M} a_i e^{-b_j t} \tag{10.29}$$

with $\partial\tilde{\bar{\Phi}}/\partial n$ replacing $\tilde{\bar{\Phi}}$ in (10.24), (10.26) and (10.27).

The choices for the $s_i$ are critical to the quality of the inversion. A few solutions of the BIEM can produce a plot of $s\tilde{\bar{\Phi}}/\Phi_s$ *vs.* $s$ as is shown in Figure 10.2. The transient response occurs in the variable part of the curve; the initial solution is at the end (large $s$) and the steady state solution is at the beginning (small $s$). Thus the values of $s$ should be chosen in the transient portion. They are usually chosen in a geometric series

$$s_i = P^{i-1}s_1 \qquad i \geqslant 1 \tag{10.30}$$

such as (for $P = 10$ and $s_1 = 1$) 1, 10, 100, . . . . The number $M$ of $s$ values must be large enough for (10.23) to fit the true curve. On the other hand the matrix of (10.28) becomes ill-conditioned for increasing $M$, creating difficulty in solving the equations and producing unstable solutions. The compromise seems to be to choose from 5 to 12 terms.

Also in choosing the form of (10.23) we have assumed the form of the solution. That is, we have assumed in this case that the solution begins at an initial value and proceeds with time to a steady state. If the solution is oscillatory, or behaves in some other manner, (10.23) should be chosen to reflect that behavior. The analyst's knowledge of the physical problem must be used to advantage. Other methods of inverting the Laplace transform are found in Cost (1964) and Bellman *et al.* (1966).

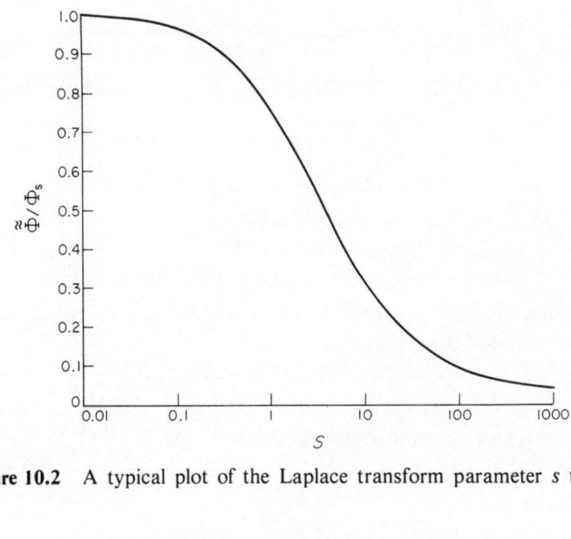


**Figure 10.2** A typical plot of the Laplace transform parameter $s$ *vs.* $\tilde{\bar{\Phi}}/\bar{\Phi}_s$.

### 10.1.3 The direct solution

A Green function which solves (10.4) directly is (Morse & Feshbach 1953)

$$G(r, \tau) = \frac{1}{\tau} \exp(-r^2/4\tau) H(\tau) \tag{10.31}$$

in which $\tau = t - t_0$ and $H(\tau)$ is the unit step function

$$\begin{aligned} H(\tau) &= 1 \qquad \tau > 0 \\ H(\tau) &= 0 \qquad \tau < 0 \end{aligned} \tag{10.32}$$

Thus $G$ is the counterpart of $\ln r$ which was used in Chapter 2. After applying Green's theorem the counterpart of (2.15) is (Morse & Feshbach 1953, p. 859; Shaw 1974)

$$2\alpha\Phi(x, y, t) = \int_0^t \mathrm{d}t_0 \int_\Gamma \left[ G \frac{\partial \Phi}{\partial n} - \Phi \frac{\partial G}{\partial n} \right] \mathrm{d}s + \int_D \Phi_0 G(r, t)\, \mathrm{d}A \tag{10.33}$$

where

$$\alpha = \begin{cases} 2\pi \text{ if the point } (x, y) \text{ is in the interior} \\ \pi \text{ if the point } (x, y) \text{ is on a smooth boundary} \\ \text{the interior boundary angle if the point } (x, y) \\ \text{is at a boundary corner} \end{cases} \tag{10.34}$$

and where $\Phi_0$ represents the initial ($t = 0$) conditions. The boundary angle $\alpha$ is shown in Figure 2.2(b).

Equation (10.33) can be treated similarly to the corresponding case for Laplace's equation. Additional features are as follows:

(1) The volume integration must be performed initially, thus requiring a discretization of the solution domain (see Sec. 10.4.2). If the initial conditions are homogeneous, $\Phi_0 = 0$, this integral is zero; otherwise, it remains in the equation and must be evaluated for each time the solution is desired. Note that the Green function appearing in this term is to be evaluated at $t_0 = 0$.

(2) The first term on the right of (10.33) involves both a line integral and a time integral. The straightforward method of numerical integration is first to integrate around the boundary (as in Ch. 2) and then to advance the time integration by an increment $\Delta t$ using the Euler method of a simple forward difference scheme. Because of the inherent

inaccuracy of the Euler method, small time steps must be taken, especially at the beginning of the calculation where the Green function changes rapidly.

(3) Time appears both in the integrand and in the limits. Thus if the solution is desired at some time, say $t = t_1$, this value must be used in the integrand from the beginning of the calculation. If, after this initial calculation, the solution is wanted at $t = t_2 > t_1$, the analyst must either begin another calculation using as initial conditions the solution at $t_1$ (which requires finding the values at enough interior points to perform the area integration) or begin again from the original initial conditions using $t_2$ in the integrand. Either method is expensive if solutions are needed for many points in time.

A discrete form of (10.33) is found in Liggett and Liu (1979b).

### *10.1.4 Numerical examples*

Two simple examples are taken to illustrate the theory. They are chosen as generalizations of one-dimensional problems because of their simplicity and because the BIEM solutions can be compared to exact results. The first is

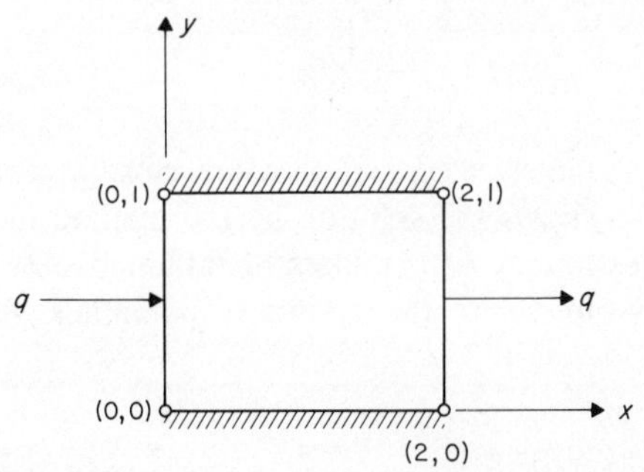


**Figure 10.3** The solution region for the examples.

the simple rectangular region shown in Figure 10.3. It will be solved as a Neumann problem with boundary conditions as follows:

$$\frac{\partial}{\partial n}\,\Phi(x, 0, t) = \frac{\partial}{\partial n}\,\Phi(x, 1, t) = 0 \tag{10.35}$$

$$\frac{\partial}{\partial n}\,\Phi(0, y, t) = 10 \qquad \frac{\partial}{\partial n}\,\Phi(1, y, t) = -10 \tag{10.36}$$

We take the potential as zero initially,

$$\Phi(x, y, 0) = 0 \tag{10.37}$$

Using the Laplace transform method the transformed boundary conditions are

$$\frac{\partial}{\partial n}\,\tilde{\tilde{\Phi}}(0, y; s) = \frac{1}{s} \qquad \frac{\partial}{\partial n}\,\tilde{\tilde{\Phi}}(1, y; s) = -\frac{1}{s} \tag{10.38}$$

Eight values of the parameter $s$ are chosen which are

$$s = 0.0001,\ 0.001,\ 0.01, \ldots, 1000 \tag{10.39}$$

and (10.20) is solved eight times for the boundary potential using the BIEM. The solutions were used to produce the plot of Figure 10.2. The figure indicates that the important range of $s$ is between 0.01 and 1000.

Next (10.26) was solved for the eight unknown values of $a_i$ at each node point. The results appear in Figure 10.4 where they are compared to the exact solution

$$\Phi = 10(x-1) + \frac{80}{\pi^2}\sum_{k=1}^{\infty}\frac{1}{(2k-1)^2}\exp\left[-(2k-1)^2\pi^2 t\right]\cos\left[(2k-1)\pi x/2\right] \tag{10.40}$$

The results of the numerical calculation are close to exact.

Figure 10.4 also shows the solution by the method of (10.33). Since all the $\partial\Phi/\partial n$ are known, they can be integrated immediately (zero in this case) and (10.23) leads directly to the boundary potentials. A trapezoidal time

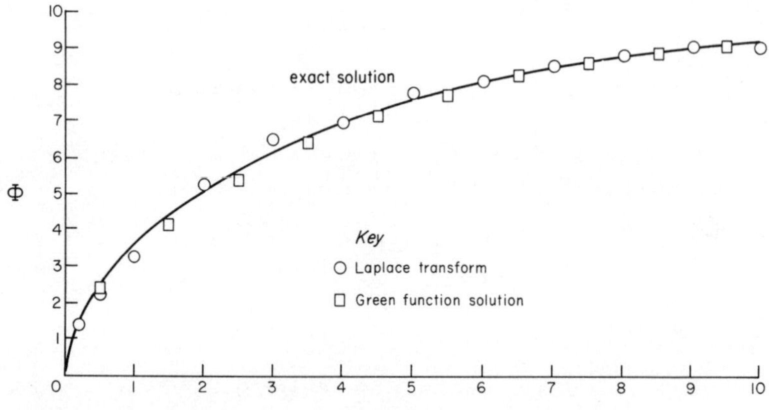


**Figure 10.4** A comparison of $\Phi$ *vs.* $t$ for the two methods and the exact solution for the example of the Neumann problem.

integration can be used for the potentials. Even with this simplicity the calculations to produce a complete curve are rather lengthy since each time point requires a start from the initial condition.

As the next example we take a square region, $0 \leqslant x \leqslant 1$ and $0 \leqslant y \leqslant 1$, and the same configuration but change the boundary condition so that the potential is prescribed on the ends. The initial condition is still given by (10.37) but the boundary conditions become

$$\Phi(0, y, t) = 1 \qquad \Phi(1, y, t) = 0$$
$$\frac{\partial}{\partial n}\Phi(x, 0, t) = \frac{\partial}{\partial n}\Phi(x, 1, t) = 0 \tag{10.41}$$

The exact solution to this problem is

$$\Phi = 1 - x + \frac{2}{\pi}\sum_{n=1}^{\infty}\frac{(-1)^n}{n}\,\mathrm{e}^{-n^2\pi^2 t}\sin(1-x)n\pi \tag{10.42}$$

For the Laplace transform method the transformed boundary conditions are

$$\tilde{\tilde{\Phi}}(0, y; s) = \frac{1}{s} \qquad \tilde{\tilde{\Phi}}(1, y; s) = 0 \tag{10.43}$$

Five values of $s$ were used and the BIEM solution is shown in Figure 10.5 for $x = 0.25$. An exact solution for $\tilde{\tilde{\Phi}}$ is available which is

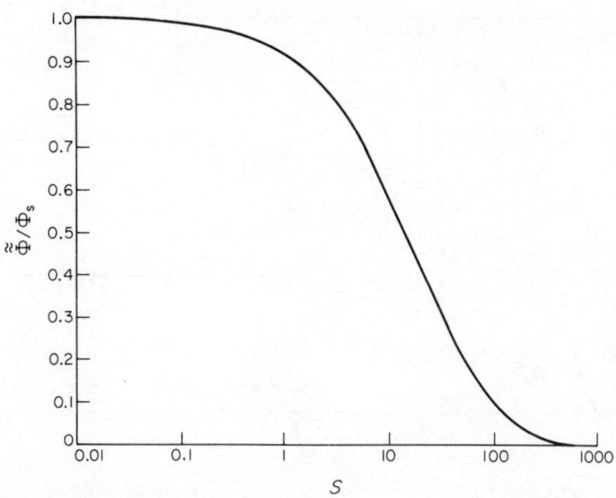


**Figure 10.5** The transformed solution for the mixed boundary problem.

$$\frac{s\tilde{\tilde{\Phi}}(x, s)}{\Phi_s} = \frac{e^{-s^{1/2}x} - e^{-2s^{1/2}x}e^{s^{1/2}x}}{1 - e^{-2s^{1/2}}} \tag{10.44}$$

Equation (10.44) checks well with the BIEM result.

The numerical Laplace transform inversion of the BIEM result produces the curves shown in Figure 10.6 for various values of $x$. Again the accuracy is excellent.

In using the direct integration method there is an immediate difficulty. The initial normal derivatives must be known in order to begin the time integration, yet these values are part of the solution. Further, the exact solution (or simply the result of a close examination of the boundary conditions) indicates that at time zero the normal derivatives are infinite. In any general problem of this type the normal derivative along a smooth boundary can be written as

$$\frac{\partial \Phi}{\partial n} = (\Phi_1 - \Phi_0)(1 + 2 \sum_{n=1}^{\infty} e^{-n^2\pi^2 t}) \qquad t < \Delta t \tag{10.45}$$

on which $\Phi_1 - \Phi_0$ is the jump in the potential in a very small time $\Delta t$. Equation (10.45) checks with the exact solution (10.40). The series is convergent for $t > 0$ but not for $t = 0$. Equation (10.45) serves in a general case to begin the time integration up to some small value $\Delta t$. From there it can be carried

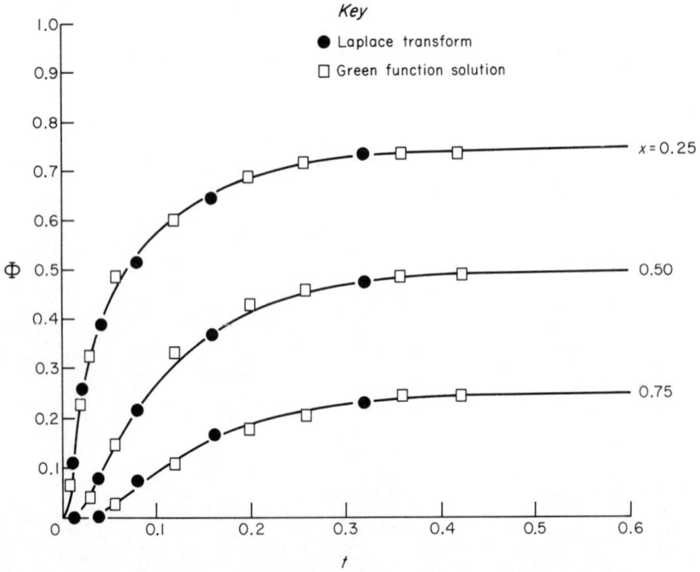


**Figure 10.6** A comparison of $\Phi$ *vs.* $t$ for the two methods and the exact solution for the mixed boundary problem.

on in the usual manner, care being necessary for small time because $\partial\Phi/\partial t$ is still changing rapidly.

The results of the direct integration are shown in Figure 10.6.

## 10.2 Nearly horizontal flows

In a great many groundwater aquifers the flow takes place in a 'horizontal' plane; there is little flow in the 'vertical'. The words 'horizontal' and 'vertical' are used here to define the plane of the primary flow and the normal to that plane. Nothing in the equations or the analysis to follow requires that the plane of flow actually be horizontal.

If we assume that the flow is nearly horizontal, we are neglecting the pressure gradient in the vertical direction. Consider a situation such as that shown in Figure 10.7. Although vertical flow may take place, we assume that the vertical head gradients are concentrated in the semi-pervious aquitards. The vertical head drop across a layer (aquifer) or aquitard is, respectively,

$$\Delta\Phi_{\mathrm{v}} = \frac{q_{\mathrm{v}} b}{K_{\mathrm{z}}} \qquad \Delta\Phi_{\mathrm{lv}} = \frac{q_{\mathrm{lv}} b_{\mathrm{l}}}{K_{\mathrm{l}}} \tag{10.46}$$

in which $q_{\mathrm{v}}$ is the vertical flow per unit area (vertical velocity) through the aquifer, $q_{\mathrm{lv}}$ is the corresponding quantity in the aquitard, $b$ is the depth of aquifer, $b_{\mathrm{l}}$ is the depth of the aquitard and $K_{\mathrm{l}}$ is the permeability of the aquitard. Thus the nearly horizontal flow approximation requires that

$$\frac{\Delta\Phi_{\mathrm{v}}}{\Delta\Phi_{\mathrm{lv}}} = \frac{bK_{\mathrm{l}}}{b_{\mathrm{l}}K_{\mathrm{z}}} \ll 1 \tag{10.47}$$

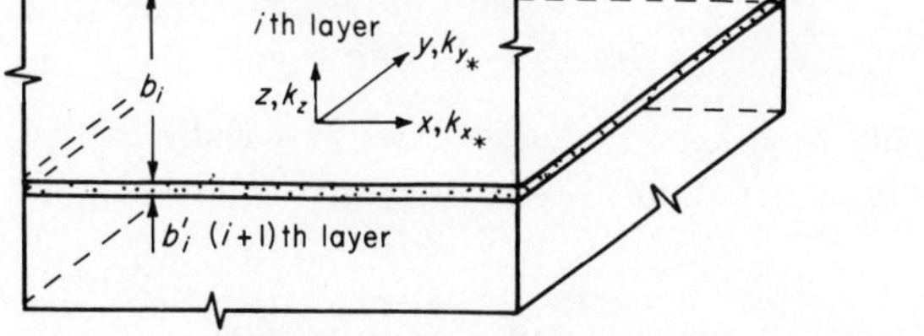


**Figure 10.7** A section of a groundwater basin showing three layers (aquifers) separated by semi-permeable aquitards.

Obviously, fulfillment of (10.47) is not a sufficient condition to make the approximation valid. If a comparatively strong horizontal flow does not exist, a computation based on the horizontal flow approximation will not compute the essential features. The ratio of vertical velocity to horizontal velocity should also be small:

$$\frac{q_v}{q_h} = \frac{K_z \Delta\Phi_v / b}{[K_x^2(\partial\Phi/\partial x)^2 + K_y^2(\partial\Phi/\partial y)^2]^{\frac{1}{2}}} \ll 1 \qquad (10.48)$$

The conditions (10.47) and (10.48) are usually (but not always) satisfied if the vertical scale of the problem (i.e. the aquifer thickness) is much less than the horizontal scale and if there are no steep gradients in an unconfined flow (i.e. the Dupuit approximation is valid; Bear 1972). Even if these conditions are violated in a local area (e.g. in the vicinity of a partially penetrating well or at the edge of an aquitard), the computation should give good results except in that local region. Another program (e.g. the three-dimensional or axisymmetric program) can be used in such a region. The boundary conditions for a program applied locally can be taken from the overall calculation.

Since the head difference across the aquifer is small, we can define an average head

$$\phi = \frac{1}{b}\int_0^b \Phi \, dz \qquad (10.49)$$

Equations which use the approximation of nearly horizontal flow will use this average head as the dependent variable.

## 10.3 Leaky aquifers

In this section we consider nearly horizontal flow in aquifers that are confined by leaky aquitards. A single aquifer is considered initially. Subsequently, we treat the case of several aquifers which are coupled through the leakage terms (Fig. 10.7).

### *10.3.1 Single layers*

The governing equation for steady flows in a leaky aquifer with constant transmissivity is given in (1.61). It is rewritten, using the definition of (10.49), as

$$\frac{\partial^2\phi}{\partial x^2} + \frac{\partial^2\phi}{\partial y^2} + \frac{K_1}{b_1 T}(\phi_1 - \phi) = 0 \qquad (10.50)$$

in which $K_1$ and $b_1$ are the permeability and the thickness of the aquitard, respectively, and $\phi_1$ is the head outside of the confining layer (see Fig. 1.4). When $\phi_1$ is constant and there are pumping or recharge wells in the aquifer, (10.50) can be modified as follows:

$$\frac{\partial^2\tilde{\phi}}{\partial x^2}+\frac{\partial^2\tilde{\phi}}{\partial y^2}-\frac{1}{\lambda^2}\tilde{\phi}=-\frac{1}{T}\sum_{k=1}^{N_W} Q_k\delta(x-x_k)\delta(y-y_k) \tag{10.51}$$

in which

$$\lambda^2=\frac{b_1T}{K_1} \qquad \tilde{\phi}=\phi-\phi_1 \tag{10.52}$$

and $Q_k$ is the volume rate of flow of the $k$th well.

Using the concept of superposition (see Sec. 3.6), we decompose the solution of (10.51) into two parts:

$$\tilde{\phi}=\tilde{\phi}_a+\tilde{\phi}_b \tag{10.53}$$

where $\tilde{\phi}_a$ is the non-singular portion of the solution and $\tilde{\phi}_b$ is given by

$$\tilde{\phi}_b=\frac{1}{T}\sum_{k=1}^{N_W}\frac{Q_k}{2\pi}K_0(r_k/\lambda) \tag{10.54}$$

In the preceding equation $N_w$ denotes the number of wells and $r_k$ represents the distance from the well. The objective is to solve for $\tilde{\phi}_a$ using the BIEM and then add $\tilde{\phi}_b$, in order to obtain the complete solution. Since the governing equation for $\tilde{\phi}_a$, which is the homogeneous part of (10.51), is exactly the same as (10.11) after replacing $1/\lambda^2$ by $s$, the BIEM described in Section 10.1.1 can be readily applied here.

In order to specify the problem fully, the boundary conditions must be formulated in terms of the complete solution. Assume that on the boundaries

$$\tilde{\phi}=\tilde{\phi}_c \qquad \text{and/or} \qquad \partial\tilde{\phi}/\partial n=\partial\tilde{\phi}_c/\partial n \tag{10.55}$$

Then in terms of $\tilde{\phi}_a$

$$\tilde{\phi}_a=\tilde{\phi}_c-\frac{1}{T}\sum_{k=1}^{N_W}\frac{Q_k}{2\pi}K_0(r_k/\lambda)$$

and/or (10.56)

$$\frac{\partial\tilde{\phi}_a}{\partial n}=\frac{\partial\tilde{\phi}_c}{\partial n}+\frac{1}{T}\sum_{k=1}^{N_W}\frac{Q_k}{2\pi\lambda}K_1(r_k/\lambda)\frac{\partial r_k}{\partial n}$$

in which $K_0$ and $K_1$ are modified Bessel functions of the second kind (see Sec. 10.1.1). In the case that the boundary is a dividing line between two zones with different transmissivities, compatibility equations similar to (4.5) and (4.7) should be used.

### 10.3.2 A numerical example

Consider an aquifer of two concentric zones, 1 and 2, of radii $R_1$ and $R_2$, respectively (Fig. 10.8), in which zone 1 is leaky and zone 2 is not. A well with strength $Q$ is located at the center of the aquifer. The transmissivity is different in each of the zones. As shown in Figure 10.8, the exterior

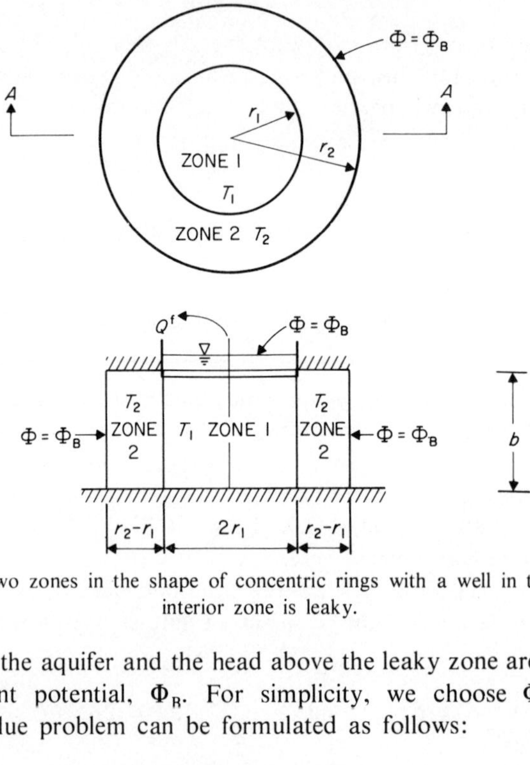


**Figure 10.8** Two zones in the shape of concentric rings with a well in the center. The interior zone is leaky.

boundary of the aquifer and the head above the leaky zone are maintained at a constant potential, $\Phi_B$. For simplicity, we choose $\Phi_B = 0$. The boundary value problem can be formulated as follows:

$$\frac{\partial^2\phi_1}{\partial r^2} + \frac{1}{r}\frac{\partial\phi_1}{\partial r} + \frac{1}{r^2}\frac{\partial^2\phi_1}{\partial\theta^2} - \frac{\phi_1}{\lambda^2} = \frac{Q}{T_1}\delta(r) \qquad 0 < r \leqslant R_1 \qquad (10.57)$$

$$\frac{\partial^2\phi_2}{\partial r^2} + \frac{1}{r}\frac{\partial\phi_2}{\partial r} + \frac{1}{r^2}\frac{\partial^2\phi_2}{\partial\theta^2} = 0 \qquad R_1 \leqslant r \leqslant R_2 \qquad (10.58)$$

where $Q$ is positive and $\lambda^2$ is defined as

$$\lambda^2 = b_1 T_1/K_1 \tag{10.59}$$

The boundary conditions are

$$\phi_1 = \phi_2 \qquad r = R_1 \tag{10.60}$$

$$T_1 \frac{\partial \phi_1}{\partial r} = T_2 \frac{\partial \phi_2}{\partial r} \qquad r = R_1 \tag{10.61}$$

$$\phi_2 = 0 \qquad r = R_2 \tag{10.62}$$

The physical problem is an axisymmetric one. The analytical solution to (10.57)–(10.62) can be readily obtained and is:

$$\phi_1 = \frac{Q}{2\pi T_1} \left\{ \left[ K_0(R_1/\lambda) + \left(\frac{R_1}{\lambda}\right)\left(\frac{T_1}{T_2}\right) \ln (R_1/R_2) K_1(R_1/\lambda) \right] I_0(r/\lambda)/J + K_0(r/\lambda) \right\} \qquad 0 < r \leqslant R_1 \tag{10.63}$$

$$\phi_2 = \frac{Q}{2\pi T_2} \left(\frac{R_1}{\lambda}\right) [I_0(R_1/\lambda) K_1(R_1/\lambda) + I_1(R_1/\lambda) K_0(R_1/\lambda)] \ln (r/R_2)/J \qquad R_1 \leqslant r \leqslant R_2 \tag{10.64}$$

where

$$J = \left(\frac{T_1}{T_2}\right)\left(\frac{R_1}{\lambda}\right) \ln \left(\frac{R_1}{R_2}\right) I_1(R_1/\lambda) - I_0(R_1/\lambda) \tag{10.65}$$

and $I_0$ and $I_1$ are modified Bessel functions of the first kind of order zero and one, respectively (Janke & Emde 1945).

The BIEM was used to solve the problem with the following numerical data:

$$\begin{aligned} Q &= 0.007885 \text{ m}^3/\text{s} & K_1/b_1 &= 0.1337 \text{ d}^{-1} \\ T_1 &= 10.38 \text{ m}^3/\text{m/d} & R_1 &= 263 \text{ m} \\ T_2 &= 20.77 \text{ m}^3/\text{m/d} & R_2 &= 528 \text{ m} \end{aligned} \tag{10.66}$$

The leakage factor $\lambda$ can be computed from (10.59) and is 528 m. In the numerical computations only a half circular region is used (Fig. 10.9a). The no flux boundary condition is employed along the line $\theta = 0, \pi$. The exterior boundary, interzonal boundary, and the solid boundary are divided into linear segments as shown in Figure 10.9b (8 elements on each semi-circle

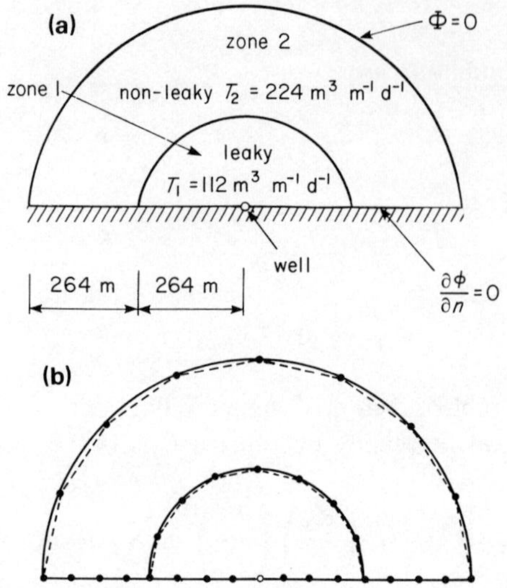


**Figure 10.9** Discretization for the solution of the case of Figure 10.8.

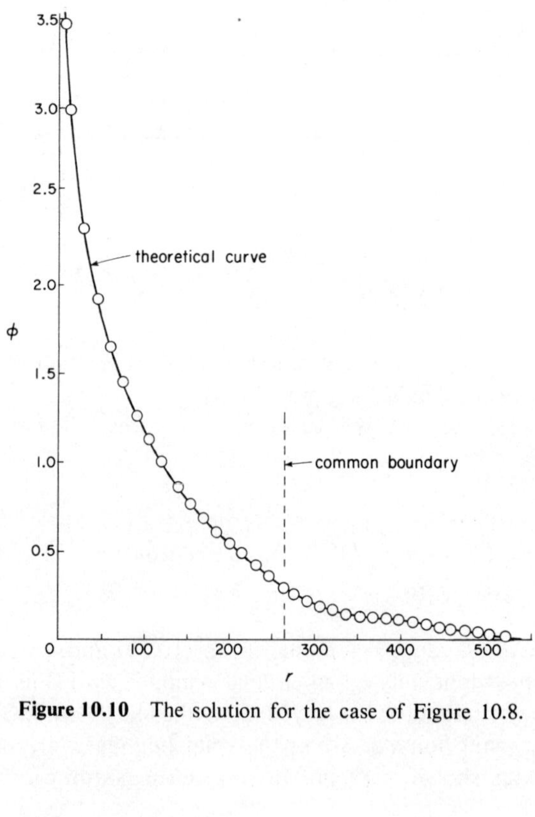


**Figure 10.10** The solution for the case of Figure 10.8.

and 19 elements on the solid boundary). The comparison between the numerical results and theoretical solutions are shown in Figure 10.10. Excellent agreement is observed.

### *10.3.3 Layered, leaky aquifers*

Consider a general situation such as that shown in Figure 10.7 in which an aquifer is separated from the aquifers above and below by leaky aquitards. The leakage through the aquitards depends on the pressure across the aquitards, and that pressure is a part of the solution in each aquifer. Thus (10.50) is modified to give

$$\frac{\partial^2\phi}{\partial x^2}+\frac{\partial^2\phi}{\partial y^2}-\frac{K_1^{i-1}}{Tb_1^{i-1}}\phi-\phi^{i-1})-\frac{K_1^{i}}{Tb_1^{i}}(\phi-\phi^{i+1})$$

$$=-\frac{1}{T}\sum_{k=1}^{N_W}Q_k\delta(x-x_k)\delta(y-y_k) \tag{10.67}$$

in which the superscript $i-1$ refers to the aquifer above and $i+1$ to the aquifer below. The superscripted terms form the coupling between aquifers. Collecting terms gives

$$\frac{\partial^2\phi}{\partial x^2}+\frac{\partial^2\phi}{\partial y^2}-\frac{1}{T}\left(\frac{K_1^{i-1}}{b_1^{i-1}}+\frac{K_1^{i}}{b_1^{i}}\right)\phi=-\frac{K_1^{i}}{Tb_1^{i}}\phi^{i+1}-\frac{K_1^{i-1}}{Tb_1^{i-1}}\phi^{i-1} \tag{10.68}$$

$$-\frac{1}{T}\sum_{k=1}^{N_W}Q_k\delta(x-x_k)\delta(y-y_k)$$

The latter equation is in the same form as (10.51) with two added terms on the right-hand side and

$$\lambda=\left[\frac{1}{T}\left(\frac{K_1^{i-1}}{b_1^{i-1}}+\frac{K_1^{i}}{b_1^{i}}\right)\right]^{-\frac{1}{2}} \tag{10.69}$$

Thus the same solution techniques apply to the differential operator, $\nabla^2-1/\lambda^2$. The dependent variables on the right-hand side, $\phi^{i+1}$ and $\phi^{i-1}$, must come from the solution of the aquifers above and below the $i$th aquifer. The layered solution proceeds in an iterative manner. For example, the solution $\phi^1$ is found for the topmost aquifer assuming no leakage to the aquifer below. Next the solution $\phi^2$ is found for the second aquifer using the solution $\phi^1$ to the topmost aquifer to define the leakage from above and assuming no leakage below. Each successive aquifer in the stack is treated similarly until the solution is found for the lowest aquifer. The solution $\phi^1$ is then redone for the topmost aquifer using the solution $\phi^2$ of the second aquifer to define the leakage. The solution $\phi^2$ of the second aquifer uses $\phi^1$ and $\phi^3$ to define the leakage. After reaching the bottom of the stack, the program again returns to the topmost layer. When there is no significant change in the solution of any layer the calculation stops.

As in (10.12) the free space Green function for (10.68) is

$$G = -K_0(r/\lambda) \tag{10.70}$$

If the values of $K_1^i/b_1^i$ and $K_1^{i-1}/b_1^{i-1}$ are small, the leakage is small and the vertical coupling is weak. In such a case the solution should converge quickly. As the leakage of the aquitard gets very small, $\lambda \to \infty$, the Bessel function begins to behave as $\ln r$ (see (10.13)), and the solution becomes the solution to Laplace's equation.

As indicated by (10.20) the solution must involve an area integration

$$I = \int_D \frac{K_0(r/\lambda)}{T}\left[-\frac{K_1^i}{b_1^i}\phi^{i+1} - \frac{K_1^{i-1}}{b_1^{i-1}}\phi^{i-1} - \sum_{k=1}^{N_W} Q_k\delta(x-x_k)\delta(y-y_k)\right] \mathrm{d}A \tag{10.71}$$

The technique for doing the area integrations is found in Section 10.4.2.

## 10.4 Numerical calculations

For the general groundwater basin such as that shown in Figure 10.1, all the equations and integrations are confined to the horizontal plane even though the solution region is divided into a three-dimensional grid of zones. In the vertical direction the zones are coupled by the leakage terms as in (10.68). In the present section we indicate how the solution can be carried out economically, even for quite large problems. The area integrations present a special problem, but they also can be done economically and without intervention from the user of the computer program.

### *10.4.1 Matrix substructuring*

At the zonal level we can write the equations which result from the BIEM integrals as

$$[E^k]\{u^k\} = \{r^k\} \tag{10.72}$$

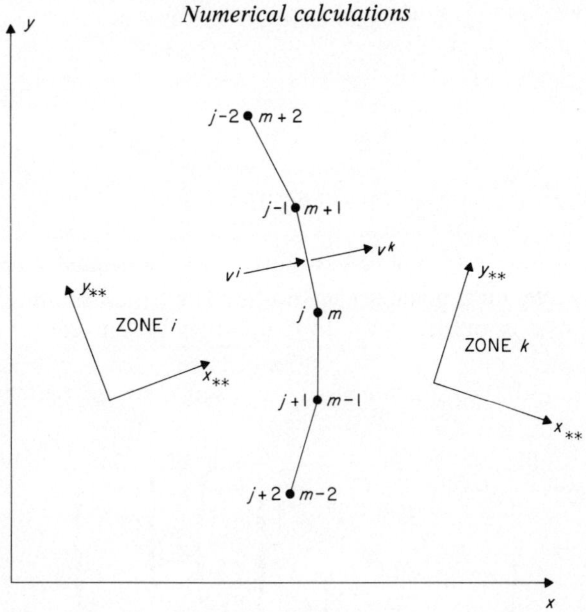


**Figure 10.11** The interzonal boundary between zone $i$ and zone $k$.

in which the superscript refers to the $k$th zone, $\{u\}$ is a vector of unknowns, $[E]$ is a combination of the assembled $[R]$ and $[L]$ matrices as defined by (2.34), and $\{r\}$ is the assembled right-hand side. The unknowns consist of either $\Phi$ or $\partial\Phi/\partial n$ at the nodes on the exterior boundaries (if the $k$th zone has an exterior boundary) and both $\Phi$ and $\partial\Phi/\partial n$ of the nodes on the boundaries which divide zones (interior or interzonal boundaries). A typical zone is pictured in Figure 10.11 in which there are $N_1^k$ exterior boundary nodes and $N_2^k$ interior boundary nodes.

It is convenient to partition (10.72) as

$$\begin{bmatrix} E_{11}^k & E_{12}^k \\ E_{21}^k & E_{22}^k \end{bmatrix} \begin{Bmatrix} u_1^k \\ u_2^k \end{Bmatrix} = \begin{Bmatrix} r_1^k \\ r_2^k \end{Bmatrix} \tag{10.73}$$

in which $u_1^k$ represents those $N_1^k$ unknowns on the exterior boundary and $u_2^k$ represents the $N_2^k$ unknowns on the common boundaries of all the adjacent zones. On the exterior boundaries there is one unknown and one equation per node; thus $E_{11}^k$ is a square matrix of size $N_1^k \times N_1^k$. Symbolically we can solve for $u_1$ as

$$\{u_1^k\} = [E_{11}^k]^{-1}(\{r_1^k\} - [E_{12}^k]\{u_2^k\}) \tag{10.74}$$

even though $u_2^k$ is still unknown. Substitution of (10.74) into (10.73) yields

$$[G^k]\{u_2^k\} = \{w^k\} \tag{10.75}$$

in which

$$|G^k| = |E_{22}^k| - |E_{21}^k||E_{11}^k|^{-1}|E_{12}|^k \tag{10.76}$$

$$\{w^k\} = \{r_2^k\} - [E_{21}^k][E_{11}^k]^{-1}\{r_1^k\} \tag{10.77}$$

The exterior boundary unknowns have been eliminated from (10.75), leaving only interzonal boundary unknowns. The coefficient matrix, $[G^k]$, is not square but is of size $N_2^k \times 2N_2^k$; thus the system cannot be solved directly.

Equations (10.75) are assembled into a single set for the entire basin:

$$\begin{bmatrix} |G^1| & & & \\ & |G^2| & & \\ & & \ddots & \\ & & & |G^n| \end{bmatrix} \begin{Bmatrix} \{u_2^1\} \\ \{u_2^2\} \\ \vdots \\ \{u_2^n\} \end{Bmatrix} = \begin{Bmatrix} \{w^1\} \\ \{w^2\} \\ \vdots \\ \{w^n\} \end{Bmatrix} \tag{10.78}$$

in which $n$ is the number of zones. In shorter notation equation (10.78) becomes

$$|G^*|\{u_2\} = \{w\} \tag{10.79}$$

Equation (10.79) contains four unknowns for each interzonal node. The $u_2$ consists of $\phi$ and $\partial\phi/\partial n$ in each of the bordering zones (i.e. $\phi^i$, $\partial\phi^i/\partial n$, $\phi^k$, and $\partial\phi^k/\partial n$).

The interzonal boundary conditions must be used to complete the system of equations. These conditions are expressed in (4.5) and (4.7) after replacing (1) with $i$ and (2) with $k$. The flow condition (4.7) can be written

$$a^i \frac{\partial\phi^i}{\partial n} = -a^k \frac{\partial\phi^k}{\partial n} \tag{10.80}$$

in which the values of $a^i$ and $a^k$ are the transmissivities if both zones are isotropic and, in general, are given by (4.49). Equation (4.5) together with (10.80) are combined so that the unknowns along the common boundary in zone $i$ are eliminated in favour of the unknowns on the same nodes in zone $k$,

$$\{u_2^i\} = [\chi]\{u_2^k\} \tag{10.81}$$

Using (10.81) in (10.80) produces

$$\begin{bmatrix} |G^{1i} \quad G^{1k}| & & & \\ & |G^{2i} \quad G^{2k}| & & \\ & & \ddots & \\ & & & |G^{ni} \quad G^{nk}| \end{bmatrix} \begin{Bmatrix} \{u_2^{1i}\} \\ |\chi^1|\{u_2^{1k}\} \\ \{u_2^{2i}\} \\ |\chi^2|\{u_2^{2k}\} \\ \vdots \\ \{u_2^{ni}\} \\ |\chi^n|\{u_2^{nk}\} \end{Bmatrix}$$

$$= \begin{Bmatrix} \{w^1\} \\ \{w^2\} \\ \vdots \\ \{w^n\} \end{Bmatrix} \tag{10.82}$$

in which the superscripts $i$ and $k$ have the same meaning as in (10.81), i.e. some of the unknowns in zone $i$ have been eliminated in favor of the corresponding unknowns in zone $k$ through the compatibility conditions. Thus (10.82) can be written as

$$\begin{bmatrix} |G^{1i}| + |G^{1k}||\chi^1| & & & \\ & |G^{2i}| + |G^{2k}||\chi^2| & & \\ & & \ddots & \\ & & & |G^{ni}| + |G^{nk}||\chi^n| \end{bmatrix} \begin{Bmatrix} \{u_2^{1i}\} \\ \{u_2^{2i}\} \\ \vdots \\ \{u_2^{ni}\} \end{Bmatrix} = \begin{Bmatrix} \{w^1\} \\ \{w^2\} \\ \vdots \\ \{w^n\} \end{Bmatrix} \tag{10.83}$$

Equations (10.83) form a solvable system for the unknowns, $u_2$, at the interzonal nodes. After their solution (10.74) is used to find the exterior boundary unknowns.

If there are $N_1$ exterior boundary nodes in each zone and $N_2$ interzonal boundary nodes between each zone, the entire system begins with

$$N = \sum_{k=1}^{n} (N_1^k + 4N_2^k) \tag{10.84}$$

unknowns which consist of $\phi$ or $\partial\phi/\partial n$ on the $\Sigma N_1^k$ exterior nodes plus both $\phi$ and $\partial\phi/\partial n$ for both adjacent zones on the $\Sigma N_2^k$ interior nodes. The number of simultaneous equations (10.83) is

$$S = 2 \sum_{k=1}^{n} N_2^k \tag{10.85}$$

In forming the equations it is necessary to invert $n_e$ of the $E_{11}$ matrices (equation (10.73)), where $n_e$ is the number of zones with exterior boundaries. After solution for the interior unknowns, $u_2$, the exterior unknowns are found from a matrix multiplication (equation (10.74)).

The coefficient matrix of (10.83) is block diagonal so that a banded equation solver can be used. The band width can be minimized by judicious numbering of the zones in much the same way that finite element band widths are minimized (Gallagher 1975). If a computer of limited storage capacity is used, the coefficient matrix of (10.83) is the largest matrix that has to be stored (assuming that the number of interior nodes is at least one-half the number of exterior nodes in any zone). Thus the calculation can be easily staged with most of the matrices stored externally and brought in as needed.

Internal solutions are found in the normal manner using zonal calculations only. That is, once the complete boundary data (both $\phi$ and $\partial\phi/\partial n$ at each node) are known, the interior solutions require only an integration around the zonal boundaries.

### 10.4.2 Area integrations

The fact that the governing equations are not homogeneous means that the solution cannot be written entirely in terms of a boundary integration. In (10.20) the initial conditions must be integrated over the domain. In (10.68) the terms which couple aquifers form the area integral of (10.71). The fact that the BIEM ceases to operate entirely on the boundary spoils some of its advantages. The goal of the area integration is that it should be done efficiently and without the explicit intervention of the analyst. Thus the BIEM remains efficient and the analyst needs to supply only boundary data as though the program were operating only on the boundaries. Of course, in some cases, such as with initial condition, the interior data must be given in any method in order that the problem is well posed mathematically.

The method we have used to perform the area integration is shown in Figure 10.12. A grid is generated by selecting a central point ($P_M$ in Fig. 10.12a) which is at the midpoint of the longest chord between boundary nodes. Lines are then extended from this point to each of the boundary nodes (Fig. 10.12b). These lines are divided into segments which are joined to form a spider web (Fig. 10.12c). At this point the interior of the zone may be sufficiently discretized to perform an accurate integration. If further discretization is necessary, the midpoints of the connecting lines are joined (Fig. 10.12d), thus doubling the number of elements. Finally, the rectangular elements are divided into triangles (Fig. 10.12e).

Quadrature formulas for areas are well known (Zienkiewicz 1971, Cowper 1973). In the present case there is an obvious tradeoff between the number of elements and the number of quadrature points within an element. Both one-point quadrature (point located at the centroid of the element) and

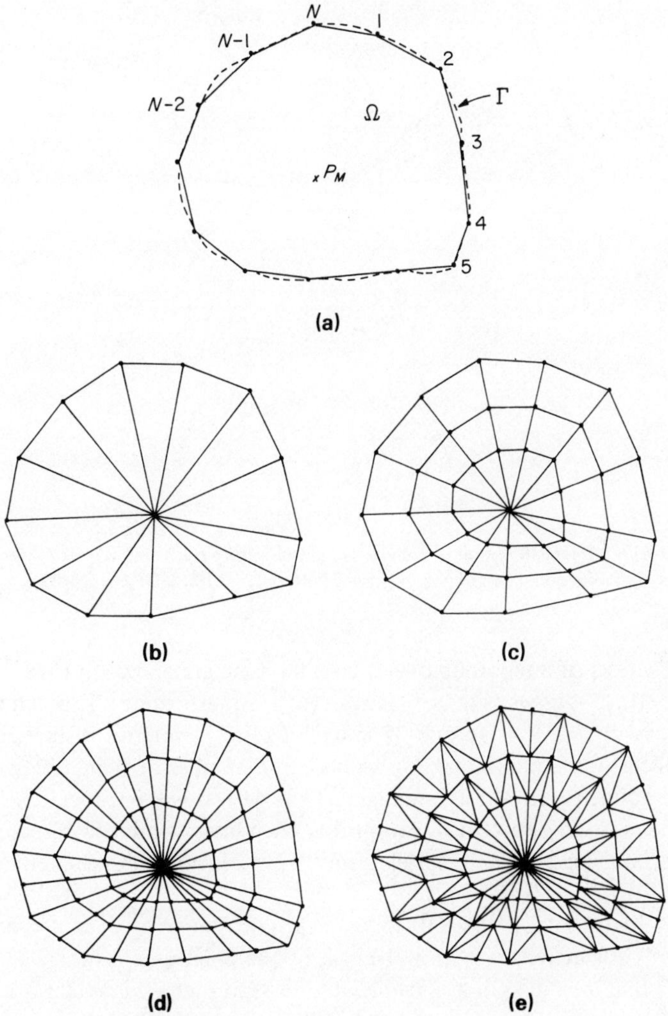


**Figure 10.12** Steps in the discretization for the area integration.

four-point quadrature have been used (Fig. 10.13). The symmetric Gaussian formulas of Cowper (1973) are employed in the four-point quadrature (Fig. 10.13b). The quadrature points in the latter case are located according to

$$\begin{bmatrix} x_1 & y_1 \\ x_2 & y_2 \\ x_3 & y_3 \\ x_4 & y_4 \end{bmatrix} = \begin{bmatrix} \frac{1}{3} & \frac{1}{3} & \frac{1}{3} \\ 0.6 & 0.2 & 0.2 \\ 0.2 & 0.6 & 0.2 \\ 0.2 & 0.2 & 0.6 \end{bmatrix} \begin{bmatrix} x_A & y_A \\ x_B & y_B \\ x_C & y_C \end{bmatrix} \tag{10.86}$$

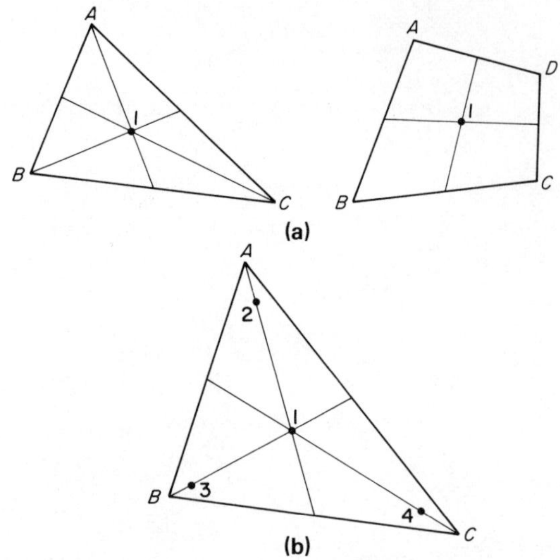


**Figure 10.13** Quadrature points: (a) one-point quadrature rule, weight = 1; Cowper's four-point quadrature rule, $W_1 = -27/48$, $W_2 = W_3 = W_4 = 25/48$.

A few tests of integration over a circular zone are shown in Figure 10.14; Figure 10.15 shows similar tests over a square zone. The number of boundary nodes, $N$, is 16 and 32 in each case. The interior discretization is controlled by the parameter $M$, which is the number of points between the central point, $P_M$, and the boundary. Thus $M = 0$ will produce a mesh like that shown in Figure 10.12b except that each interior angle is bisected once more producing $2N$ triangular elements; $M = 2$ leads to the discretization of Figure 10.12e.

Both Figures 10.14 and 10.15 indicate that the error does not decrease uniformly for increases in $M$ in the case of one-point quadrature. The reason may be that a single point is inadequate to represent a triangular area if the triangle departs greatly from isosceles. The one point rule, however, requires fewer functional evaluations and thus may be more efficient. The error plots indicate that the quadrature can be done as accurately as needed.

The mesh generation scheme assumes that the boundary is sufficiently convex in order that $P_M$ and the lines connecting $P_M$ to the nodes do not fall outside of a zone. The analyst can create additional zones if a problem exists. Alternatively, additional logic in the mesh generation scheme could solve the problem of oddly shaped zones. Further, there is no doubt that the area discretization could be optimized for maximum efficiency; however, the chief criterion is that the operator should not have to interfere in the automatic process.

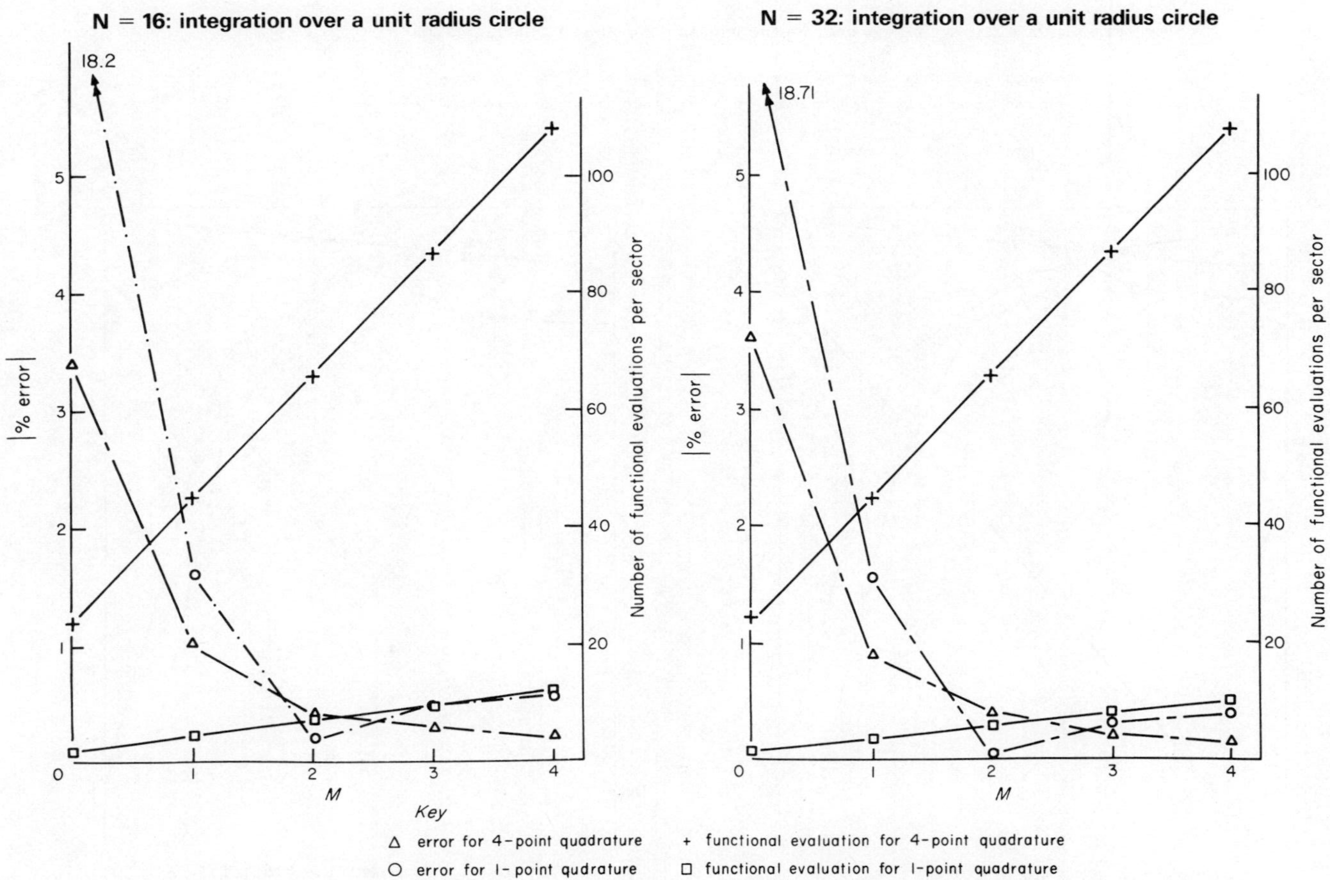


**Figure 10.14** Error and number of functional evaluations for area integration over a circular region.

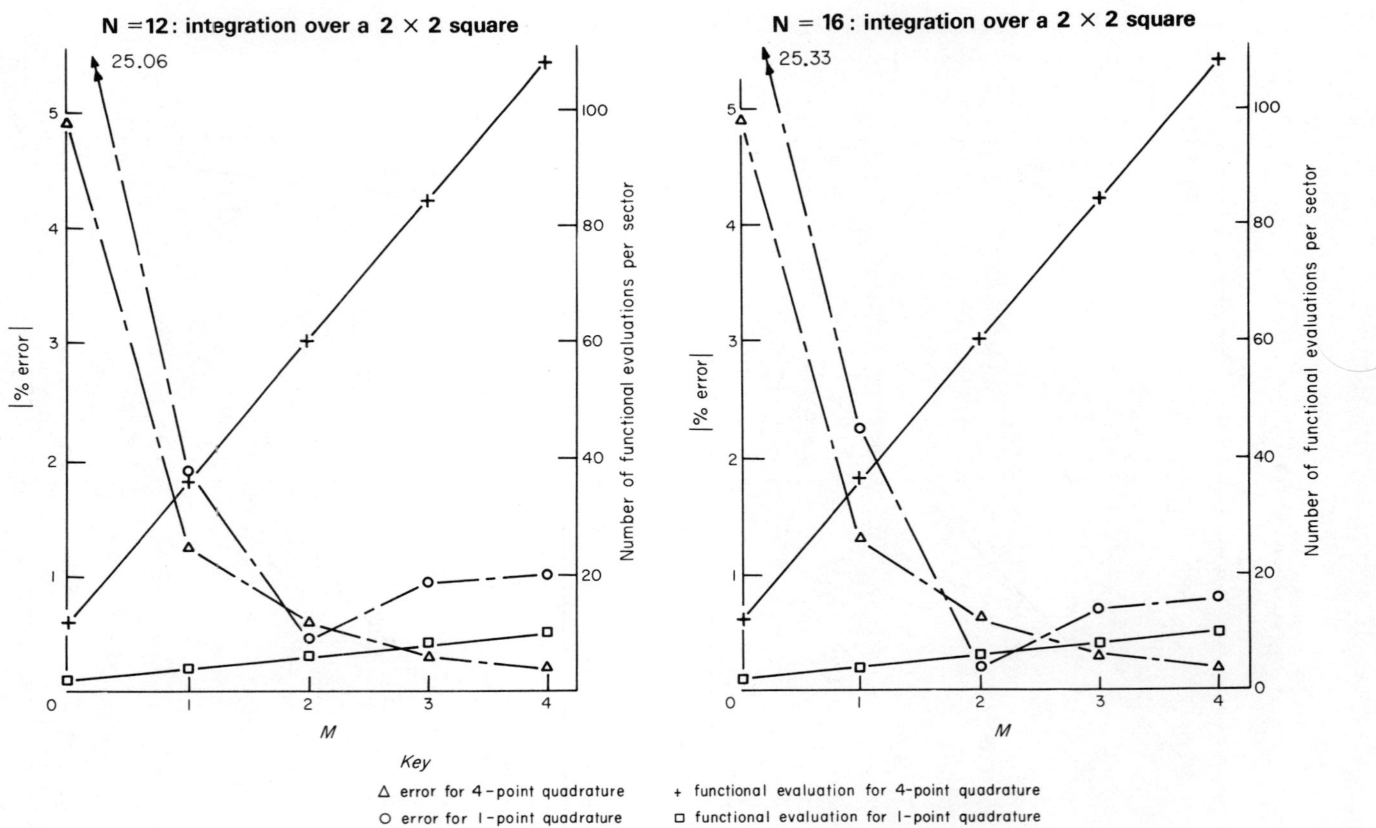


**Figure 10.15** Error and number of functional evaluations for area integration over a square region.

## 10.5 Unconfined aquifers

The basic equation for two-dimensional flow in unconfined aquifers under the Dupuit assumption is developed in Section 1.7. In the form of (10.68) it is

$$\frac{\partial^2\phi^2}{\partial x^2}+\frac{\partial^2\phi^2}{\partial y^2}-\frac{K_1^i}{Kb_1^i}\phi$$

$$=-\frac{K_1^i}{Kb_1^i}\phi^{i+1}+\frac{W}{K}-\frac{1}{K}\sum_{k=1}^{N_w} Q_k\delta(x-x_k)\delta(y-y_k) \qquad (10.87)$$

The third and fourth terms represent the leakage to the aquifer below. The leakage to the aquifer above is replaced by the recharge, $W$. The differential operator in (10.87) is non-linear and must be linearized in order to apply the BIEM. Three possible forms have been studied; they are

$$L(\phi)=\frac{\partial^2\phi}{\partial x^2}+\frac{\partial^2\phi}{\partial y^2}-\phi/\lambda^2\bar{\phi}=R/\bar{\phi} \qquad (10.88)$$

$$L(\phi^2)=\frac{\partial^2\phi^2}{\partial x^2}+\frac{\partial^2\phi^2}{\partial y^2}-\phi^2/\lambda^2\bar{\phi}=R \qquad (10.89)$$

$$L(\phi^2)=\frac{\partial^2\phi^2}{\partial x^2}+\frac{\partial^2\phi^2}{\partial y^2}=F(\phi^2) \qquad (10.90)$$

in which $R$ represents the right-hand side of (10.87) and $\lambda=(K_1^i/b_1^iK)^{-\frac{1}{2}}$. In (10.88) and (10.89) $\bar{\phi}$ is a zonal average and $\phi^2$ is approximated in some terms by $\phi\bar{\phi}$; thus (10.88) is linear in $\phi$ whereas equation (10.89) is linear in $\phi^2$. However, (10.88) has been rejected as it does not reduce to (1.56) in the case that $R\equiv 0$ and there is no leakage.

### *10.5.1 The Poisson equation method*

No approximations have been made in (10.90) if $F(\phi^2)$ includes the third and all succeeding terms of (10.87). The differential operator is linear in $\phi^2$; thus, following (10.18),

$$\int_\Gamma\left(\frac{\phi^2}{r}\frac{\partial r}{\partial n}-\ln r\frac{\partial\phi^2}{\partial n}\right)\mathrm{d}s=\int_D F\ln r\,\mathrm{d}A \qquad (10.91)$$

The solution procedure is to first guess $\phi$ everywhere in the solution domain. Then $F(\phi^2)$ can be calculated and the area integration done. The BIEM can then be applied in the usual way with the results of the area integration added to the right-hand side of the simultaneous equations for $\phi^2$ and $\partial\phi^2/\partial n$. Next the interior solutions are obtained at enough points for the area integration to be done accurately. The sequence of calculations is repeated until there is no change from one iteration to the next.

Even though the solution is iterative, the boundary integrations do not have to be repeated. The BIEM coefficients are dependent only on the geometry; thus, the coefficient matrix can be formed initially, inverted, then used for each subsequent iteration. The program can be arranged so that the analyst does not have to intervene to find an initial guess of $\phi$ in $D$. The guess can be the average of the boundary values of $\phi$, or simply an arbitrary constant.

Various strategies can be employed to speed the iteration process. For example, the area integration can be done crudely at the outset, then more accurately as the result is refined. However, there is some advantage in using the same interior points in each iteration. The interior solution is

$$\phi^2 = \frac{1}{2\pi}\int_\Gamma \left(\frac{\phi^2}{r}\frac{\partial r}{\partial n} - \ln r\frac{\partial \phi^2}{\partial n}\right) \mathrm{d}s - \frac{1}{2\pi}\int_D F \ln r \, \mathrm{d}A \quad (10.92)$$

For a given interior point (10.92) becomes

$$\phi^2(P) = \frac{1}{2\pi}\sum_{i=1}^{n}\left(a_i\phi_i^2 - b_i\frac{\partial\phi_i^2}{\partial n}\right) - \frac{c}{2\pi} \quad (10.93)$$

in which $\phi^2$ and $\partial\phi_i^2/\partial n$ represent nodal values on the boundary, $a_i$ and $b_i$ represent the parts of the integrals of $(1/r)\partial r/\partial n$ and $\ln r$, respectively, which contribute to node $i$, and $c$ is the value of the area integral. The $a_i$ and $b_i$ are dependent only on the geometry of the boundary and the location of point $P$. Thus they need to be found only once. In the entire process we need to do the following calculations:

(a) The boundary integral is done $m + 1$ times where $m$ is the number of interior points to be used in the area integration.
(b) One inversion (or decomposition) of the BIEM coefficient matrix is necessary.
(c) One area integration is performed for each iteration.
(d) The arithmetic implied by (10.93) solves for $\phi^2$ at each interior point in each iteration.

The solution can be obtained to any desired accuracy by using enough boundary and interior nodes.

### *10.5.2 The averaging method*

In using (10.89) we assume that $R$ is known but we need an estimate for $\bar{\phi}$. In order that the equation be valid, $\bar{\phi}$ must be a constant, but also it should be representative of the solution in the entire zone. We have generally used

$$\bar{\phi} = \frac{1}{n} \sum_{i=1}^{n} \phi_i \tag{10.94}$$

That is, $\bar{\phi}$ is the average of the boundary nodes. If $\lambda$ is small (large leakage, in which case the Dupuit approximation is in doubt), satisfactory accuracy may not be achieved by this method.

As in the previous case the solution is iterative. An approximation is made for $\bar{\phi}$. Then, analogous to Section 10.1.1, the Green function is again (see (10.12) and (10.70))

$$G = -K_0\left(\frac{r}{\lambda(\bar{\phi})^{\frac{1}{2}}}\right) \tag{10.95}$$

and following (10.20)

$$\int_\Gamma \left[\frac{\phi^2}{\lambda(\bar{\phi})^{\frac{1}{2}}} K_1\left(\frac{r}{\lambda(\bar{\phi})^{\frac{1}{2}}}\right)\frac{\partial r}{\partial n} + K_0\left(\frac{r}{\lambda(\bar{\phi})^{\frac{1}{2}}}\right)\frac{\partial \phi^2}{\partial n}\right] \mathrm{d}s - \int_D RK_0\left(\frac{r}{\lambda(\bar{\phi})^{\frac{1}{2}}}\right) \mathrm{d}A = 0 \tag{10.96}$$

with interior points evaluated from (see (10.21))

$$2\pi\phi^2(P) = \int_\Gamma \left[\frac{\phi^2}{\lambda(\bar{\phi})^{\frac{1}{2}}} K_1\left(\frac{r}{\lambda(\bar{\phi})^{\frac{1}{2}}}\right)\frac{\partial r}{\partial n} + K_0\left(\frac{r}{\lambda(\bar{\phi})^{\frac{1}{2}}}\right)\frac{\partial \phi^2}{\partial n}\right] \mathrm{d}s - \int_D RK_0\left(\frac{r}{\lambda(\bar{\phi})^{\frac{1}{2}}}\right) \mathrm{d}A \tag{10.97}$$

The evaluation of $K_0$ and $K_1$ can be done using a polynomial approximation (Abramowitz & Stegun 1974, p. 379) or a library program (e.g. an IMSL routine).

### *10.5.3 Comparison of methods*

Both of the above methods require iteration. The Poisson equation method always requires area integration whereas the averaging method does not in

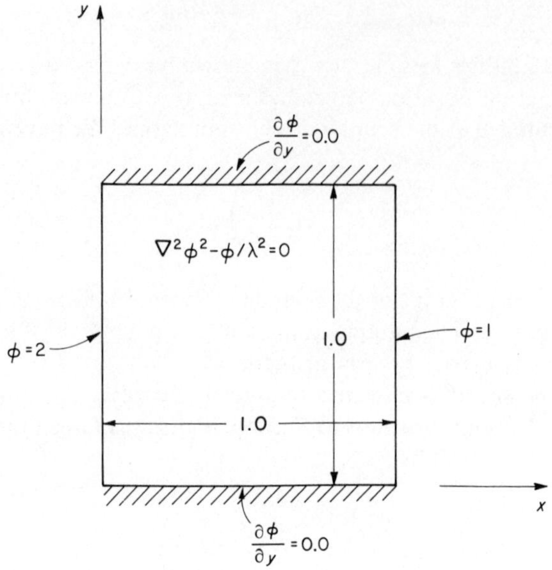


**Figure 10.16** Configuration of the example problem.

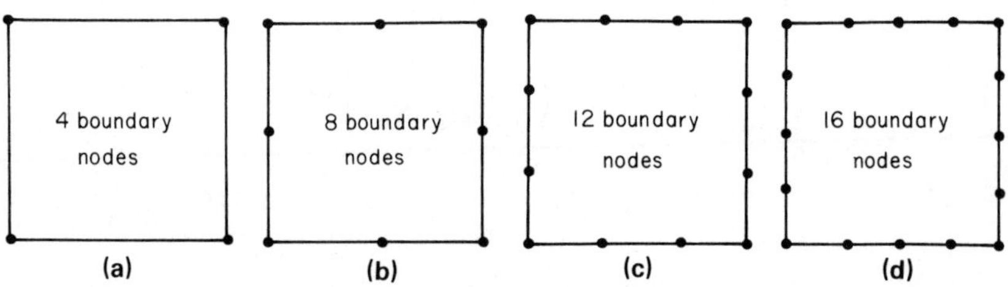


**Figure 10.17** Boundary discretization for the case of Figure 10.16.

the case that $R \equiv 0$. The Poisson equation method generally requires more computer time due to the interior evaluations that are necessary for the area integration. On the other hand it can solve the equation to arbitrary accuracy whereas the accuracy is limited with the averaging method.

As an example, consider the problem shown in Figure 10.16. Four different boundary discretizations were used as shown in Figure 10.17. Because $R \equiv 0$ the area integral of (10.96) does not need to be done.

The accuracy and time plots for both methods are shown in Figure 10.18. For this particular example, the averaging method gives more accurate results than the Poisson equation method if a crude area integration is used for the latter. An accurate area integration leads to more accurate results for the Poisson equation method. There is no ambiguity concerning computing time; the averaging method is clearly faster. That result might change, however, in the case where an area integration is necessary for the averaging method.

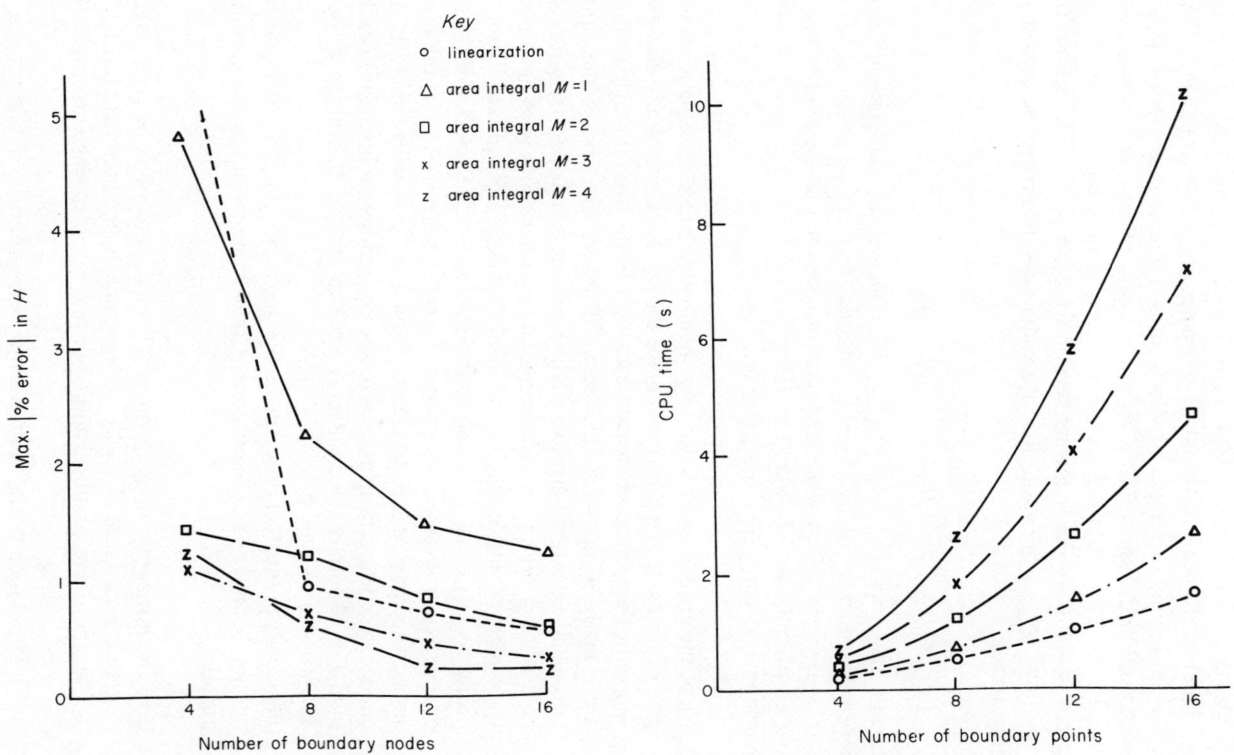


**Figure 10.18** Accuracy and computer time plots for the case of Figure 10.16.

## 10.6 Calculations in complex basins

The two-dimensional calculation involving the nearly horizontal flow approximation has been coded into a computer program named GM5. This program is designed to solve problems in combinations of zoned, leaky, layered, confined, unconfined, isotropic and anisotropic groundwater basins. Figure 10.1 shows many of the features of a basin in which this program is useful. In the following section we describe the calculations for steady flow. A modification for unsteady flow is briefly described in Section 10.6.2.

### *10.6.1 Steady flow*

Although the program performs a two-dimensional calculation, it is designed to retain as many three-dimensional features as possible. Some attempt was made to design the computer program so that the analyst needs to supply a minimum of data. The following illustrates the steps of data preparation and entry into the program.

(1) First, the groundwater basin must be divided into a three-dimensional grid of zones. In the horizontal plane the zones can be of any size and shape. In the vertical plane the basin must be divided into layers which are roughly of constant thickness. All layers must be of the same horizontal extent, but artifices such as zero or infinite permeability can be used. The primary consideration in choosing zones is to have constant permeability in each zone in all layers. If an aquitard changes properties or ends, a zonal division should be selected along that change. Thus the physical properties of the basin determine the divisions. The only reason to select a zonal division where there is no change in physical properties is to specify an internal boundary condition. For example, the head may need to be specified along a river which penetrates to the water table.

(2) The analyst must choose the nodal locations on the exterior and interior boundaries of each zone. Considerations in choosing nodes include:

   (a) The horizontal plane geometry in each zone must be reasonably well reproduced. Usually straight line boundaries between node locations are assumed although higher order elements could be used.

   (b) Interpolation of the head and flux between nodes should provide a reasonable approximation no matter which type of element is used. This criterion implies that the nodes should be denser in areas in which rapid change in boundary values is specified or where rapid change in the solution is expected, if such changes are not those assumed by the element type (usually linear).

No consideration need be given for the accuracy of the approximation to the governing equation, as would be the case in finite element or finite difference methods. The BIEM solves the governing equation almost exactly (it does approximate the right-hand side). Thus in selecting zones and nodes the analyst needs to be concerned only with the physical properties of the solution; the mathematical properties will take care of themselves.

In supplying nodes, limited data generation is possible. If there is a long straight boundary in which the analyst wants equally spaced nodes, he needs only to supply the number of such nodes and the program will automatically compute their location.

(3) The exterior boundary data consist of specifying either the head or the flux at each exterior node. Again, there is the possibility of automatic data generation if a number of nodes has the same value.

(4) Various other properties of the basin are given in each zone. These consist of:

(a) information on the zonal connectivity is given by specifying the number of layers and the zone numbers in the vertical;
(b) the program is told if the aquifer is confined or unconfined;
(c) well locations and strengths are given;
(d) the characteristics of the aquitards are specified;
(e) recharge values are given.

(5) Finally, the analyst can specify points, off the zonal boundaries, where the interior solution is needed. If a line of such points is needed, the program can generate the intermediate points automatically.

The program will return the head and flux at each interior and exterior node and the head at the specified points off the boundaries. If additional analysis is necessary, this or another BIEM program can be used at the zonal level, taking the previously computed results as boundary conditions.

### *10.6.2 Unsteady flow*

The governing equations for unsteady flow in confined and unconfined aquifers are (Bear 1972):

$$T_{x_*}\frac{\partial^2\phi}{\partial x_*^2} + T_{y_*}\frac{\partial^2\phi}{\partial y_*^2} = S\frac{\partial\phi}{\partial t} \qquad (10.98)$$

$$K_{x_*}\frac{\partial^2\phi^2}{\partial x_*^2}+K_{y_*}\frac{\partial^2\phi^2}{\partial y_*^2}=2n_e\frac{\partial\phi}{\partial t} \quad (10.99)$$

in which $S$ is the storativity and $n_e$ is the effective porosity. The time derivative can be approximated as

$$\frac{\partial\phi}{\partial t}\approx\frac{\phi_{k+1}-\phi_k}{\Delta t} \quad (10.100)$$

in which the subscript represents the time level. Thus finite difference analogs of (10.98) and (10.99) are

$$T_{x_*}\frac{\partial^2\phi_{k+1}}{\partial x_*^2}+T_{y_*}\frac{\partial^2\phi_{k+1}}{\partial y_*^2}-\frac{S}{\Delta t}\phi_{k+1}=-\frac{S}{\Delta t}\phi_k \quad (10.101)$$

$$K_{x_*}\frac{\partial^2\phi_{k+1}^2}{\partial x_*^2}+K_{y_*}\frac{\partial^2\phi_{k+1}^2}{\partial y_*^2}-\frac{2n_e}{\Delta t}\phi_{k+1}=-\frac{2n_e}{\Delta t}\phi_k \quad (10.102)$$

Equation (10.101) is in the same form as (10.10) and can be solved by the same methods. (The leakage and recharge terms can be added with no additional complication.) Similarly, (10.102) is analogous to (10.87). The program used to solve (10.10) and (10.87) can be used with slight modification in a time stepping process for the unsteady flow problem.

### *10.6.3 A two-layer example*

As an example of a multilayer calculation the problem shown in Figure 10.19 is solved. The aquifer is divided into four zones, two in the vertical direction and two in the horizontal direction. The boundary conditions have been selected so that the solution is uniform in the $y$ direction; thus an analytical solution can be found (Lafe 1981).

The nodal arrangement for this example is shown in Figure 10.20. The required area integrations are done with $M = 2$, giving $6 \times 16 = 96$ area elements per zone, and one-point quadrature is used. Because of the vertical coupling the problem is iterative. First, the potential in the lower layer is assumed to be zero and the solution is performed for the upper layer. No area integration is necessary for this step since the zero assumption removes the nonhomogeneous terms from (10.68). Using the results for the upper layer, the solution is found for the lower layer. The program then returns to the upper layer and finds a new result. After five such iterations the exact values were found to within about three significant figures. The numerical results are shown in Table 10.1 together with the parameters used for the calculation.

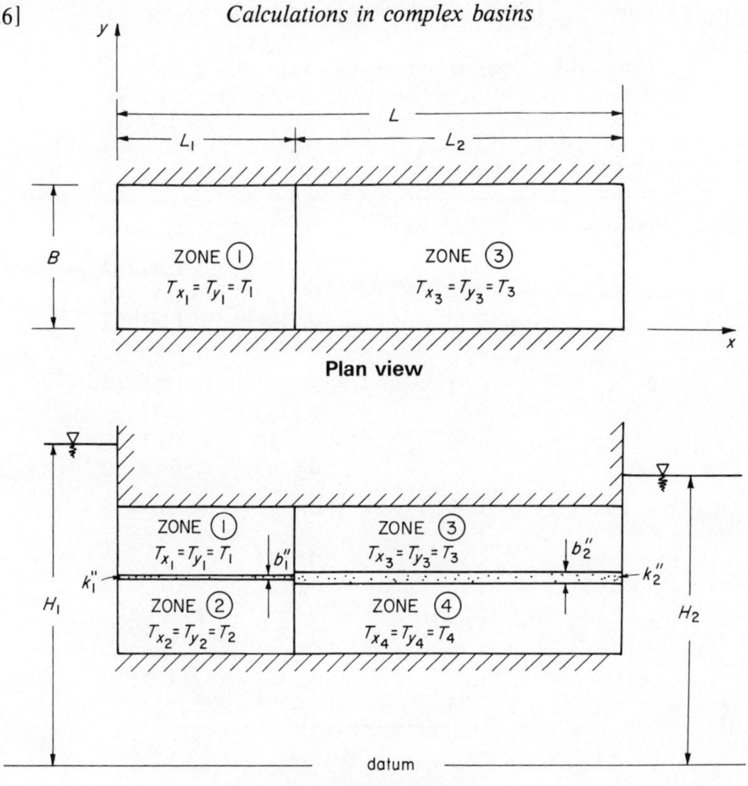


**Figure 10.19** An example of a two layer calculation.

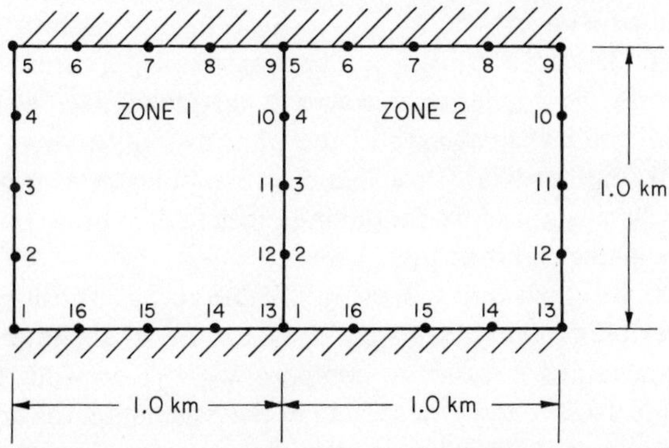


**Figure 10.20** The boundary discretization for the case of Figure 10.19.

## *10.6.4 The Yun-Lin aquifer*

A plan view of the Yun-Lin aquifer on the island of Taiwan is shown in Figure 10.21. The aquifer consists of a single layer. Both the selection of the zonal shapes and the exterior boundary were influenced by a previous finite

**Table 10.1** Comparison of the BIEM and exact solutions for the four-zone, two-layer aquifer after five iterations. Parameters $H_1 - H_2 = 1.0$ m; $L_1 = L_2 = 1000$ m; $T_1 = 4T_2 = 1440$ m²/d; $T_3 = 5T_4 = 14\,400$ m²/d; $K_1^1/b_1^1 = 1.44 \times 10^{-4}$ d⁻¹; $K_1^2/b_e^2 = 1.44 \times 10^{-5}$ d⁻¹.

| *Zone no.* | $\xi = x/1000$ | *BIEM* $\phi$ | *Exact* $\phi$ |
|---|---|---|---|
| 1 | 0.25 | 0.7729 | 0.77282 |
| | 0.50 | 0.5457 | 0.54560 |
| | 0.75 | 0.3148 | 0.31833 |
| | 1.0 | 0.0910 | 0.09097 |
| 2 | 0.25 | 0.7777 | 0.77741 |
| | 0.50 | 0.5553 | 0.55493 |
| | 0.75 | 0.3330 | 0.33268 |
| | 1.0 | 0.1109 | 0.11080 |
| 3 | 1.0 | 0.0910 | 0.09097 |
| | 1.25 | 0.0683 | 0.06823 |
| | 1.50 | 0.0455 | 0.04549 |
| | 1.75 | 0.0228 | 0.02274 |
| 4 | 1.0 | 0.1109 | 0.11080 |
| | 1.25 | 0.0832 | 0.08309 |
| | 1.50 | 0.0555 | 0.05539 |
| | 1.75 | 0.0277 | 0.02770 |

element solution. Straight line interior and exterior boundaries are not necessary, or even useful, in the BIEM program; however, a major objective was the comparison of solutions.

Several of the zones are leaky. The transmissivity varies between each zone. Although the aquifer is anisotropic, the global axes were selected to coincide with the principal axes of the transmissivity tensor. As shown in Figure 10.21, the heads are specified on some boundaries (right and left in the figure) whereas a no flux condition is specified on other boundaries (top and bottom in the figure).

The nodal locations are shown in Figure 10.22. Those points not on interior or exterior boundaries were the locations of the interior solutions, which were obtained in order to compare with the previous finite element results. Figure 10.23 shows the results of the calculation for both the BIEM and finite elements. Generally good agreement is observed.

There is no direct comparison of costs for the BIEM program and the finite element program since they were run on different computers. The BIEM program required 2.80 s of CPU time (47 cents worth at Cornell University) on the IBM 370/168 whereas the finite element solution required 84.69 s of CPU time on a Univac 1100.

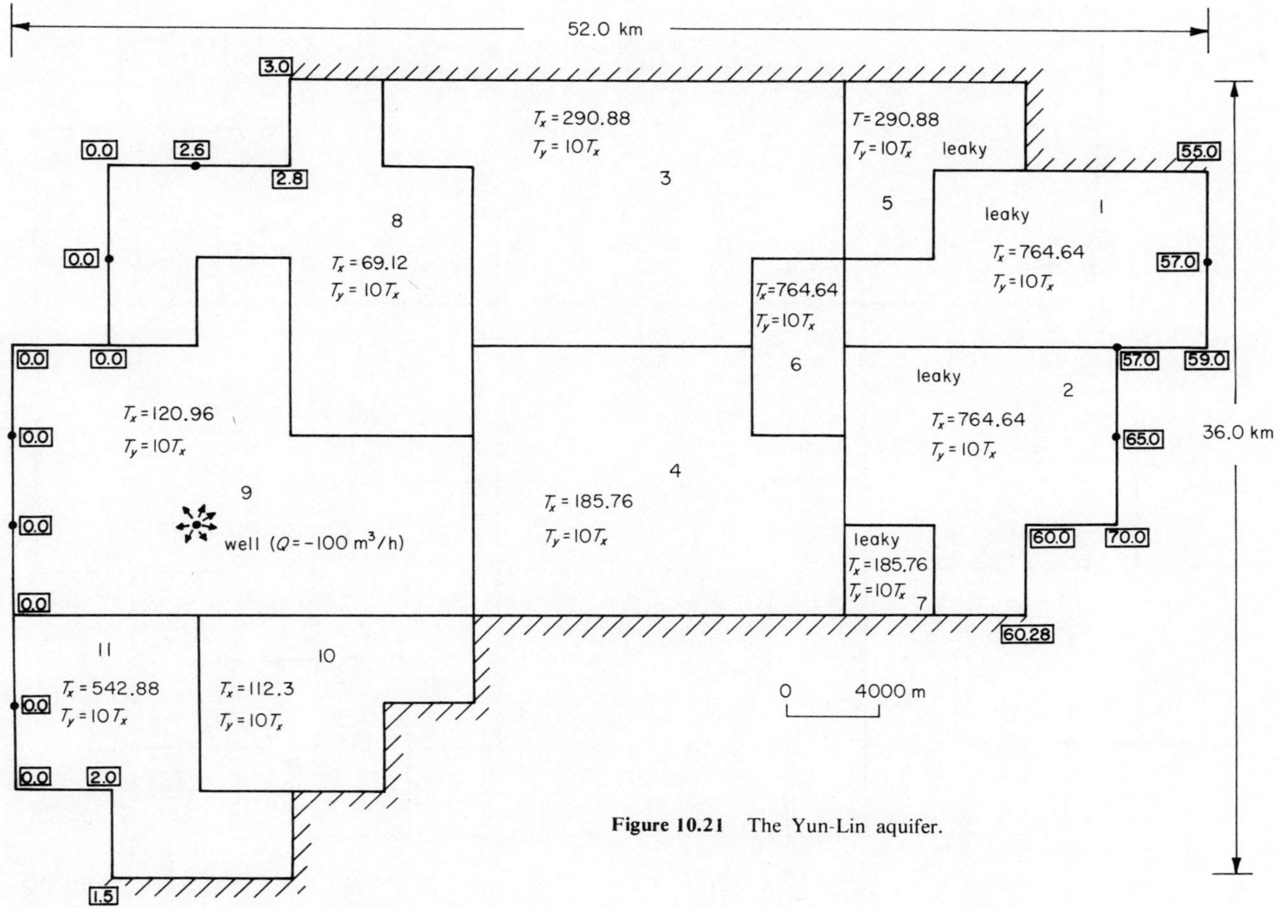


**Figure 10.21** The Yun-Lin aquifer.

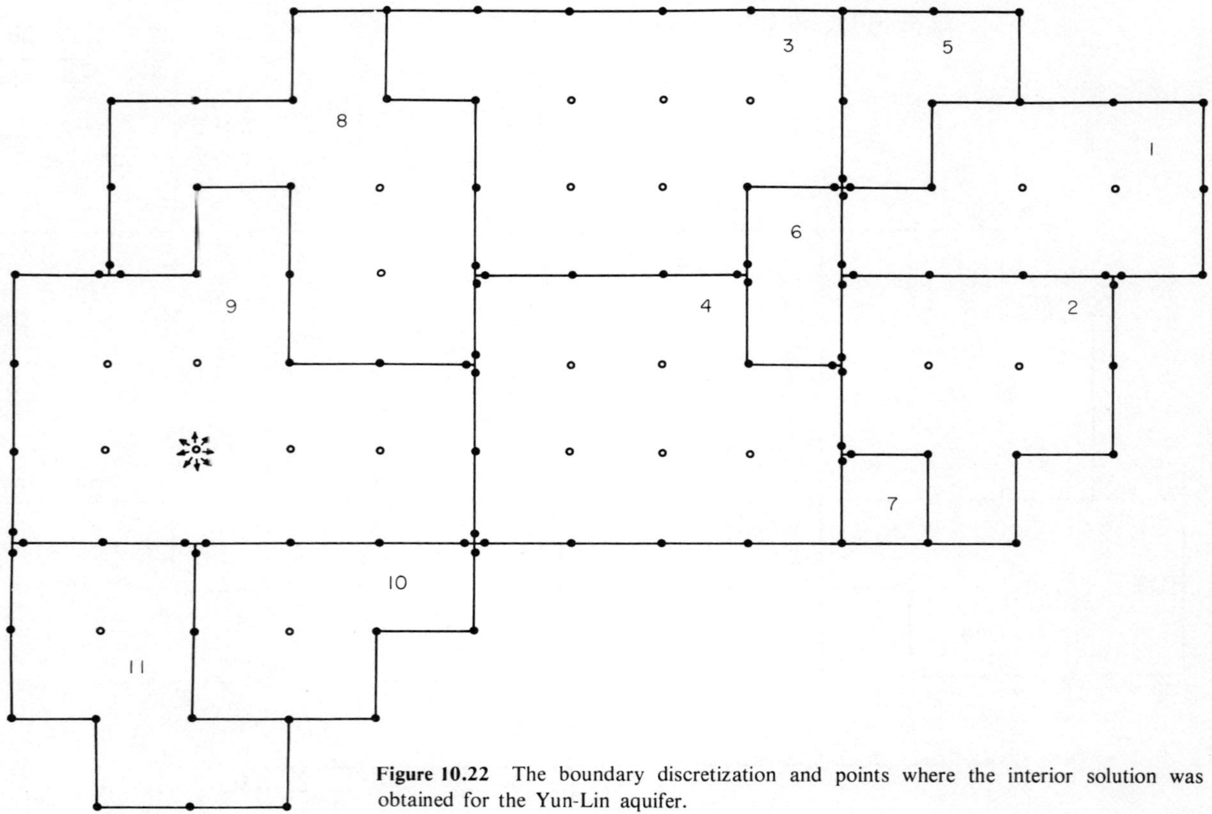


**Figure 10.22** The boundary discretization and points where the interior solution was obtained for the Yun-Lin aquifer.

| | | | | | | | | | | | | | |
|---|---|---|---|---|---|---|---|---|---|---|---|---|---|
| | | | (3.0)*<br>•<br>3.0 | (26.3)<br>•<br>25.1 | (34.9)<br>•<br>33.2 | (41.7)<br>•<br>39.8 | (48.2)<br>•<br>46.1 | (–)<br>•<br>52.2 | (58.9)<br>•<br>56.8 | (61.4)<br>•<br>59.7 | (61.6)<br>•<br>60.4 | | |
| | (0.0)*<br>•<br>0.0 | (2.6)*<br>•<br>2.6 | (2.8)*<br>•<br>2.8 | (–)<br>•<br>24.5 | (35.3)<br>•<br>33.7 | (41.6)<br>•<br>39.7 | (48.2)<br>•<br>46.1 | (54.6)<br>•<br>52.3 | (58.9)<br>•<br>56.8 | (61.4)<br>•<br>59.7 | (–)<br>•<br>60.3 | (58.9)<br>•<br>58.5 | (55.0)*<br>•<br>55.0 |
| | (0.0)*<br>•<br>0.0 | (3.3)<br>•<br>3.2 | (–)<br>•<br>7.0 | (23.0)<br>•<br>22.3 | (35.8)<br>•<br>34.3 | (41.4)<br>•<br>39.7 | (48.2)<br>•<br>46.1 | (55.1)<br>•<br>52.7 | (58.9)<br>•<br>57.0 | (61.2)<br>•<br>59.6 | (61.5)<br>•<br>60.3 | (58.3)<br>•<br>58.2 | (57.0)*<br>•<br>57.0 |
| (0.0)*<br>•<br>0.0 | (0.0)*<br>•<br>0.0 | (–)<br>•<br>3.7 | (8.6)<br>•<br>8.3 | (21.9)<br>•<br>21.1 | (35.4)<br>•<br>34.1 | (41.4)<br>•<br>39.6 | (–)<br>•<br>46.1 | (55.3)<br>•<br>52.9 | (58.8)<br>•<br>57.0 | (60.9)<br>•<br>59.5 | (62.3)<br>•<br>60.8 | (57.0)*<br>•<br>57.0 | (59.0)*<br>•<br>59.0 |
| (0.0)*<br>•<br>0.0 | (–)<br>•<br>1.0 | (4.0)<br>•<br>3.9 | (9.7)<br>•<br>9.3 | (21.0)<br>•<br>20.1 | (34.2)<br>•<br>32.4 | (41.3)<br>•<br>39.4 | (48.1)<br>•<br>46.1 | (55.2)<br>•<br>52.9 | (59.0)<br>•<br>57.0 | (–)<br>•<br>59.4 | (61.1)<br>•<br>60.8 | (65.0)*<br>•<br>65.0 | |
| (0.0)*<br>•<br>0.0 | (1.4)<br>•<br>1.4 |  | (10.1)<br>•<br>9.8 | (20.5)<br>•<br>19.6 | (33.1)<br>•<br>31.3 | (41.2)<br>•<br>39.3 | (48.2)<br>•<br>46.0 | (54.7)<br>•<br>52.5 | (59.2)<br>•<br>57.4 | (61.1)<br>•<br>59.6 | (60.0)<br>•<br>60.0 | (70.0)*<br>•<br>70.0 | |
| (0.0)*<br>•<br>0.0 | (1.9)<br>•<br>1.8 | (3.7)<br>•<br>3.8 | (9.9)<br>•<br>9.4 | (19.8)<br>•<br>19.0 | (31.9)<br>•<br>30.8 | (–)<br>•<br>39.3 | (48.2)<br>•<br>46.1 | (54.5)<br>•<br>52.4 | (59.2)<br>•<br>57.2 | (61.0)<br>•<br>59.7 | (60.3)*<br>•<br>60.3 | | |
| (0.0)*<br>•<br>0.0 | (1.9)<br>•<br>1.9 | (3.7)<br>•<br>3.5 | (9.1)<br>•<br>8.5 | (19.0)<br>•<br>17.4 | (30.6)<br>•<br>27.5 | | | | | | | | |
| (0.0)*<br>•<br>0.0 | (2.0)*<br>•<br>2.0 | (3.9)<br>•<br>3.8 | (7.4)<br>•<br>6.9 | (18.2)<br>•<br>16.4 | | | | | | | | | |
| | (1.5)*<br>•<br>1.5 | (4.2)*<br>•<br>4.0 | (6.8)*<br>•<br>6.5 | | | | | | | | | | |

*Key*

* applied boundary condition

(–) finite element solution not available

**Figure 10.23** BIEM and finite element results for the Yun-Lin aquifer.

# 11 Programs

In this final chapter we describe three of the programs which are used to solve BIEM problems. The BIEM programs are modular, or can be made modular, so that sophistication is added by adding new sections or subsections. The programs described in this chapter form the basis for the simple groundwater calculations. Due to their size and complexity we have not included programs for many of the calculations described in the latter chapters. Such programs can be written with the basic units described herein plus the descriptions in those chapters.

## 11.1 GM8

The program GM8 is basic to the BIEM calculation. It is designed to solve the Laplace equation in a closed region where the boundary conditions may be of either the Neumann or Dirichlet type or a mixture of the two. The region may have arbitrary shape but is assumed to be homogeneous.

This and the other BIEM programs can be divided into the following sections:

(1) An input section which receives the data and also prints (echoes) the input data so that the user can review them for corrections.

(2) An integration loop which finds (see (2.34))

$$R_{i,j} = \int_{\xi_{j-1}}^{\xi_j} \frac{\xi - \xi_{j-1}}{\xi_j - \xi_{j-1}} \frac{1}{r} \frac{\partial r}{\partial n} \, \mathrm{d}\xi + \int_{\xi_j}^{\xi_{j+1}} \frac{\xi_{j+1} - \xi}{\xi_{j+1} - \xi_j} \frac{1}{r} \frac{\partial r}{\partial n} \, \mathrm{d}\xi \quad (11.1)$$

$$L_{i,j} = \int_{\xi_{j-1}}^{\xi_j} \frac{\xi - \xi_{j-1}}{\xi_j - \xi_{j-1}} \ln r \, \mathrm{d}\xi + \int_{\xi_j}^{\xi_{j+1}} \frac{\xi_{j+1} - \xi}{\xi_{j+1} - \xi_j} \ln r \, \mathrm{d}\xi \quad (11.2)$$

(3) An assembly which forms the equations

$$[H]\{u\} = \{r\} \quad (11.3)$$

in which $\{u\}$ is the vector of unknown $\Phi$ and $\partial\Phi/\partial n$ values at each point $i$, $[H]$ is the matrix of coefficients from $R_{i,j}$ and $L_{i,j}$ which multiplies the unknowns, and $\{r\}$ is the product of the known $\Phi$ and $\partial\Phi/\partial n$ with the proper coefficients of $R_{i,j}$ and $L_{i,j}$.

(4) An equation solver which solves for $\{u\}$ in (11.3).

(5) A de-assembly which sorts out the $\{u\}$ vector into the $\Phi$ and $\partial\Phi/\partial n$ vectors.

(6) An output section which prints the boundary results.

(7) An internal integration loop which solves

$$\Phi = \frac{1}{2\pi}\int_\Gamma \left(\frac{\Phi}{r}\frac{\partial r}{\partial n} - \frac{\partial \Phi}{\partial n}\ln r\right) ds \tag{11.4a}$$

$$\frac{\partial \Phi}{\partial x} = \frac{1}{2\pi}\int_\Gamma \left(\frac{\Phi}{r^2}\frac{\partial \eta}{\partial x} - 2\eta\frac{\Phi}{r^3}\frac{\partial r}{\partial x} - \frac{1}{r}\frac{\partial r}{\partial x}\frac{\partial \Phi}{\partial n}\right) ds \tag{11.4b}$$

$$\frac{\partial \Phi}{\partial y} = \frac{1}{2\pi}\int_\Gamma \left(\frac{\Phi}{r^2}\frac{\partial \eta}{\partial y} - 2\eta\frac{\Phi}{r^3}\frac{\partial r}{\partial y} - \frac{1}{r}\frac{\partial r}{\partial y}\frac{\partial \Phi}{\partial n}\right) ds \tag{11.4c}$$

(8) An output section which prints the solution at the internal points.

There are no internal checks in the program except that, if the $[H]$ matrix is singular, the program prints 'NO SOLUTION' and stops. There is no warning for a nearly singular matrix. Consistent input data are assumed; no checks are made for consistency.

### *11.1.1 Input data (free format)*

Card 1: title (limit of 80 characters in one line)
Card 2: print key, $=0$ to limit output, $=1$ for full output
Card 3: number of nodal points, $N$
Card 4 to card $N+3$: nodal point data which consist of:

Item 1: $x$ co-ordinate of node
Item 2: $y$ co-ordinate of node
Item 3: $\Phi$ value (if unknown use zero)
Item 4: $\partial\Phi/\partial n$ value (if unknown use zero)
Item 5: ID code
ID $=2$ if $\Phi$ is known and $\partial\Phi/\partial n$ is to be computed
ID $=3$ if $\partial\Phi/\partial n$ is known and $\Phi$ is to be computed
Item 6: ICT code for corner designation (see definition of a 'corner' below)
ICT $=0$ if the node is not a corner

ICT = 1 if $\partial\Phi/\partial n$ is given when approaching the node and $\Phi$ is given when leaving the node
ICT = 2 if $\Phi$ is given when approaching the node and $\partial\Phi/\partial n$ is given when leaving the node
*Note*: There is no provision in the program for a corner type where $\partial\Phi/\partial n$ is given on both sides of the node but is discontinuous.

Card $N + 4$: number $M$ of interior points where $\Phi$ is to be computed.
Card $N + 5$ to card $N + 5 + M$: interior point data:

Item 1: the $x$ co-ordinate of the point
Item 2: the $y$ co-ordinate of the point

Definition of 'corner': A corner is a point where the boundary condition changes. It has nothing to do with the geometry of the region, but a corner is often located at a geometrical change in direction. There are three types of changes which create a corner: (1) a change from $\Phi$ given to $\partial\Phi/\partial n$ given, (2) a change from $\partial\Phi/\partial n$ given to $\Phi$ given, and (3) a discontinuous change in the given value of $\partial\Phi/\partial n$ (not implemented in GM8).

### *11.1.2 Program listing*

A complete listing of GM8 is given herein. The program is written in double precision FORTRAN for an IBM 370 computer. It is a basic program with no attempt to make it efficient. Consider the following notes:

(1) The integration loop renumbers the node points so that

$$(x_i, y_i) \rightarrow (x_{i+1}, y_{i+1}), (x_N, y_N) \rightarrow (x_1, y_1)$$

The renumbering enables the program to complete the boundary integration so that points 1 and $N$ do not form exceptions. This makes the capacity of the program (number of nodes) one less than the number in the dimension statement.

(2) The equation solver is a subroutine called EQSOL. It requires that the right-hand vector be added to the coefficient matrix as the $N + 1$ column. It could be replaced with any equation solver.

(3) In this and other programs given herein there should be a large amount of data checking and self diagnostics added if the program is to be used commercially. Also, computer graphics would lend itself to the preprocessing of input data and the display of results.

Program listing:

```
      IMPLICIT REAL *8(A-H,O-Z)
      DIMENSICN RLN(50,50),RN(50,50),X(50),Y(50),XX(100),YY(100)
     1,U(100),F(50),P(50),PN(50),ID(50),PN1(5),PN2(5),TITLE(20)
     2,ICT(50),CPDX(50),DPDY(50)
      COMMON H(100,100),RHS(100),MP
      READ 7,TITLE
    7 FORMAT(20A4)
      PRINT 7777,TITLE
 7777 FORMAT (//1X,20A4)
      READ (5,*) KEY
      READ(5,*)N
      MP=N
      IF (KEY.EQ.1) PRINT 6933,N
 6933 FORMAT('  NUMBER OF NODES  =',I3/)
      IF (KEY.EQ.1) PRINT 6938
 6938 FORMAT('..... KEY TO INPUT DATA AND IDENTIFICATION NUMBERS.....
     1'/,' X,Y  ARE THE NODAL COORDINATES '/,' P,PN ARE THE POTENTIAL AND
     2D THE NORMAL DERIVATIVE AT THE NODE')
      IF (KEY.EQ.1) PRINT 6939
 6939 FORMAT('  ID: IDENTIFIER FOR THE NODAL UNKNOWNS'/,
     1'              =2 FOR P  GIVEN'/
     2'              =3 FOR PN GIVEN'/)
      IF (KEY.EQ.1) PRINT 6941
 6941 FORMAT('  ICT:  NODAL CORNER TYPE IDENTIFIER'/,
     1'             =1 FOR PN GIVEN WHEN APPROACHING NODE'/,
     2'             =2 FCR PN GIVEN WHEN LEAVING NODE'/,
     3'...... ALL MOVEMENTS ARE IN THE CLOCKWISE SENSE.....'/)
C
C     NODAL DATA INPUT AND OUTPUT
C
 6932 L=N+1
      IF (KEY.EQ.1) PRINT 9953
 9953 FORMAT('  POINT',7X,'X-COOD',6X,'Y-COOD',5X,'POTENTIAL',3X,
     1   'NORM DERV',3X,'ID',8X,'ICT')
      DO 6946 I=2,L
      READ (5,*)X(I),Y(I),P(I),PN(I),ID(I),ICT(I)
      IM1=I-1
 6946 IF (KEY.EQ.1) PRINT 9950,IM1,X(I),Y(I),P(I),PN(I),ID(I),ICT(I)
 9950 FORMAT(I5,5X,F10.4,2X,F10.4,2X,F10.4,2X,F10.4,2X,I5,5X,I5)
C
C     CLEARING THE GLOBAL MATRIX
C
      DO 17 I=1,N
      DO 16 J=1,L
   16 H(I,J)=0.0
   17 F(I)=0.0
C
C     BOUNDARY INTEGRATION
C
      X(1)=X(L)
      X(L+1)=X(2)
      Y(1)=Y(L)
      Y(L+1)=Y(2)
      DO 200 I=2,L
      II1=1
      II2=1
      FNJP=0.0
      PJP=0.0
      DO 140 J=2,L
      IF(I.EQ.J)GO TO 130
      IF(I.EQ.J+1)GO TO 135
      IF(J.NE.L)GO TO 98
      IF(I.EQ.2)GO TO 135
   98 R1=DSQRT((X(J+1)-X(J))**2+(Y(J+1)-Y(J))**2)
      CO=(X(J+1)-X(J))/R1
      SI=(Y(J+1)-Y(J))/R1
      XIA=(Y(J)-Y(I))*SI+(X(J)-X(I))*CO
      XIB=(Y(J+1)-Y(I))*SI+(X(J+1)-X(I))*CO
      ETA=DABS((Y(I)-Y(J))*CO-(X(I)-X(J))*SI)
  100 IF(ETA.LT.1.E-5)GO TO 102
      BTN=DATAN(XIB/ETA)
```

```
      ATN=DATAN(XIA/ETA)
  105 ASQ=XIA**2+ETA**2
      ALN=DLOG(ASQ)
      BSQ=XIB**2+ETA**2
  106 BLN=DLOG(BSQ)
  107 CNE=BSQ*(BLN-1.)-ASQ*(ALN-1.)
      TWO=XIB*BLN-XIA*ALN-2.*(XIB-XIA)+2.*ETA*(BTN-ATN)
      SIGNRN=-(X(J)-X(I))*(Y(J+1)-Y(J))+(X(J+1)-X(J))*(Y(J)-Y(I))
      PJ=(XIB*(BTN-ATN)-0.5*ETA*(BLN-ALN))/(XIB-XIA)
      PJ=DSIGN(PJ,SIGNRN)
      XBXA=DABS(XIB-XIA)
      PNJ=TWO*XIB/(2.*XBXA)-ONE/(4.*XBXA)
      IF(ICT(J).NE.1)GO TO 321
      PN1(II1)=PN(J)
      F(I-1)=F(I-1)+PNJP*PN(J)
      II1=II1+1
      PNJP=0.0
  321 IF(ICT(J).NE.2)GO TO 432
      F(I-1)=F(I-1)+PNJ*PN(J)
      PN2(II2)=PN(J)
      II2=II2+1
      PNJ=0.0
  432 RN(I,J)=PJ+PJP
      RLN(I,J)=PNJ+PNJP
      PJP=(ETA*(BLN-ALN)*0.5-XIA*(BTN-ATN))/(XIB-XIA)
      PJP=DSIGN(PJP,SIGNRN)
      PNJP=ONE/(4.*XBXA)-TWO*XIA/(2.*XBXA)
      GO TO 140
  102 BTN=0.0
      ATN=0.0
      GO TO 105
  130 ALN=0.0
      BTN=0.0
      ATN=0.0
      XIB=DSQRT((X(I+1)-X(I))**2+(Y(I+1)-Y(I))**2)
      XIA=0.0
      ETA=0.0
      ASQ=0.0
      BSQ=XIB**2
      GO TO 106
  135 XIA=-DSQRT((X(I)-X(J))**2+(Y(I)-Y(J))**2)
      BTN=0.0
      ATN=0.0
      XIB=0.0
      BLN=0.0
      BSQ=0.0
      ASQ=XIA**2
      ALN=DLOG(ASQ)
      ETA=0.0
      GO TO 107
  140 CONTINUE
      RN(I,2)=RN(I,2)+PJP
      IF(ICT(2).NE.1)GO TO 544
      F(I-1)=F(I-1)+PNJP*PN1(1)
      PNJP=0.0
  544 RLN(I,2)=RLN(I,2)+PNJP
      ALPHA=(((X(I)-X(I-1))**2+(Y(I)-Y(I-1))**2+
     1(X(I)-X(I+1))**2+(Y(I)-Y(I+1))**2-(X(I+1)-
     2 X(I-1))**2-(Y(I+1)-Y(I-1))**2)/(2.*DSQRT(((X(I)
     3-X(I-1))**2+(Y(I)-Y(I-1))**2)*((X(I)-X(I+1
     4))**2+(Y(I)-Y(I+1))**2))))
      IF(DABS(ALPHA).LT.0.9999)GO TO 141
      ALPHA=3.14159265
      GO TO 170
  141 ALPHA=DARCOS(ALPHA)
      IF(DABS(X(I)-X(I-1)).GT.1.E-5)GO TO 144
      IF(Y(I).LT.Y(I-1))GO TO 142
      IF(X(I+1).LT.X(I))ALPHA=6.28318531-ALPHA
      GO TO 170
  142 IF(X(I+1).GT.X(I))ALPHA=6.28318531-ALPHA
      GO TO 170
  144 A=(Y(I-1)-Y(I))/(X(I-1)-X(I))
```

```
      B=Y(I)-A*X(I)
      YA=A*X(I+1)+B
      IF(Y(I).LT.Y(I-1))GO TO 150
      IF(X(I).LT.X(I-1))GO TO 146
      IF(YA.LT.Y(I+1))ALPHA=6.28318531-ALPHA
      GO TO 170
  146 IF(YA.GT.Y(I+1))ALPHA=6.28318531-ALPHA
      GO TC 170
  150 IF(X(I).GT.X(I-1))GO TO 152
      IF(YA.GT.Y(I+1))ALPHA=6.28318531-ALPHA
      GO TO 170
  152 IF(YA.LT.Y(I+1))ALPHA=6.28318531-ALPHA
  170 RN(I,I)=-ALPHA
  200 CONTINUE
   13 FORMAT(//)
C
C     ASSEMBLYING THE GLOBAL EQUATIONS
C
      DC 1C00 I=1,N
      DO 2000 J=1,N
      IF(IC(J+1).EQ.2)H(I,J)=-RLN(I+1,J+1)
      IF(IC(J+1).EQ.3)H(I,J)=RN(I+1,J+1)
 20C0 CCNTINUE
C
C     COMPUTING R-H-S VECTOR
C
      A=0.0
      CO 2001 J=1,N
      IF(ICT(J).NE.0)PN(J+1)=0.0
 2001 A=A-P(J+1)*RN(I+1,J+1)+PN(J+1)*RLN(I+1,J+1)
      H(I,L)=A+F(I)
 1000 CONTINUE
C
C     SOLVING GLOBAL EQUATIONS FOR THE NODAL UNKNOWNS
C
      CALL EQSOL
C
C     ALLOCATING EACH SOLUTION TO THE PROPER NODAL UNKNOWN
C
      JJ1=1
      DO 8003 J=2,L
      IF(ID(J).EQ.2)GO TO 8005
      P(J)=RHS(JJ1)
      JJ1=JJ1+1
      GO TO 8C03
 8005 PN(J)=RHS(JJ1)
      JJ1=JJ1+1
 8003 CCNTINUE
C
C     OUTPUT CF THE BOUNDARY SOLUTION
C
      PRINT 13
      PRINT 30
   30 FORMAT(' ..... COMPUTED RESULTS .....'/)
      MMAX=0
   36 MMIN=MMAX+1
      MMAX=MMIN+10
      MN1=N-MMIN
      IF(MN1.LT.11)MMAX=N
      PRINT 31,(J,J=MMIN,MMAX)
   31 FORMAT('  POINT  ',11I9)
      PRINT 37,(X(J+1),J=MMIN,MMAX)
      PRINT 38,(Y(J+1),J=MMIN,MMAX)
   37 FCRMAT(' X-COOD.   ',11F9.4)
   38 FCRMAT(' Y-COOD.   ',11F9.4)
      PRINT 32,(PN(J+1),J=MMIN,MMAX)
      PRINT 33,(P(J+1),J=MMIN,MMAX)
   32 FCRMAT(' NORM DERV.',11F9.4)
   33 FORMAT(' POTENTIAL ',11F9.4)
      PRINT 13
      IF(MMAX.LT.N)GO TO 36
C
```

```
C
C     INPUT FOR INTERNAL SOLUTIONS
C
      READ(5,*)LMAX
      IF(LMAX.EQ.0)GO TO 4400
      DO 1219 I=1,LMAX
 1219 READ(5,*)XX(I),YY(I)
C
C     PERFORMING BOUNDARY INTEGRATIONS TO OBTAIN INTERNAL SOL.
C
      DO 4200 I=1,LMAX
      II1=1
      II2=1
      U(I)=0.0
      DPDX(I) = 0.0
      DPDY(I) = 0.0
      PNJP=0.0
      PNXJP = 0.0
      PNYJP = 0.0
      PJP=0.0
      PXJP = 0.0
      PYJP = 0.0
      DO 4140 J=2,L
      R2=DSQRT((X(J+1)-X(J))**2+(Y(J+1)-Y(J))**2)
      CO=(X(J+1)-X(J))/R2
      SI=(Y(J+1)-Y(J))/R2
      ETA=DABS((YY(I)-Y(J))*CO-(XX(I)-X(J))*SI)
      XIA=(Y(J)-YY(I))*SI+(X(J)-XX(I))*CO
      XIB=(Y(J+1)-YY(I))*SI+(X(J+1)-XX(I))*CO
      XBXA = XIB - XIA
      SIGNRN =-(X(J)-XX(I))*(Y(J+1)-Y(J))+(X(J+1)-X(J))*(Y(J)-YY(I))
      IF (SIGNRN .LT. 0.) ETA = - ETA
      IF (DABS(ETA) .LT. 1.E-5) GO TO 4102
      BTN=DATAN(XIB/ETA)
      ATN=DATAN(XIA/ETA)
      GO TO 4105
 4102 BTN=0.0
      ATN=0.0
 4105 ASQ=XIA**2+ETA**2
      ALN=DLOG(ASQ)
      BSQ=XIB**2+ETA**2
      BLN=DLOG(BSQ)
      ONE=BSQ*(BLN-1.)-ASQ*(ALN-1.)
      TWO=XIB*BLN-XIA*ALN-2.*(XIB-XIA)+2.*ETA*(BTN-ATN)
      PJ=(XIB*(BTN-ATN)-0.5*ETA*(BLN-ALN))/(XIB-XIA)
      PXJ = ( - 0.5*SI*(BLN - ALN + 2.) - CO*(BTN - ATN) + SI*(ETA**2 +
     $ XIA*XIB)/ASQ)/(XIB - XIA) + ETA*CO/ASQ
      PYJ = (0.5*CO*(BLN - ALN + 2.) - SI*(BTN - ATN) - CO*(ETA**2 + XIA
     $ *XIB)/ASQ)/(XIB - XIA) + ETA*SI/ASQ
      PNJ=TWO*XIB/(2.*XBXA)-ONE/(4.*XBXA)
      PNXJ = - CO + ((ETA*CO - XIB*SI)*(BTN - ATN) + 0.5*(XIB*CO + ETA
     $ *SI)*(BLN - ALN))/XBXA
      PNYJ = - SI + ((ETA*SI + XIB*CO)*(BTN - ATN) + 0.5*(XIB*SI - ETA
     $ *CO)*(BLN - ALN))/XBXA
      IF(ICT(J).NE.1)GO TO 4321
      U(I)=U(I)-PNJP*PN1(II1)
      DPDX(I) = DPDX(I) + PNXJP*PN1(II1)
      DPDY(I) = DPDY(I) + PNYJP*PN1(II1)
      II1=II1+1
      PNJP=0.0
      PNXJP = 0.0
      PNYJP = 0.0
 4321 IF(ICT(J).NE.2)GO TO 4322
      U(I)=U(I)-PNJ*PN2(II2)
      DPDX(I) = DPDX(I) + PNXJ*PN2(II2)
      DPDY(I) = DPDY(I) + PNYJ*PN2(II2)
      II2=II2+1
      PNJ=0.0
      PNXJ = 0.0
      PNYJ = 0.0
```

```
4322 U(I)=U(I)+(PJ+PJP)*P(J)-(PNJ+PNJP)*PN(J)
     DPDX(I) = DPDX(I) + (PXJ + PXJP)*P(J) + (PNXJ + PNXJP)*PN(J)
     DPDY(I) = DPDY(I) + (PYJ + PYJP)*P(J) + (PNYJ + PNYJP)*PN(J)
     PJP=(ETA*(BLN-ALN)*0.5-XIA*(BTN-ATN))/(XIB-XIA)
     PXJP = (0.5*SI*(BLN - ALN - 2.) + CO*(BTN - ATN) + SI*(ETA**2 +
    $ XIA*XIB)/BSQ)/(XIB - XIA) - ETA*CO/BSQ
     PYJP = ( -0.5*CO*(BLN - ALN - 2.) + SI*(BTN - ATN) - CO*(ETA**2 +
    $ XIA*XIB)/BSQ)/(XIB - XIA) - ETA*SI/BSQ
     PNJP=CNE/(4.*XBXA)-TWO*XIA/(2.*XBXA)
     PNXJP = CO + (( -ETA*CO + XIA*SI)*(BTN - ATN) - 0.5*(XIA*CO + ETA*
    $ SI)*(BLN - ALN))/XBXA
     PNYJP = SI + ( -(ETA*SI + XIA*CO)*(BTN - ATN) - 0.5*(XIA*SI - ETA*
    $ CO)*(BLN - ALN))/XBXA
4140 CONTINUE
     U(I)=U(I)+PJP*P(2)
     DPDX(I) = DPDX(I) + PXJP*P(2)
     DPDY(I) = DPDY(I) + PYJP*P(2)
     IF(ICT(2).NE.1)GO TO 4195
     U(I)=U(I)-PNJP*PN1(1)
     DPDX(I) = DPDX(I) + PNXJP*PN1(1)
     DPDY(I) = DPDY(I) + PNYJP*PN1(1)
     PNJP=0.0
     PNXJP = 0.0
     PNYJP = 0.0
4195 CONTINUE
     U(I)=U(I)-(PNJP*PN(2))
     DPDX(I) = DPDX(I) + PNXJP*PN(2)
     DPDY(I) = DPDY(I) + PNYJP*PN(2)
4200 CONTINUE
     DO 4300 I=1,LMAX
     U(I)=U(I)/6.283185308
     DPDX(I) = DPDX(I)/6.283185308
4300 DPDY(I) = DPDY(I)/6.283185308
C
C       OUTPUT OF INTERNAL SOLUTIONS
C
     PRINT 4700
4700 FORMAT(/' ..... INTERNAL SOLUTIONS ....')
     MMAX=0
4736 MMIN=MMAX+1
     MMAX=MMIN+10
     MN1=LMAX-MMIN
     IF(MN1.LT.11)MMAX=LMAX
     PRINT 4500,(K,K=MMIN,MMAX)
4500 FORMAT(/' POINT   ',11I9)
     PRINT 37,(XX(K),K=MMIN,MMAX)
     PRINT 38,(YY(K),K=MMIN,MMAX)
     PRINT 33,(U(K),K=MMIN,MMAX)
     PRINT 40, (DPDX(K), K = MMIN, MMAX)
     PRINT 42, (DPDY(K), K = MMIN, MMAX)
  40 FORMAT (' X-COMP VEL', 11F9.4)
  42 FORMAT (' Y-COMP VEL', 11F9.4)
     IF(MMAX.EQ.LMAX)GO TO 4400
     GO TO 4736
4400 CONTINUE
     STOP
     END
C
C       LINEAR EQUATION SOLVER
C
     SUBROUTINE EQSOL
     IMPLICIT REAL *8(A-H,O-Z)
     COMMON A(100,100),X(100),N
     DIMENSION M(100)
     DO 5 I=1,N
     M(I)=1
     AMAX=A(I,1)
     DO 2 J=2,N
     IF(DABS(A(I,J)).LE.DABS(AMAX))GO TO 2
 100 AMAX=A(I,J)
```

```
      M(I)=J
    2 CCNTINUE
      IF(AMAX.EQ.0)GO TO 98
    3 NN=N+1
      DO 4 J=1,NN
    4 A(I,J)=A(I,J)/AMAX
      DC 5 IP=1,N
      IF(IP.EQ.I)GO TO 5
      MMM=M(I)
      ZMULT=A(IP,MMM)
      CO 6 J=1,NN
      IF(J.NE.MMM)GO TO 9
    8 A(IP,J)=0.0
      GO TC 6
    9 A(IP,J)=A(IP,J)-ZMULT*A(I,J)
    6 CCNTINUE
    5 CCNTINUE
      DO 7 I=1,N
      NO=M(I)
    7 X(NO)=A(I,NN)
      RETURN
   98 PRINT 1C00
 1000 FORMAT(15H NO SOLUTION    )
      STOP
      END
```

### *11.1.3 Example*

As an example calculation we use the wedge in a box (Fig. 3.3). As there is no provision for special elements, none is used.

Input data:

```
  90-DEGREE WEDGE IN 2 X 1 BOX
1
14
-1. -.5 -1. 0. 2 1
-1. 0. -1. 0. 2 0
-1. .5 -1. 0. 2 2
-.5 .5 0. 0. 3 0
-.25 .25 0. 0. 3 0
0. 0. 0. 0. 3 0
.25 .25 0. 0. 3 0
.5 .5 0. 0. 3 0
1. .5 1. 0. 2 1
1. 0. 1. 0. 2 0
1. -.5 1. 0. 2 2
.25 -.5 0. 0. 3 0
0. -.5 0. 0. 3 0
-.25 -.5 0. 0. 3 0
6
0. -.02
0. -.05
0. -.1
0. -.2
0. -.3
0. -.4
```

The output is shown in Figure 11.1.

```
 90-DEGREE WEDGE IN 2 X 1 BOX
 NUMBER OF NODES   = 14

.... KEY TO INPUT DATA AND IDENTIFICATION NUMBERS.....
X,Y  ARE THE NODAL COORDINATES
F,FN ARE THE POTENTIAL AND THE NORMAL DERIVATIVE AT THE NODE
 ID: IDENTIFIER FOR THE NODAL UNKNOWNS
            =2 FOR F  GIVEN
            =3 FOR FN GIVEN

 ICT:  NODAL CORNER TYPE IDENTIFIER
             =1 FOR FN GIVEN WHEN APPROACHING NODE
             =2 FOR FN GIVEN WHEN LEAVING NODE
..... ALL MOVEMENTS ARE IN THE CLOCKWISE SENSE.....

 POINT       X-COOD      Y-COOD      POTENTIAL   NORM DERV   ID       ICT
   1        -1.0000     -0.5000     -1.0000      0.0000      2        1
   2        -1.0000      0.0000     -1.0000      0.0000      2        0
   3        -1.0000      0.5000     -1.0000      0.0000      2        2
   4        -0.5000      0.5000      0.0000      0.0000      3        0
   5        -0.2500      0.2500      0.0000      0.0000      3        0
   6         0.0000      0.0000      0.0000      0.0000      3        0
   7         0.2500      0.2500      0.0000      0.0000      3        0
   8         0.5000      0.5000      0.0000      0.0000      3        0
   9         1.0000      0.5000      1.0000      0.0000      2        1
  10         1.0000      0.0000      1.0000      0.0000      2        0
  11         1.0000     -0.5000      1.0000      0.0000      2        2
  12         0.2500     -0.5000      0.0000      0.0000      3        0
  13         0.0000     -0.5000      0.0000      0.0000      3        0
  14        -0.2500     -0.5000      0.0000      0.0000      3        0
```

**Figure 11.1** Output from GM8.

```
..... COMPUTED RESULTS .....

 POINT          1        2        3        4        5        6        7        8        9       10       11
X-COOD.     -1.0000  -1.0000  -1.0000  -0.5000  -0.2500   0.0000   0.2500   0.5000   1.0000   1.0000   1.0000
Y-COOD.     -0.5000   0.0000   0.5000   0.5000   0.2500   0.0000   0.2500   0.5000   0.5000   0.0000  -0.5000
NORM DERV.  -0.8919  -0.7956  -0.6547   0.0000   0.0000   0.0000   0.0000   0.0000   0.6547   0.7956   0.8919
POTENTIAL   -1.0000  -1.0000  -1.0000  -0.7200  -0.5045   0.0000   0.5045   0.7200   1.0000   1.0000   1.0000


 POINT         12       13       14
X-COOD.      0.2500   0.0000  -0.2500
Y-COOD.     -0.5000  -0.5000  -0.5000
NORM DERV.   0.0000   0.0000   0.0000
POTENTIAL    0.2821   0.0000  -0.2821


..... INTERNAL SOLUTIONS ....

POINT           1        2        3        4        5        6
X-COOD.      0.0000   0.0000   0.0000   0.0000   0.0000   0.0000
Y-COOD.     -0.0200  -0.0500  -0.1000  -0.2000  -0.3000  -0.4000
POTENTIAL    0.0000   0.0000   0.0000   0.0000   0.0000   0.0000
X-COMP VEL   1.9285   1.6456   1.4438   1.2679   1.1890   1.1520
Y-COMP VEL  -0.0000  -0.0000  -0.0000  -0.0000   0.0000   0.0000
```

**Figure 11.1** – *continued.*

# 11.2 GM9

The program GM9 is an extension of GM8. This program allows the insertion of special elements (see Ch. 3) which are intended to conform to the functional behavior of $\Phi$ and/or $\partial\Phi/\partial n$ along the boundaries. For the integration over normal elements the program functions just as GM8. The integration over the special elements is done with 16-point Gaussian quadrature in 6 panels as shown in Figure 11.2. For most problems the quadrature is far more accurate than needed and unnecessarily time consuming. However, the program is designed for special elements which may have singular behavior at either end or in the middle. It is assumed that any one problem will not have many such elements and thus the inefficiency of the integration does not greatly affect the overall run time. At the cost of generality the program could be designed to operate more efficiently, and perhaps more accurately, for any specific problem.

The primary sections of the program are those listed for GM8. In addition there are:

(9) A subroutine which performs the Gaussian quadrature.
(10) A subroutine which sets the weights and abscissas for Gaussian quadrature.
(11) A function subprogram which specifies the form of the special elements.

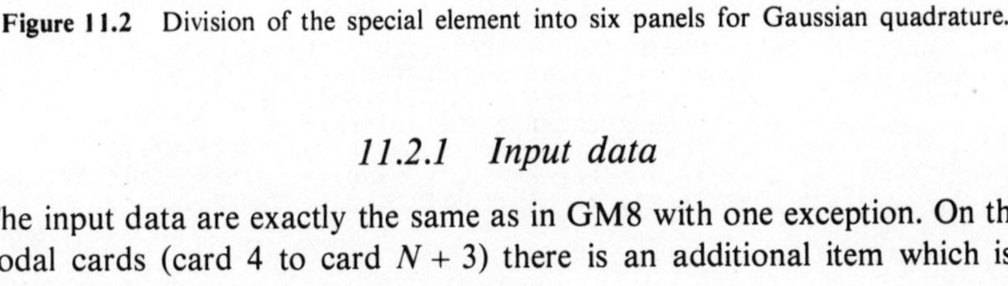


**Figure 11.2** Division of the special element into six panels for Gaussian quadrature.

### *11.2.1 Input data*

The input data are exactly the same as in GM8 with one exception. On the nodal cards (card 4 to card $N + 3$) there is an additional item which is:

Item 7: LS code for special elements
LS = 0 if the element following the node is not special
LS = 1 if the element following the node is the first special element
LS = $m$ if the element following the node is the $m$th special element

### *11.2.2 The function subprogram*

Any special element must have two interpolating functions, $N_1$ and $N_2$. At the first node of the element $N_1$ must be unity and it must be zero at the second node of the element. The opposite is the case for $N_2$; it is zero at the first node and unity at the second node. Within these restrictions the interpolation functions are arbitrary. They may be discontinuous.

For any special element four functions must be constructed, two for the $\Phi$ interpolation and two for the $\partial\Phi/\partial n$ interpolation. These are designated as follows:

$F_{11}$ is used for the $\Phi$ interpolation and is unity at the first node and zero at the second node.
$F_{12}$ is used for the $\partial\Phi/\partial n$ interpolation and is unity at the first node and zero at the second node.
$F_{21}$ is used for the $\Phi$ interpolation and is zero at the first node and unity at the second node.
$F_{22}$ is used for the $\partial\Phi/\partial n$ interpolation and is zero at the first node and unity at the second node.

The parameters for these functions are: X = distance from the first node, XL = length of the element, and LS = the special element number (see input data, Sec. 11.2.1). The LS is used to direct the function subprogram to the correct function. The other parameters are used to define the function.

### *11.2.3 Program listing*

Only those changes from GM8 are shown, surrounded by enough statements for them to be placed in the GM8 listing. In the input section:

```
          .
          .
          .

      CALL GAUSS
 6941 FORMAT('  ICT:  NODAL CORNER TYPE IDENTIFIER'/,
     1'               =1 FOR PN GIVEN WHEN APPROACHING NODE'/,
     2'               =2 FOR PN GIVEN WHEN LEAVING NODE'//,
     3        '  LS: SINGULAR ELEMENT IDENTIFIER'/,
     4'               =0 FOR NO SINGULARITY UR SPECIAL ELEMENT'/,
     5'               =J IF THE ELEMENT FOLLOWING THE JTH NODE IS SPECIAL'/
     3,'...... ALL MOVEMENTS ARE IN THE CLOCKWISE SENSE.....'/)
C
C     NODAL DATA INPUT AND OUTPUT
C
 6932 L=N+1
      IF (KEY.EQ.1) PRINT 9953
 9953 FORMAT('  POINT',7X,'X-COCD',6X,'Y-COOD',5X,'POTENTIAL',3X,
     1   'NCFM DERV',3X,'ID',8X,'ICT',7X,'LS')
      DO 6946 I=2,L
      READ (5,*)X(I),Y(I),P(I),PN(I),ID(I),ICT(I),LS(I)
      IM1=I-1

          .
          .
          .
```

In the boundary integration section:

```
C      BOUNDARY INTEGRATION
C
       LS(1)=LS(L)
       X(1)=X(L)
       X(L+1)=X(2)
              .
              .
              .
       XBXA=CABS(XIB-XIA)
       PNJ=THC*XIB/(2.*XBXA)-ONE/(4.*XBXA)
       IF (LS(J) .NE. 0) CALL INT1(PJ,PNJ,I,J,LS(J))
       PJ=CSIGN(PJ,SIGNRN)
       IF(ICT(J).NE.1)GO TO 321
              .
              .
              .
       PJP=(ETA*(BLN-ALN)*0.5-XIA*(BTN-ATN))/(XIB-XIA)
       PNJP=ONE/(4.*XBXA)-TWO*XIA/(2.*XBXA)
       IF (LS(J) .NE. 0) CALL INT2(PJP,PNJP,I,J,LS(J))
       PJP=DSIGN(PJP,SIGNRN)
       GO TC 140
```

In the integration for internal points, similar changes must be made in order to send the program to the numerical integration subroutine where needed. As in GM8 the integration sections for boundary points and internal points are separated. (They could easily be combined at a saving in program length, but the result would contain statements which are unnecessary in the case of internal points.) The subroutine for numerical integration for boundary points is listed below:

```
       SUBROUTINE INT1(PJ,PNJ,II,JJ,LS)
       IMPLICIT REAL*8(A-H,O-Z)
       COMMON /INT/XIA,XIB,ETA
       COMMON /GAUS/XI(16),W(16)
       SL=XIB-XIA
       IF (ETA .GT. 1.E-5) GO TO 110
       PJ=0.
       GC TC 1010
  11C  SUM=C.
       CO 120 I=1,16
       S=SL*(1.+XI(I))/2.
  120  SUM=SUM+W(I)*F11(S,SL,LS)/((XIA+S)**2+ETA**2)
       PJ=SUM*SL*ETA/2.
 1C1C  T=C.
       DO 1030 K=1,6
       SUM=0.
       DO 1020 I=1,16
       IF (K .EC. 1) S=SL*(1.+XI(I))*.025
       IF (K. EQ. 2) S=SL*(1.+XI(I))*.2+SL*.05
       IF (K .EQ. 3) S=SL*(1.+XI(I))*.025+SL*.45
       IF (K .EC. 4) S=SL*(1.+XI(I))*.025+SL*.5
       IF (K .EC. 5) S=SL*(1.+XI(I))*.2+SL*.55
       IF (K .EQ. 6) S=SL*(1.+XI(I))*.025+SL*.95
 102C  SUM=SUM+W(I)*F12(S,SL,LS)*DLOG((XIA+S)**2+ETA**2)
       IF (K .EC. 1) T=.05*SUM
       IF (K .EC .2) T=T+.4*SUM
       IF (K .EQ. 3) T=T+.05*SUM
       IF (K .EQ. 4) T=T+.05*SUM
       IF (K .EQ. 5) T=T+.4*SUM
```

```
      IF (K .EQ. 6) T=T+.05*SUM
 1030 CONTINUE
      PNJ=T*SL/4.
      RETURN
      ENTRY INT2(PJ,PNJ,II,JJ,LS)
      SL=XIB-XIA
      IF (ETA .GT. 1.E-5)  GO TO 3010
      PJ=0.
      GO TO 4010
 3010 SUM=0.
      DO 3020 I=1,16
      S=SL*(1.+XI(I))/2.
 3020 SUM=SUM+W(I)*F21(S,SL,LS)/((XIA+S)**2+ETA**2)
      PJ=SUM*SL*ETA/2.
 4010 T=0.
      DO 4030 K=1,6
      SUM=0.
      DO 4020 I=1,16
      IF (K .EQ. 1) S=SL*(1.+XI(I))*.025
      IF (K .EQ. 2) S=SL*(1.+XI(I))*.2+SL*.05
      IF (K .EQ. 3) S=SL*(1.+XI(I))*.025+SL*.45
      IF (K .EQ. 4) S=SL*(1.+XI(I))*.025+SL*.5
      IF (K .EQ. 5) S=SL*(1.+XI(I))*.2+SL*.55
      IF (K .EQ. 6) S=SL*(1.+XI(I))*.025+SL*.95
 4020 SUM=SUM+W(I)*F22(S,SL,LS)*DLOG((XIA+S)**2+ETA**2)
      IF (K .EQ. 1) T=.05*SUM
      IF (K .EQ. 2) T=T+.4*SUM
      IF (K .EQ. 3) T=T+.05*SUM
      IF (K .EQ. 4) T=T+.05*SUM
      IF (K .EQ. 5) T=T+.4*SUM
      IF (K .EQ. 6) T=T+.05*SUM
 4030 CONTINUE
      PNJ=T*SL/4.
      RETURN
      END
      SUBROUTINE GAUSS
      IMPLICIT REAL*8(A-H,O-Z)
      COMMON /GAUS/ X(16),W(16)
   12 X(9)=0.0950125098
      X(10)=0.2816035507
      X(11)=0.4580167776
      X(12)=0.6178762444
      X(13)=0.7554044083
      X(14)=0.8656312023
      X(15)=0.9445750230
      X(16)=0.9894009349
      W(9)=0.1894506104
      W(10)=0.1826034150
      W(11)=0.1691565193
      W(12)=0.1495959888
      W(13)=0.1246289712
      W(14)=0.0951585116
      W(15)=0.0622535239
      W(16)=0.0271524594
      X(1)=-X(16)
      X(2)=-X(15)
      X(3)=-X(14)
      X(4)=-X(13)
      X(5)=-X(12)
      X(6)=-X(11)
      X(7)=-X(10)
      X(8)=-X(9)
      W(1)=W(16)
      W(2)=W(15)
      W(3)=W(14)
      W(4)=W(13)
      W(5)=W(12)
      W(6)=W(11)
      W(7)=W(10)
      W(8)=W(9)
 9999 RETURN
      END
```

```
C
C     ******************  SPECIAL ELEMENT INSERTION  **************
C     THE FOLLOWING FUNCTIONS ARE TO BE USED TO INSERT SPECIAL INTERPOLATION
C   FUNCTIONS WHERE THERE IS A KNOWN VARIATION OF THE POTENTIAL AND/OR ITS
C   NORMAL DERIVATIVE (AS IN THE NEIGHBORHOOD OF SINGULARITIES).  SUCH
C   ELEMENTS ARE FLAGGED BY SETTING LS(J)=TO THE NUMBER OF THE SPECIAL
C   ELEMENT; THAT IS:
C         LS(J)=0 -> REGULAR, LINEAR ELEMENT
C         LS(J)=1 -> THE JTH ELEMENT IS THE FIRST SPECIAL ELEMENT
C         LS(J)=N -> THE JTH ELEMENT IS THE NTH SPECIAL ELEMENT
C   F11 AND F21 ARE USED IN THE POTENTIAL INTEGRAL
C   F12 AND F22 ARE USED IN THE NORMAL DERIVATIVE INTEGRAL
C   F11 AND F12 MUST BE UNITY AT THE FIRST NODE OF THE ELEMENT AND
C         ZERO AT THE SECOND (LAST) NODE OF THE ELEMENT
C   F21 AND F22 MUST BE UNITY AT THE SECOND (LAST) NODE OF THE ELEMENT
C         AND ZERO AT THE FIRST NODE
C     THE PROPER SPECIAL FUNCTION IS TO BE INSERTED AFTER THE CONTINUE
C   STATEMENT CONTAINING ITS ELEMENT NUMBER
```

The instructions for four function subprograms which define the special elements are given in the above comment statements. A standard form has been adopted for the function subprograms so that the operator needs only to add a line with the proper functional representation. The example is illustrative.

### *11.2.4 Example*

The wedge problem of Figure 3.3 is used to illustrate the program. The point of the wedge is a singularity in that the velocity goes to infinity at that point. If $r$ is the distance from the point,

$$\Phi \sim r^{\frac{2}{3}} \tag{11.5}$$

Along the wall $\partial\Phi/\partial n = 0$ from the boundary conditions. The proper interpolating functions for $\Phi$ are

$$\begin{aligned} F_{11} &= (1-\xi)^{\frac{2}{3}} \\ F_{21} &= 1-(1-\xi)^{\frac{2}{3}} \end{aligned} \tag{11.6}$$

for element 5 (between nodes 5 and 6) in which $\xi =$ the normalized distance from node 5 ($\xi = 0$ at 5 and $\xi = 1$ at 6). For element 6

$$\begin{aligned} F_{11} &= 1-\xi^{\frac{2}{3}} \\ F_{21} &= \xi^{\frac{2}{3}} \end{aligned} \tag{11.7}$$

where $\xi$ is now the normalized distance along element 6. The functions $F_{12}$ and $F_{22}$ are taken as zero for both elements since they would be multiplied by zero (the $\partial\Phi/\partial n$ boundary condition) in any case. The listing of the function subprogram follows:

```
C         SUBPROGRAM WDG9
      FUNCTION F11(X,XL,LS)
      IMPLICIT REAL*8(A-H,O-Z)
      GO TO (1,2,3,4,5,6),LS
    1 CONTINUE
      F11=(1.-(X/XL))**.666666666666667
      RETURN
    2 CONTINUE
      F11=1.-(X/XL)**.666666666666667
      RETURN
    3 CONTINUE
      RETURN
    4 CONTINUE
      RETURN
    5 CONTINUE
      RETURN
    6 CONTINUE
      RETURN
      END
      FUNCTION F12(X,XL,LS)
      IMPLICIT REAL*8(A-H,O-Z)
      GO TO (1,2,3,4,5,6),LS
    1 CONTINUE
      F12=0.
      RETURN
    2 CONTINUE
      F12=0.
      RETURN
    3 CONTINUE
      RETURN
    4 CONTINUE
      RETURN
    5 CONTINUE
      RETURN
    6 CONTINUE
      RETURN
      END
      FUNCTION F21(X,XL,LS)
      IMPLICIT REAL*8(A-H,O-Z)
      GO TO (1,2,3,4,5,6),LS
    1 CONTINUE
      F21=1.-(1.-X/XL)**.666666666666667
      RETURN
    2 CONTINUE
      F21=(X/XL)**.666666666666667
      RETURN
    3 CONTINUE
      RETURN
    4 CONTINUE
      RETURN
    5 CONTINUE
      RETURN
    6 CONTINUE
      RETURN
      END
      FUNCTION F22(X,XL,LS)
      IMPLICIT REAL*8(A-H,O-Z)
      GO TO (1,2,3,4,5,6),LS
    1 CONTINUE
      F22=0.
      RETURN
    2 CONTINUE
      F22=0.
      RETURN
    3 CONTINUE
      RETURN
    4 CONTINUE
      RETURN
    5 CONTINUE
      RETURN
    6 CONTINUE
      RETURN
      END
```

```
 90-DEGREE WEDGE IN 2 X 1 BOX
 NUMBER OF NODES   = 14

.... KEY TO INPUT DATA AND IDENTIFICATION NUMBERS.....
X,Y  ARE THE NODAL COORDINATES
P,PN ARE THE POTENTIAL AND THE NORMAL DERIVATIVE AT THE NODE
 ID: IDENTIFIER FOR THE NODAL UNKNOWNS
          =2 FOR P  GIVEN
          =3 FOR PN GIVEN

 ICT:  NODAL CORNER TYPE IDENTIFIER
           =1 FOR PN GIVEN WHEN APPROACHING NODE
           =2 FOR PN GIVEN WHEN LEAVING NODE

 LS: SINGULAR ELEMENT IDENTIFIER
           =0 FOR NO SINGULARITY OR SPECIAL ELEMENT
           =J IF THE ELEMENT FOLLOWING THE JTH NODE IS SPECIAL
..... ALL MOVEMENTS ARE IN THE CLOCKWISE SENSE.....

 POINT        X-COOD       Y-COOD       POTENTIAL    NORM DERV   ID        ICT        LS
   1         -1.0000      -0.5000      -1.0000       0.0000      2          1          0
   2         -1.0000       0.0000      -1.0000       0.0000      2          0          0
   3         -1.0000       0.5000      -1.0000       0.0000      2          2          0
   4         -0.5000       0.5000       0.0000       0.0000      3          0          0
   5         -0.2500       0.2500       0.0000       0.0000      3          0          1
   6          0.0000       0.0000       0.0000       0.0000      3          0          2
   7          0.2500       0.2500       0.0000       0.0000      3          0          0
   8          0.5000       0.5000       0.0000       0.0000      3          0          0
   9          1.0000       0.5000       1.0000       0.0000      2          1          0
  10          1.0000       0.0000       1.0000       0.0000      2          0          0
  11          1.0000      -0.5000       1.0000       0.0000      2          2          0
  12          0.2500      -0.5000       0.0000       0.0000      3          0          0
  13          0.0000      -0.5000       0.0000       0.0000      3          0          0
  14         -0.2500      -0.5000       0.0000       0.0000      3          0          0
```

**Figure 11.3** Output from GM9.

```
..... COMPUTED RESULTS .....
 POINT             1          2          3          4          5          6          7          8          9         10         11
X-COOD.     -1.0000    -1.0000    -1.0000    -0.5000    -0.2500     0.0000     0.2500     0.5000     1.0000     1.0000     1.0000
Y-COOD.     -0.5000     0.0000     0.5000     0.5000     0.2500     0.0000     0.2500     0.5000     0.5000     0.0000    -0.5000
NORM DERV.  -0.8762    -0.7783    -0.6353     0.0000     0.0000     0.0000     0.0000     0.0000     0.6353     0.7783     0.8762
POTENTIAL   -1.0000    -1.0000    -1.0000    -0.7280    -0.5197    -0.0000     0.5197     0.7280     1.0000     1.0000     1.0000

 POINT            12         13         14
X-COOD.      0.2500     0.0000    -0.2500
Y-COOD.     -0.5000    -0.5000    -0.5000
NORM DERV.   0.0000     0.0000     0.0000
POTENTIAL    0.2875     0.0000    -0.2875


..... INTERNAL SOLUTIONS ....
POINT              1          2          3          4          5          6
X-COOD.      0.0000     0.0000     0.0000     0.0000     0.0000     0.0000
Y-COOD.     -0.0200    -0.0500    -0.1000     0.2000    -0.3000    -0.4000
POTENTIAL    0.0000     0.0000     0.0000     0.0000     0.0000     0.0000
X-COMP VEL   2.6371     1.9752     1.6055     1.3364     1.2285     1.1802
Y-COMP VEL   0.0000    -0.0000    -0.0000     0.0000    -0.0000    -0.0000
```

**Figure 11.3** – *continued.*

The input data for the wedge problem is the same as that shown for GM9 except that the LS parameter is added to each nodal card. That is, cards 4–7 and 10–17 have a zero added, card 8 has a 1 added (the first special element), and card 9 has a 2 added. The output is shown in Figure 11.3.

## 11.3 DAM

The program DAM is written for the calculation of unsteady, free surface flow through an earth dike of constant permeability. The dike is sitting on an impermeable foundation. The water level on the upstream face can rise with time in a prescribed manner, but the level cannot decrease since there is no provision for a seepage surface on the upstream face. The water level on the downstream face remains constant throughout the calculation. The free surface points can be moved at arbitrary angles which are specified by the user (see Fig. 11.4). A few, but not all, of the details in the program are listed below.

### *11.3.1 Input data*

In addition to those items specified for GM8, DAM needs the following:

(1) Some specification of the headwater level with time.
(2) The time step and weighting ($\Delta t$ and $\theta$, see (5.14)).
(3) The permeability.
(4) The angle, $\gamma$, along which each free surface point will move (Figs 5.1 & 11.4).
(5) Codes which identify points on the free surface and the seepage surface.

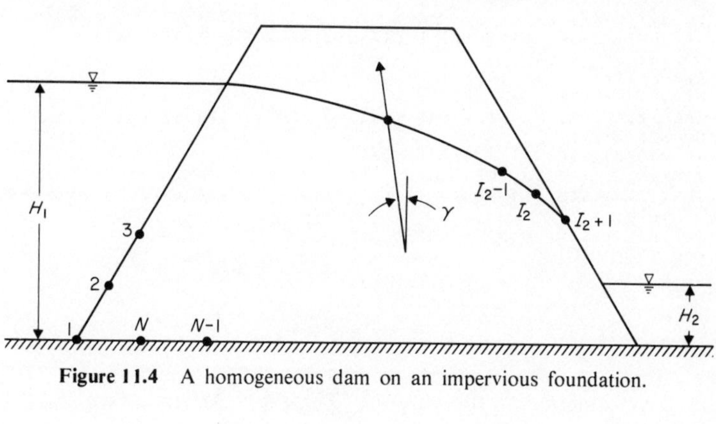


**Figure 11.4** A homogeneous dam on an impervious foundation.

## 11.3.2 Program details

Only a few program statements are given herein. There are four nodal point identifiers instead of two:

```
      .
      .
      .
      PRINT 6939
6939  FORMAT('  ID: IDENTIFIER FOR THE NODAL UNKNOWNS'/,
     1'           =2 FOR P  GIVEN'/
     2'           =3 FOR PN GIVEN'/
     3'           =4 FOR A FREE SURFACE POINT'/
     4'           =5 FOR A SEEPAGE SURFACE POINT'/)
      PRINT 6941
6941  FORMAT('  ICT:  NODAL CORNER TYPE IDENTIFIER'/,
     1'          =1 FOR PN GIVEN WHEN APPROACHING NODE'/,
     2'          =2 FOR PN GIVEN WHEN LEAVING NODE'/,
     3'...... ALL MOVEMENTS ARE IN THE CLOCKWISE SENSE.....'/)
      .
      .
      .
```

These identifiers, together with the angle $\gamma$, are read in on the cards which specify node location, etc. The basic integration loop can stand unchanged. The identifiers are used in the assembly and de-assembly:

```
      .
      .
      .
  200 CONTINUE
      DO 604 I=1,N
      DO 602 J=1,N
      IF (ID(J+1) .EQ. 2 .OR. ID(J+1) .EQ. 5) H(I,J)=-RLN(I+1,J+1)
      IF (ID(J+1) .EQ. 3) H(I,J)=RN(I+1,J+1)
      IF(ID(J+1).EQ.4)H(I,J)=-RLN(I+1,J+1)-RN(I+1,J+1)*DELTA(J+1)*THETA
  602 CONTINUE
      H(I,I2+1) = RN(I+1,I2+2)
      A=0.
      DO 596 J=1,I1
  596 A=A-RN(I+1,J+1)*P(J+1)
      DO 597 J=KP,I2
  597 A=A-RN(I+1,J+1)*(P(J+1)-DELTA(J+1)*(1.-THETA)*PN(J+1))
      DO 598 J=KCP,I4
  598 A=A-RN(I+1,J+1)*P(J+1)
  604 H(I,L)=A+F(I)
  605 CALL ECSCL(N, MAXSIZ, MAXSIZ+1, H, RHS, MSTATS)
      DO 620 J=2,L
      IF (ID(J) .EQ. 3) P(J)=RHS(J-1)
      IF (ID(J) .EQ. 2 .OR. ID(J) .EQ. 5) PN(J)=RHS(J-1)
      IF (ID(J) .EQ. 4) P(J)=P(J)-DELTA(J)*(THETA*RHS(J-1)+PN(J)*(1.
     1 -THETA))
      IF(ID(J).EQ.4)PN(J) = RHS(J-1)
  620 CONTINUE
      .
      .
      .
```

In the above statements the variable DELTA $= \Delta t/\cos\beta$ (see (5.14)). The adjustment of the free surface is:

```
      .
      .
      .
      DO 642 J=KP, KQ
      X(J+1) = X(J+1) + (Y(J+1)-P(J+1)) * TAN(PIOT-GAMA(J+1))
  642 Y(J+1) = P(J+1)
      .
      .
      .
```

There are, of course, other details which enter into the free surface program. They are rather obvious. The resulting program represents a quite efficient method to calculate unsteady, free surface flow.

### 11.3.3 Example

The example computes the free surface profile in a dam from an initial low water level as the head water rises. The results of the computation are shown in Figure 11.5. The time step was rather coarse (beginning at $t = 0.2$ and increasing by 10% each step), causing some inaccuracy in the results.

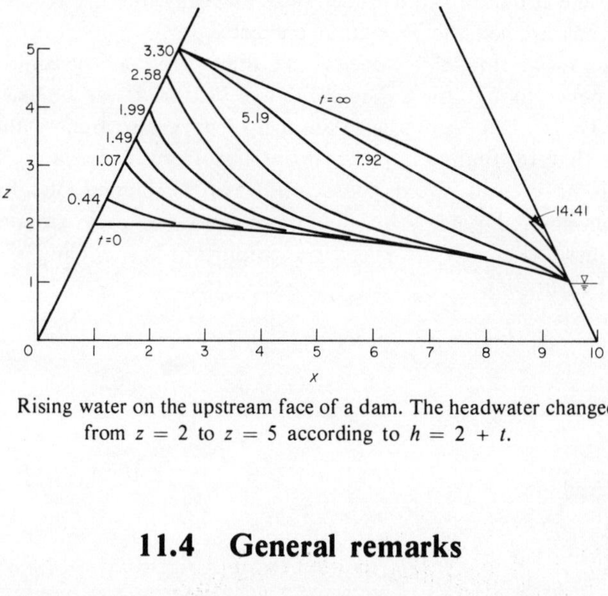


**Figure 11.5** Rising water on the upstream face of a dam. The headwater changed elevation from $z = 2$ to $z = 5$ according to $h = 2 + t$.

## 11.4 General remarks

Three basic BIEM programs have been illustrated in considerable detail in this chapter. A further extension to illustrate zoning, axisymmetric, three-dimensional, and the Chapter 10 programs would require far too much space. These programs can be constructed from the fundamental considerations and the information in those chapters.

There is a wide gap between the fundamental programs illustrated herein and programs which are suitable for industrial application. The latter must be made as foolproof and easy to operate as possible. There should be a section which checks the input data for consistency and issues appropriate error and warning messages. The calculation should be checked for loss of accuracy (e.g. a nearly singular coefficient matrix, internal solution points too near the boundaries, nearly singular integrands, etc.). Moreover, the input data should be made simple with as much as possible generated internally. A well designed program often spends as much computational time in internal checks as it does in the computation for which it was designed. Such checks are necessary, however, to insure that the program is not misused and does not return incorrect answers.

We have not addressed the very complex problems of groundwater modeling. The algorithms, the BIEM, proposed in this book constitute only one aspect of groundwater modeling. We have assumed that the geology and aquifer properties are known. Obviously, the major part of a modeling project consists of the collection of data and the final result cannot be more accurate than those data. Of equal importance is the verification of the model. Even though the numerical algorithm is efficiently and correctly written, it must be verified for any field situation. This step ensures that, at least for some conditions, the modeler has correctly interpreted the available data and has applied the algorithm correctly.

In groundwater modeling, even more than in other scientific activities, there are never enough data. The numerical algorithm can be used to fill in the gaps. There have been a large number of papers written on the 'inverse problem', that of finding aquifer properties from well data. Since this problem is never well posed, there is always the danger of inferring too much from limited data. Although these considerations are outside the scope of this book, they are at least as important in modeling as the computational algorithm.

# Appendix

## A.1 Notation

We have attempted to use the uniform notation throughout, i.e. any symbol carries the same meaning. Unfortunately, we have occasionally been forced to use a symbol for more than one meaning. Most of the symbols are listed below; some which have a clear, local meaning are omitted (e.g. the dummy variable in an integral). A bold roman symbol such as **q** indicates a vector. A bold italic symbol such as ***K*** indicates a tensor. See Appendix A.2 for matrix notation.

| | |
|---|---|
| $A$ | area |
| $A$ | coefficient of the Fourier series |
| $A_e$ | area of a surface element |
| $B$ | half-width of the rectangular recharge area |
| $b$ | thickness of a confined aquifer |
| $b_1$ | thickness of an aquitard |
| $D$ | flow domain |
| $E$ | error term |
| $E(m)$ | complete elliptic integral of the second kind |
| e | 2.71828 |
| $G$ | axisymmetric Green function |
| $g$ | three-dimensional Green function |
| $g$ | acceleration due to gravity |
| $H_{i,j}$ or $[H]$ | coefficient matrix for the unknown $\mathbf{\Phi}$ and/or $\partial\Phi/\partial n$ |
| $h$ | head or potential |
| $h$ | water depth |
| $h_s$ | water level along the perimeter of a well |
| $h_w$ | water level inside a well |
| $I$ | indicates an integral |
| $J$ | Jacobian of a co-ordinate transformation |
| $K$ | permeability |
| $K_1$ | permeability of an aquitard |
| $\boldsymbol{K}$ or $[K]$ | permeability tensor |
| $K_0$ | modified Bessel function of the second kind and order zero |
| $K(m)$ | complete elliptic integral of the second kind |
| $k$ | number of the time step |
| $k$ | wave number |
| $L$ | length |
| $L_{i,j}$ or $[L]$ | matrix multiplying the boundary values of $\partial\Phi/\partial n$ |
| $M$ | number of Gaussian quadrature points |
| $m$ | modulus of the elliptic integral |
| $m_1$ | complementary modulus of the elliptic integral of the first kind |
| $N$ | number of node points |
| $N_e^{(1)}$ | number of external node points |
| $N_1$, $N_2$, $N_3$ | interpolating functions |
| $N_i$ | number of nodes in each zone |

| | |
|---|---|
| $N_w$ | number of wells |
| $N_z$ | number of zones |
| $n$ | porosity |
| $n^*$ | number of unknowns |
| $\mathbf{n}$ | outward unit normal vector of the boundary of the flow domain |
| $n_p$ | number of sources/sinks |
| $n_r$, $n_z$ | direction cosines in the $r$, $z$ co-ordinate system |
| $P$ | point designation (the 'base point') |
| $P_i$ | singular point; base point |
| $p$ | pore water pressure |
| $Q$ | point designation (the 'field point') |
| $Q$ | flow rate |
| $Q_D$ | flow rate calculated from the Dupuit approximation |
| $Q_I$ | flow rate into a well from an aquifer |
| $Q_W$ | pumping rate from a well |
| $\mathbf{q}$ | specific discharge; seepage velocity |
| $q$ | $k \Delta t \tanh kh$ |
| $q_b$ | boundary flux |
| $q_l$ | leakage across an aquitard |
| $q_x$, $q_y$ | component flow rates |
| $R$ | distance between two points in three-dimensional space |
| $R_{i,j}$ or $[R]$ | matrix multiplying the boundary values of $\Phi$ |
| $r$ | distance from a base point to a field point |
| $\{r\}$ | right-hand side of matrix equation |
| $r$, $\theta$, $z$ | cylindrical co-ordinates |
| $R$, $\theta$, $\phi$ | spherical co-ordinates |
| $r_p$ | radius of the production pipe of a well |
| $r_w$ | well radius |
| $S$ | degree of saturation |
| $S$ | surface area |
| $S_e$ | specific storage |
| $s$ | length along a boundary |
| $s$ | Laplace transform parameter |
| $T$ | transmissivity |
| $t$ | time |
| $U$, $V$ | general functions |
| $W$ | recharge intensity |
| $u$, $v$, $w$ | component velocities |
| $x$, $y$, $z$ | co-ordinates |
| $x_i$, $y_i$, $z_i$ | co-ordinates of $P_i$ |
| $x_*$, $z_*$ | rotated (transformed) co-ordinates or dimensionless co-ordinates |
| $z_{WT}$ | vertical distance between pond surface and the undisturbed water table |
| $z_d$ | difference in the $z$ co-ordinates of two points |
| $\alpha$ | vertical compressibility |
| $\alpha$ | exterior boundary angle |
| $\alpha$ | angle in the zoned problem |
| $\beta$ | angle between the free surface and horizontal |
| $\beta$ | fluid compressibility (bulk modulus)$^{-1}$ |

| | |
|---|---|
| $\delta$ | a small quantity |
| $\delta(\ )$ | the Dirac delta function |
| $\delta_{i,j}$ | the Kronecker delta |
| $\boldsymbol{\delta}_x$, $\boldsymbol{\delta}_y$ | component unit vectors |
| $\gamma$ | angle between the $x$–$z$ co-ordinate system (with $z$ vertical) and the $x_*$–$z_*$ co-ordinate system |
| $\Delta t$ | time step |
| $\varepsilon$ | a small quantity |
| $\varepsilon$ | difference between $\sigma$ and $\sigma_N$ |
| $\eta$ | the normal distance between a base point and a linear element |
| $\eta$ | free surface elevation |
| $\eta_*$ | $z_*$ co-ordinate of the free surface in the $x_*$–$z_*$ co-ordinate system |
| $\theta$ | spherical or cylindrical co-ordinate |
| $\theta$ | weighting factor which positions the time derivative between times $k(\Delta t)$ and $(k + 1)(\Delta t)$ |
| $\lambda$ | eigenvalue |
| $\xi$ | distance along the boundary or a boundary element |
| $\pi$ | 3.14159265 |
| $\rho$ | fluid density |
| $\sigma$ | surface surrounding the singular point |
| $\sigma$ | analytical damping rate |
| $\sigma_N$ | numerical damping rate |
| $\sigma_z$ | vertical intergranular stress |
| $\Phi$ | velocity potential |
| $\tilde{\Phi}$ | Laplace transformed potential |
| $\Phi_s$ | steady state potential |
| $\phi$ | spherical co-ordinate |
| $\varphi$ | potential averaged across a layer |
| $\chi$ | zero of the Legendre polynomial, used as a sampling point for Gaussian quadrature |
| $\chi^s$ | sampling point for special Gaussian quadrature |
| $\omega$ | Gaussian quadrature weighting factor |
| $\omega^s$ | special Gaussian quadrature weighting factor |
| $\boldsymbol{\nabla}$ | vector operator $\left(\boldsymbol{\nabla} = \boldsymbol{\delta}_x \frac{\partial}{\partial x} + \boldsymbol{\delta}_y \frac{\partial}{\partial y} + \boldsymbol{\delta}_z \frac{\partial}{\partial z}\right)$ |
| * | indicates transformed or dimensionless quantity |

## A.2 Matrix notation

We have attempted to use standard matrix notation throughout with the following symbols:

| | |
|---|---|
| $[\ \ ]$ | a square or rectangular matrix |
| $\{\ \ \}$ | an $n \times 1$ matrix or a column vector |
| $\lfloor\ \ \rfloor$ | a $1 \times n$ matrix or a row vector |
| $[\ \ ]^{\mathrm{T}}$ | transpose of the original rectangular matrix; thus $\{\ \ \}^{\mathrm{T}}$ is the transpose of a column vector $(\{A\}^{\mathrm{T}} = \lfloor A \rfloor)$ |

| | |
|---|---|
| $[\quad]^{-1}$ | inverse of the original square matrix |
| $\lceil\quad\rfloor$ | a diagonal matrix, i.e. all elements are zero except those on the main diagonal |
| $\lceil I \rfloor$ | the unity matrix, i.e. a diagonal matrix in which the main diagonal consists solely of ones |

## A.3 Free space Green functions for Laplace's equation

The free space Green functions are constructed so that they satisfy the governing equation in an infinite domain except at a singular point $P_i(\mathbf{x}_i)$. In the case of the Laplacian operator, the free space Green function $V$ satisfies the following equation:

$$\nabla^2 V = \begin{cases} -4\pi\delta(x-x_i)\delta(y-y_i)\delta(z-z_i) & \text{for three-dimensional problems} \\ 2\pi\delta(x-x_i)\delta(z-z_i) & \text{for two-dimensional problems} \end{cases} \tag{A.1}$$

where $\delta(\quad)$ is the Dirac delta function. One can interpret the free space Green function as the velocity potential corresponding to a point source with strength $4\pi$ at $P_i(\mathbf{x}_i)$ for three-dimensional problems and corresponding to a line sink with intensity $2\pi$ for two-dimensional cases.

Consider the three-dimensional problems. We can express the Laplacian operator in terms of spherical co-ordinates $(R, \phi, \theta)$ with the origin at the singular point $(x_i, y_i, z_i)$. Laplace's equation is

$$\nabla^2 V = \frac{1}{R^2}\frac{\partial}{\partial R}\left(R^2\frac{\partial V}{\partial R}\right) + \frac{1}{R^2 \sin\phi}\frac{\partial}{\partial\phi}\left(\sin\phi\frac{\partial V}{\partial\phi}\right) + \frac{1}{R^2\sin^2\phi}\frac{\partial^2 V}{\partial\theta^2} = 0 \tag{A.2}$$

Since $V$ is the potential induced by a point source at the origin, it is reasonable to seek $V$ as a function of only the radial distance, $R$. Thus, for all $R > 0$, we have

$$\nabla^2 V = \frac{1}{R^2}\frac{\mathrm{d}}{\mathrm{d}R}\left(R^2\frac{\mathrm{d}V}{\mathrm{d}R}\right) = 0 \tag{A.3}$$

Integration of (A.3) yields

$$V = \frac{A}{R} + B \tag{A.4}$$

where $A$ and $B$ are constants of integration. Equation (A.4) is valid only for $R > 0$. To determine $A$ and $B$, we integrate (A.1) over an infinitesimal sphere of radius $\varepsilon$, centered at the singular point,

$$\iiint \nabla^2 V \, d\tau = \iiint -4\pi\delta \, d\tau = -4\pi \tag{A.5}$$

Applying the divergence theorem (2.1) to the left-hand integral, we have

$$\iint_\sigma \frac{\partial V}{\partial R} \, dA = -4\pi \tag{A.6}$$

where σ denotes the surface of the sphere. Equation (A.6) is a statement of conservation of mass. Substitution of (A.4) into (A.6) yields $A = 1$. Setting $B = 0$ for convenience, we have

$$V = \frac{1}{R} \tag{A.7}$$

For two-dimensional problems Laplace's equation can be expressed in terms of polar co-ordinates $(r, \theta)$ with the origin at the singular point $(x_i, z_i)$. Thus

$$\nabla^2 V = \frac{1}{r}\frac{\partial}{\partial r}\left(r\frac{\partial V}{\partial r}\right) + \frac{1}{r^2}\frac{\partial^2 V}{\partial \theta^2} = 0 \tag{A.8}$$

For $r > 0$, so that we are away from the singularity, we have

$$\frac{1}{r}\frac{d}{dr}\left(r\frac{dV}{dr}\right) = 0 \tag{A.9}$$

Integrating twice yields

$$V = A \ln r + B \tag{A.10}$$

where $A$ and $B$ are constants of integration. Integrating (A.1) over a circle of radius ε centered at $(x_i, y_i)$ gives

$$\iint \nabla^2 V \, dA = 2\pi \iint \delta \, dA = 2\pi \tag{A.11}$$

Applying the divergence theorem to the left-hand integral and using (A.10), we find $A = 1$. Again, we can set $B = 0$ for convenience. The two-dimensional free space Green function becomes

$$V = \ln r \tag{A.12}$$

## A.4 The integrals using linear interpolation

First consider (2.28). The distance $r_i$ (see Fig. 2.5 or 2.7) is given by (2.32). The derivative $\partial r_i / \partial n$ is the cosine of the angle between $r_i$ and the normal to the boundary, which is simply $\eta_i/r_i$. Thus the integral of (2.28) becomes

$$\int \frac{\eta_i}{\xi^2 + \eta_i^2}\, \xi \, d\xi = \frac{\eta_i}{2} \ln(\xi^2 + \eta_i^2) \tag{A.13}$$

The integral of (2.29) is

$$\int \frac{\eta_i}{\xi^2 + \eta_i^2}\, d\xi = \tan^{-1} \frac{\xi}{\eta_i} \tag{A.14}$$

The integral of (2.30) is

$$\int \tfrac{1}{2} \ln(\xi^2 + \eta_i^2)\xi \, d\xi = \tfrac{1}{4}(\xi^2 + \eta_i^2)[\ln(\xi^2 + \eta_i^2) - 1] \tag{A.15}$$

The integral of (2.31) is

$$\int \tfrac{1}{2} \ln(\xi^2 + \eta_i^2)\, d\xi = \tfrac{1}{2}\left[\xi \ln(\xi^2 + \eta_i^2) - 2\xi + 2\eta_i \tan^{-1} \frac{\xi}{\eta_i}\right] \tag{A.16}$$

In the process of obtaining the interior solutions, (2.71) and (2.72), the following integrals are encountered in addition to some of those given above:

$$\int \frac{d\xi}{(\xi^2 + \eta^2)^2} = \frac{\xi}{2\eta^2(\xi^2 + \eta^2)} + \frac{1}{2\eta^3} \tan^{-1} \frac{\xi}{\eta} \tag{A.17}$$

$$\int \frac{\xi \, d\xi}{(\xi^2 + \eta^2)^2} = -\frac{1}{2(\xi^2 + \eta^2)} \tag{A.18}$$

$$\int \frac{\xi^2 \, d\xi}{\xi^2 + \eta^2} = \xi - \eta \tan^{-1} \frac{\xi}{\eta} \tag{A.19}$$

$$\int \frac{\xi^2 \, d\xi}{(\xi^2 + \eta^2)^2} = -\frac{\xi}{2(\xi^2 + \eta^2)} + \frac{1}{2\eta} \tan^{-1} \frac{\xi}{\eta} \tag{A.20}$$

## A.5 Gaussian quadrature over contours

Although there are many methods of numerical integration, all of our computer programs use Gaussian quadrature. The expensive or time consuming part of numerical integration is usually in the functional evaluations. Gaussian quadrature appears to be the most accurate method for a given number of functional evaluations. Efficiency in integration is important in the BIEM. Since the number of simultaneous equations which are generated is small (compared to domain methods), the time a computer program spends in evaluating the integrals is not insignificant compared to the total time of the calculation.

Derivation of the Gaussian quadrature method can be found in Lanczos (1956), Ralston (1965) or Milne (1949). The weights and zeros (sampling points) to be used in the following equations are tabulated in Abramowitz and Stegun (1974) and Stroud and Secrest (1966). We simply present the method without derivation.

### A.5.1 Standard (Legendre) Gaussian quadrature

Assume that we wish to integrate

$$I = \int_a^b f(x)\,\mathrm{d}x \tag{A.21}$$

in which $f(x)$ is some known function and $a$ and $b$ represent the limits of integration. Then

$$I \approx \frac{b-a}{2} \sum_{i=1}^{M} \omega_i f(x_i) + E(M) \tag{A.22}$$

in which

$$x_i = \frac{b-a}{2}\chi_i + \frac{b+a}{2} \tag{A.23}$$

$$E(M) = \frac{(b-a)^{2M+1}(M!)^4}{(2M+1)[(2M)!]^3} f^{(2M)}(x) \tag{A.24}$$

where $f^{(2M)}$ represents the $2M$th derivative of the function. The $\omega_i$ are the weights and the $\chi_i$ are the zeros of the Legendre polynomial. The error term is $E(M)$ which depends on the $2M$th derivative of the function. We note that for a polynomial in which the largest exponent is $2M-1$ or smaller, the $f^{(2M)}(x) = 0$ and thus Gaussian quadrature is exact.

### A.5.2 Example

Consider the integral

$$I = \int_2^3 \frac{x}{(x^2+4)^{\frac{1}{2}}}\,\mathrm{d}x \tag{A.25}$$

Using $M = 4$ the weights and zeros are:

| $\omega_i$ | $\chi_i$ |
|---|---|
| 0.6521451549 | $\pm 0.3399810436$ |
| 0.3478548451 | $\pm 0.8611363116$ |

Using (A.23) the functional evaluations should occur at:

$$x_1 = \frac{3-2}{2}(-0.8611363116) + \frac{3+2}{2} = 2.069431844$$

$$x_2 = 2.33009478$$

$$x_3 = 2.669990522$$

$$x_4 = 2.930568156$$

The corresponding values of $f(x)$ are:

$$f(x_1) = \frac{2.069431844}{((2.069431844)^2 + 4)^{\frac{1}{2}}} = 0.719066643$$

$$f(x_2) = 0.758797962$$

$$f(x_3) = 0.800358547$$

$$f(x_4) = 0.825979933$$

From (A.22) the value of the integral, neglecting $E(M)$, is

$$I \approx \frac{3-2}{2}[(0.3478548451)(0.719066643) + (0.6521451549)(0.758797962)$$
$$+ (0.6521451549)(0.800358547) + (0.3478548451)(0.825979933)]$$
$$= 0.77712415$$

The exact value is

$$I = (13)^{\frac{1}{2}} - 8^{\frac{1}{2}} = 0.777124150$$

Thus the Gaussian quadrature is extremely accurate with only four functional evaluations.

The weights and zeros tend to concentrate the functional evaluations near the ends of the interval but weight more heavily the evaluations nearer the center.

### *A.5.3 Special Gaussian quadrature*

Frequently the BIEM produces an integral of the form

$$I = \int_a^b f(x) \ln(x-a)\,dx \tag{A.26}$$

This integral can be evaluated by the process outlined above, but accuracy will suffer due to the fact that

$$\lim_{x \to a} \ln(x-a) = -\infty \tag{A.27}$$

The accuracy can be recovered by taking a sufficiently large number of quadrature points and/or breaking the interval of integration into parts, with the part near the singularity small. A more efficient method is to use special Gaussian quadrature, where the weights and zeros automatically take care of the logarithmic singularity.

First, let

$$\xi = \frac{x-a}{b-a} \tag{A.28}$$

then

$$\begin{aligned} I &= (b-a)\int_0^1 \ln\,[\xi(b-a)]\, f[\xi(b-a)+a]\,\mathrm{d}\xi \\ &= I_1 + I_2 \end{aligned} \tag{A.29}$$

where

$$I_1 = (b-a)\int_0^1 \ln \xi\, f[\xi(b-a)+a]\,\mathrm{d}\xi \tag{A.30}$$

$$I_2 = (b-a)\int_0^1 \ln\,(b-a)\, f[\xi(b-a)+a]\,\mathrm{d}\xi \tag{A.31}$$

The integral $I_2$ has a non-singular integrand and can be evaluated by standard Gaussian quadrature. The formula for $I_1$ is

$$I_1 = (b-a)\sum_{i=1}^{M} \omega_i^{\mathrm{s}} f(x_i^{\mathrm{s}}) \tag{A.32}$$

in which

$$x_i^{\mathrm{s}} = a + (b-a)\chi_i^{\mathrm{s}} \tag{A.33}$$

The $\omega_i^{\mathrm{s}}$ and $\chi_i^{\mathrm{s}}$ are the special weights and zeros.

A more general case is

$$I = \int_a^b f(x) \ln|x-c|\,\mathrm{d}x \qquad a \leqslant c \leqslant b \tag{A.34}$$

in which the integral is divided into two parts:

$$\begin{aligned} I &= \int_a^c f(x)\ln\,(c-x)\,\mathrm{d}x + \int_c^b f(x)\ln\,(x-c)\,\mathrm{d}x \\ &= I_3 + I_4 \end{aligned}$$

For $I_3$ we use the transformation

$$\xi = \frac{c-x}{c-a} \tag{A.35}$$

The integral $I_4$ is the same as the one in (A.26) with $a$ replaced by $c$.

### A.5.4 Example

In order to illustrate the special quadrature, we use the integral

$$I = \int_1^3 \frac{\ln (x-1)}{(x^2+4)^{\frac{1}{2}}}\, dx \tag{A.36}$$

From (A.30) and (A.31)

$$I_1 = 2\int_0^1 \frac{\ln \xi \, d\xi}{[(2\xi+1)^2+4]^{\frac{1}{2}}} \tag{A.37}$$

$$I_2 = 2\ln 2\int_0^1 \frac{d\xi}{[(2\xi+1)^2+4]^{\frac{1}{2}}} \tag{A.38}$$

Using either analytical methods or Gaussian quadrature,

$$I_2 = 0.49460 \tag{A.39}$$

In order to integrate $I_1$ we use 4-point special Gaussian quadrature. The weights and zeros are:

| $\omega_i^s$ | $\chi_i^s$ |
|---|---|
| −0.3834640681 | 0.0414484802 |
| −0.3868753177 | 0.2452749143 |
| −0.1904351269 | 0.5561654535 |
| −0.0392254871 | 0.8489823945 |

The values of $f(\xi)$ are

$$f(\chi_1) = \frac{1}{\{[2(0.0414484802)+1]^2+4\}^{\frac{1}{2}}} = 0.439686$$

$$f(\chi_2) = 0.400907$$

$$f(\chi_3) = 0.343768$$

$$f(\chi_4) = 0.297759$$

and the value of the integral is

$$I_1 = -2[(0.439686)(0.3834640681) + (0.400907)(0.3868753177) + (0.343768)(0.1904351269) + (0.297759)(0.0392254871)$$
$$= -0.80170$$

Finally,

$$I = I_1 + I_2 = -0.30710$$

## A.6 Gaussian integration over triangular areas

In a three-dimensional BIEM problem the integration is performed over a surface which is divided into small areas (elements) of a prescribed shape. We confine the discussion to integration of triangular elements. It is important to use an efficient integration method since a substantial portion of the computer time may be used in evaluating integrals.

Gaussian quadrature weights and zeros for triangular elements are expressed in terms of triangular co-ordinates. Any point, $P$, can be expressed in terms of three co-ordinates, $L^{(1)}$, $L^{(2)}$, $L^{(3)}$ which are defined with reference to Figure A.1 as

$$L^{(1)} = \frac{A_{P23}}{A_e} \qquad L^{(2)} = \frac{A_{P13}}{A_e} \qquad L^{(3)} = \frac{A_{P12}}{A_e} \tag{A.40}$$

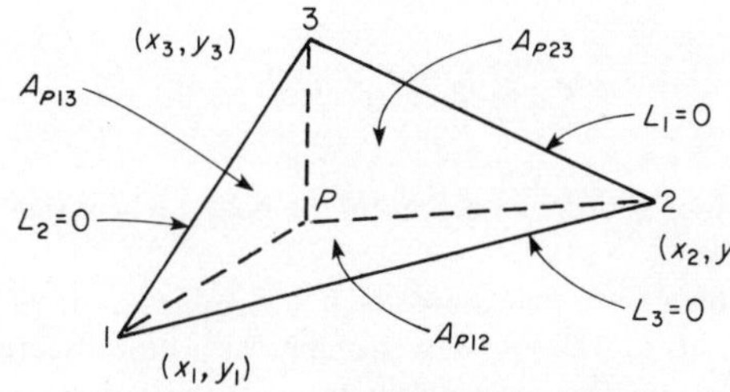


**Figure A.1** Triangular co-ordinates.

The area $A_e$ is the total area of the triangle and $A_{P23}$, for example, represents the area of the sub-triangle $P23$. The sum of these areas equals the total, thus $L^{(1)}$, $L^{(2)}$, and $L^{(3)}$ are not independent but are related through

$$L^{(1)} + L^{(2)} + L^{(3)} = 1 \tag{A.41}$$

Specifically, the triangular co-ordinates are defined by

$$\begin{aligned} L^{(1)} &= [x_2y_3 - x_3y_2 + (y_2 - y_3)x + (x_3 - x_2)y]/2A_e \\ L^{(2)} &= [x_3y_1 - x_1y_3 + (y_3 - y_1)x + (x_1 - x_3)y]/2A_e \\ L^{(3)} &= [x_1y_2 - x_2y_1 + (y_1 - y_2)x + (x_2 - x_1)y]/2A_e \end{aligned} \tag{A.42}$$

in which $x_i, y_i$ are the Cartesian co-ordinates of the $i$th node. It is interesting to note that the linear interpolation functions of (7.7) are simply the triangular co-ordinates

$$N_1 = L^{(1)} \qquad N_2 = L^{(2)} \qquad N_3 = L^{(3)} \tag{A.43}$$

since $L^{(i)} = 1$ at node $i$ and zero at the other nodes.

The integrals over the element are

$$I = \iint_{A_e} f(x, y)\, \mathrm{d}A \tag{A.44}$$

The function $f$ can always be expressed in triangular co-ordinates by inverting (A.42):

$$\begin{aligned} x &= L^{(1)}x_1 + L^{(2)}x_2 + L^{(3)}x_3 \\ y &= L^{(1)}y_1 + L^{(2)}y_2 + L^{(3)}y_3 \end{aligned} \tag{A.45}$$

Then

$$I = \iint_{A_e} f(L^{(1)}, L^{(2)}, L^{(3)})\, \mathrm{d}A \tag{A.46}$$

For numerical integration we write (A.46) as

$$I = A_e \sum_{i=1}^{M} \omega_i f(L_i^{(1)}, L_i^{(2)}, L_i^{(3)}) \tag{A.47}$$

in which $\omega_i$ is the weight and the $L_i$ represent the points where the function is to be evaluated.

The weights and quadrature points are given in Zienkiewicz (1971) up to $M = 25$ and in Stroud (1971) up to $M = 64$. The problem with those references is that the points are not symmetric with respect to the three vertices of the triangle, and thus give slightly different results depending on the orientation of the triangle with respect to the quadrature points. Cowper (1973) gives symmetric weights and sampling points up to $M = 13$.

Numerical integration over areas or volumes has not developed to the extent of the one-dimensional counterpart. In the BIEM the Green function spoils the polynomial type of integrand so that such quadrature formulas are badly needed. Especially important is the special quadrature, either in one or two dimensions, which integrates accurately near a singularity. Fortunately, the singularity that occurs in the BIEM can be separated and treated analytically as explained in Section 7.1.

### *A.6.1 Example*

Suppose the vertices of the triangle of Figure A.1 are located at

$$(x_1, y_1) = (0, 0) \qquad (x_2, y_2) = (1.5, 0.5) \qquad (x_3, y_3) = (1, 1)$$

The area of the triangle is

$$A_e = \tfrac{1}{2}\begin{vmatrix} 1 & x_1 & y_1 \\ 1 & x_2 & y_2 \\ 1 & x_3 & y_3 \end{vmatrix} = \tfrac{1}{2}$$

and the triangular co-ordinates are (i.e. (A.42))

$$\begin{aligned} L^{(1)} &= 1 - 0.5x - 0.5y \\ L^{(2)} &= x - y \\ L^{(3)} &= -0.5x + 1.5y \end{aligned}$$

From equation (A.45)

$$x = 1.5L^{(2)} + L^{(3)}$$

$$y = 0.5L^{(2)} + L^{(3)}$$

Now consider the integral

$$I = \iint_{A_e} x^2 y \, \mathrm{d}A$$

Using the area co-ordinates

$$I = \iint_{A_e} (1.125L^{(2)^3} + 3.75L^{(2)^2}L^{(3)} + 3.5L^{(2)}L^{(3)^2} + L^{(3)^3})\mathrm{d}A$$

Cowper (1973) gives the positions and weights for a three point formula as

$\omega_i = 1/3, \; i = 1, 2, 3$

| | *i = 1* | *i = 2* | *i = 3* |
|---|---|---|---|
| $L^{(1)}$: | 2/3 | 1/6 | 1/6 |
| $L^{(2)}$: | 1/6 | 2/3 | 1/6 |
| $L^{(3)}$: | 1/6 | 1/6 | 2/3 |

From equation (A.47)

$$\begin{aligned} I = (1/2)\{&[1/3][1.125(1/6)^3 + 3.75(1/6)^2(1/6) + 3.5(1/6)(1/6)^2 + (1/6)^3] \\ &+ [1/3][1.125(2/3)^3 + 3.75(2/3)^2(1/6) + 3.5(2/3)(1/6)^2 + (1/6)^3] \\ &+ [1/3][1.125(1/6)^3 + 3.75(1/6)^2(2/3) + 3.5(1/6)(2/3)^2 + (2/3)^3]\} \\ = 0.22569& \end{aligned}$$

The exact result is $I = 0.22708$, a 0.6% error.

# Bibliography

Abel, N. H. 1823. Oplösning af et Par Opgaver ved Hjelp af bestemte Integraler. *Mag. Naturvidensk.* **2**, 55–68, 205–15.

Abramowitz, M. and I. A. Stegun 1974. *Handbook of mathematical functions.* New York: Dover.

Banerjee, P. K. and R. Butterfield 1977. Boundary element methods in geomechanics. In *Finite elements in geomechanics*, G. Gudehus (ed.), Ch. 16. New York: Wiley.

Bear, J. 1972. *Dynamics of fluids in porous media.* New York: American Elsevier.

Bear, J. and G. Dagan 1964. Moving interface in coastal aquifers. *J. Hyd. Div., ASCE* **90**(HY4), 193–216.

Bellman, R. E., R. E. Kalaba and J. Lockett 1966. *Numerical inversion of the Laplace transform.* New York: American Elsevier.

Bouwer, H. 1962. Analyzing groundwater mounds by resistance network. *J. Irrig. Drainage Div., ASCE* **88**(IR3), 15–36.

Charmonman, S. 1966. A numerical method of solution of free surface problems. *J. Geophys. Res.* **71**(16), 3861–8.

Cost, T. L. 1964. Approximate Laplace transform inversions in viscoelastic stress analysis. *AIAA J.* **2**, 2157–66.

Courant, R. and D. Hilbert 1953. *Methods of mathematical physics.* New York: Interscience.

Cowper, G. R. 1973. Gaussian quadrature formulas for triangles. *Int. J. Num. Meth. Engng.* **7**(3), 405–8.

Dagan, G. 1967a. Linearized solutions of free surface groundwater flow with uniform recharge. *J. Geophys. Res.* **72**(4), 1183–93.

Dagan G. 1967b. A method of determining the permeability and effective porosity of unconfined anisotropic aquifers. *Water Resources Res.* **3**(4), 1059–71.

Darcy, H. 1856. *Les fontaines publique de la Ville de Dijon.* Paris: Dalmont.

de Josselin de Jong, G. 1960. Singularity distributions for the analysis of multiple-fluid flow through porous media. *J. Geophys. Res.* **65**(11), 3739–58.

Durell, R. and E. Arnold 1932. *New solid geometry.* New York: Merrill.

Dwight, H. B. 1961. *Tables of integrals and other mathematical data.* New York: Macmillan.

Eagleson, P. S. 1970. *Dynamic hydrology.* New York: McGraw-Hill.

Forchheimer, P. 1901. Wasserbewegung durch Boden. *Z. Ver. Deutsch. Ing.* **45**, 1782–8.

Franklin, P. 1944. *Methods of advanced calculus.* New York: McGraw-Hill.

Gallagher, R. H. 1975. *Finite element analysis: fundamentals.* Englewood Cliffs, NJ: Prentice-Hall.

Gambolati, G. 1976. Transient free surface flow to a well: an analysis of theoretical solutions. *Water Resources Res.* **12**(1), 27–39.

Gardner, G. H. F., J. Downie and H. A. Kendall 1962. Gravity segregation of miscible fluids in linear model. *Soc. Pet. Engrs J., Trans AIME* **225**, 95–104.

Glover, R. E. 1959. The pattern of freshwater flow in a coastal aquifer. *J. Geophys. Res.* **64**(4), 457–9.

Greenberg, M. D. 1971. *Applications of Green's functions in science and engineering.* Englewood Cliffs, NJ: Prentice-Hall.

Hall, H. P. 1955. An investigation of steady flow toward a gravity well. *La Houille Blanche* **10**(8), 8–35.

Hart, J. F., E. W. Cheney, C. L. Lawson, H. J. Machly, C. K. Mesztenyi, J. R. Rice, H. G. Thatcher, Jr and C. Witzgall 1968. *Computer approximations.* New York: Wiley.

Hewson-Browne, R. C. 1974. An analytical solution for the problem of a confined aquifer in which a fresh-water canal runs along the middle of a mud flat. *Q. J. Mech. Appl. Math.* **XXVII**, 143–8.

Hornbeck, R. W. 1975. *Numerical methods.* New York: Quantum.

Janke, E. and F. Emde 1945. *Tables of functions.* New York: Dover.

Jaswon, M. A. and G. T. Symm 1977. *Integral equation methods in potential theory and elastostatics.* New York: Academic Press.

Jeng, G. and A. Wexler 1977. Isoparametric, finite element, variational solution of integral equations for three-dimensional fields. *Int. J. Num. Meth. Engng* **11**(9), 1455–71.

Keulegan, G. H. 1954. *Ninth progress report on model laws for density currents; An example of density current flow in permeable media.* Report No. 3411. Washington, DC: National Bureau of Standards.

Kumar, V. and S. Mukherjee 1977. A boundary-integral equation formulation for time-dependnt inelastic deformation in metals. *Int. J. Mech. Sci.* **19**, 713–24.

Kupradze, V. D. 1965. *Potential methods in the theory of elasticity.* Jerusalem: Israel Program for Scientific Translations.

Lafe, O. E. 1981. *Boundary integral solutions to nearly horizontal flows in multiply zoned aquifers.* PhD Thesis. Cornell University.

Lafe, O. E., J. S. Montes, A. H-D. Cheng, J. A. Liggett and P. L-F. Liu 1980. Singularities in Darcy flow through porous media. *J. Hyd. Div., ASCE* **106**(HY6), 977–97.

Lanczos, C. 1956. *Applied analysis.* Englewood Cliffs, NJ: Prentice-Hall.

Lass, H. 1950. *Vector and tensor analysis.* New York: McGraw-Hill.

Lennon, G. P., P. L-F. Liu and J. A. Liggett 1979a. Boundary integral equation solution to axisymmetric potential flows, 1, Basic formulation. *Water Resources Res.* **15**(5), 1102–6.

Lennon, G. P., P. L-F. Liu and J. A. Liggett 1979b. Boundary integral equation solution to axisymmetric potential flows, 2, Recharge and well problems in porous media. *Water Resources Res.* **15**(5), 1107–15.

Liggett, J. A. 1977. Locations of free surface in porous media. *J. Hyd. Div., ASCE* **103**(HY4), 353–65.

Liggett, J. A. and J. A. Cunge 1975. Numerical methods of solution of the unsteady flow equations. In *Unsteady flow in open channels*, Ch. 4. Fort Collins, Colorado: Water Resources Publications.

Liggett, J. A. and C. Hadjitheodorou 1965. Initial motion problem in porous media. *J. Hyd. Div., ASCE* **91**(HY3), 61–80.

Liggett, J. A. and P. L-F. Liu 1979a. Unsteady interzonal flow in porous media. *Water Resources Res.* **15**(2), 240–6.

Liggett, J. A. and P. L-F. Liu 1979b. Unsteady flow in confined aquifers – a comparison of two boundary integral methods. *Water Resources Res.* **15**(4), 861–6.

Liu, P. L-F., A. H-D. Cheng, J. A. Liggett and J. H. Lee 1981. Boundary integral equation solutions to moving interface between two fluids in porous media. *Water Resources Res.* **17**(5), 1445–52.

Liu, P. L-F. and J. A. Liggett 1978. An efficient numerical method of two dimensional steady groundwater problems. *Water Resources Res.* **14**(3), 385–90.

Liu, P. L-F. and J. A. Liggett 1979. Boundary solutions to two problems in porous media. *J. Hyd. Div., ASCE* **105**(HY3), 171–83.

Long, R. R. 1961. *Mechanics of solids and fluids*. Englewood Cliffs, NJ: Prentice-Hall.

Marino, M. A. 1967. Hele–Shaw model study of the growth and decay of groundwater ridges. *J. Geophys. Res.* **72**(4), 1195–1205.

Milne, W. E. 1949. *Numerical calculus*. Princeton University Press.

Morse, P. M. and H. Feshbach 1953. *Methods of theoretical physics*. New York: McGraw-Hill.

Neuman, S. P. 1973. Saturated–unsaturated seepage by finite elements. *J. Hyd. Div., ASCE* **99**(HY12), 2233–51.

Neuman, S. P., T. N. Narasimhan and P. A. Witherspoon 1976. *Application of mixed explicit–implicit finite element method to nonlinear diffusion-type problems*. International Conference on Finite Elements in Water Resources, Princeton University.

Neuman, S. P. and P. A. Witherspoon 1971. Analysis of nonsteady flow with a free surface using the finite element method. *Water Resources Res.* **7**(3), 611–23.

Pinder, G. F. and W. G. Gray 1977. *Finite element simulation in surface and subsurface hydrology*. New York: Academic Press.

Polubarinova-Kochina, P. Ya 1962. *Theory of groundwater movement* (translated from Russian by J. M. R. DeWiest). Princeton University Press.

Prager, W. 1928. Die Druckverteilung an Körpern in ebener Potentialströmung. *Physik. Z.* **XXIX**, 865.

Ralston, A. 1965. *A first course in numerical analysis*. New York: McGraw-Hill.

Remson, I., G. M. Hornberger and F. J. Molz 1971. *Numerical methods in subsurface hydrology*. New York: Wiley–Interscience.

Richards, L. A. 1931. Capillary conduction of liquids through porous medium. *Physics* **1**, 318–33.

Rizzo, F. J. and D. J. Shippy 1970. A method of solution for certain problems of transient heat conduction. *AIAA J.* **8**(11), 2004–9.

Rizzo, F. J. and D. J. Shippy 1977. An advanced boundary integral equation method for three-dimensional thermoelasticity. *Int. J. Num. Meth. Engng* **11**(11), 1753–68.

Sá da Costa, A. A. and J. L. Wilson 1979. *A numerical model of seawater intrusion in aquifers*. Report no. 247. Ralph M. Parsons Lab., MIT.

Schapery, R. A. 1962. Approximate methods of transform inversion for viscoelastic stress analysis. *Proc. Fourth US Nat. Congr. Appl. Mech.* **2**, 1075–85.

Segerlind, L. J. 1976. *Applied finite element analysis*. New York: Wiley.

Shamir, V. and G. Dagan 1971. Motion of the seawater interface in coastal aquifers: a numerical solution. *Water Resources Res.* **7**(3), 644–57.

Shaw, R. P. 1974. An integral equation approach to diffusion. *Int. J. Heat Mass Transfer* **17**, 693–9.

Shaw, R. P. 1975. Boundary integral equation method applied to water waves. *ASME Proc. AMD* (II), 23–5.

Stark, K. P. and R. E. Volker 1967. *Nonlinear flow through porous materials – some theoretical aspects*. Research Bulletin No. 1. Dept of Engineering, Univ. College of Townsville, Australia.

Stroud, A. H. 1971. *Approximate calculation of multiple integrals*. Englewood Cliffs, NJ: Prentice-Hall.

Stroud, A. H. and D. Secrest 1966. *Gaussian quadrature formulas*. Englewood Cliffs, NJ: Prentice-Hall.

Vauclin, M., D. Khanji and G. Vauchaud 1979. Experimental and numerical study of transient, two-dimensional unsaturated–saturated water table recharge problem. *Water Resources Res.* **15**(5), 1089–1101.

Verma, R. D. 1969. *Physical analysis of the outflow from an unconfined aquifer*. PhD thesis. Cornell University.

Verma, R. D. and W. H. Brutsaert 1971. Unsteady free surface groundwater seepage. *J. Hyd. Div., ASCE* **97**(HY8), 1213–29.

Verruijt, A. 1980. The rotation of a vertical interface in a porous medium. *Water Resources Res.* **16**(1), 239–40.

Voinov, V. V., O. V. Voinov and A. G. Petrov 1974. A method of computing the potential flow round a solid of revolution in an incompressible fluid. *USSR Comp. Maths. Math. Phys.* **14**, 263–8.

Volker, R. E. 1980. *Predicting the movement of sea water into a coastal aquifer*. Technical Paper No. 51. Australia Water Resources Council.

Wu, Y. S. 1976. *The boundary integral equation method using various approximation techniques for problems governed by Laplace's equation*. Thesis presented to University of Kentucky, Lexington, Ky, in partial fulfillment of the degree of Master of Science in Engineering Materials.

Zienkiewicz, O. C. 1971. *The finite element method in engineering science*, 2nd edn. New York: McGraw-Hill.

# Author index

# Subject index

Numbers in bold type refer to text sections.